Survey Sampling

Statistical Distributions
Professor N. Balakrishnan
McMaster University

Statistical Process Improvement
Professor G. Geoffrey Vining
Virginia Polytechnic Institute

Stochastic Processes
Professor V. Lakshmikantham
Florida Institute of Technology

Survey Sampling
Professor Lynne Stokes
Southern Methodist University

Time Series
Sastry G. Pantula
North Carolina State University

1. The Generalized Jackknife Statistic, *H. L. Gray and W. R. Schucany*
2. Multivariate Analysis, *Anant M. Kshirsagar*
3. Statistics and Society, *Walter T. Federer*
4. Multivariate Analysis: A Selected and Abstracted Bibliography, 1957–1972, *Kocherlakota Subrahmaniam and Kathleen Subrahmaniam*
5. Design of Experiments: A Realistic Approach, *Virgil L. Anderson and Robert A. McLean*
6. Statistical and Mathematical Aspects of Pollution Problems, *John W. Pratt*
7. Introduction to Probability and Statistics (in two parts), Part I: Probability; Part II: Statistics, *Narayan C. Giri*
8. Statistical Theory of the Analysis of Experimental Designs, *J. Ogawa*
9. Statistical Techniques in Simulation (in two parts), *Jack P. C. Kleijnen*
10. Data Quality Control and Editing, *Joseph I. Naus*
11. Cost of Living Index Numbers: Practice, Precision, and Theory, *Kali S. Banerjee*
12. Weighing Designs: For Chemistry, Medicine, Economics, Operations Research, Statistics, *Kali S. Banerjee*
13. The Search for Oil: Some Statistical Methods and Techniques, *edited by D. B. Owen*
14. Sample Size Choice: Charts for Experiments with Linear Models, *Robert E. Odeh and Martin Fox*
15. Statistical Methods for Engineers and Scientists, *Robert M. Bethea, Benjamin S. Duran, and Thomas L. Boullion*
16. Statistical Quality Control Methods, *Irving W. Burr*

17. On the History of Statistics and Probability, *edited by D. B. Owen*
18. Econometrics, *Peter Schmidt*
19. Sufficient Statistics: Selected Contributions, *Vasant S. Huzurbazar (edited by Anant M. Kshirsagar)*
20. Handbook of Statistical Distributions, *Jagdish K. Patel, C. H. Kapadia, and D. B. Owen*
21. Case Studies in Sample Design, *A. C. Rosander*
22. Pocket Book of Statistical Tables, *compiled by R. E. Odeh, D. B. Owen, Z. W. Birnbaum, and L. Fisher*
23. The Information in Contingency Tables, *D. V. Gokhale and Solomon Kullback*
24. Statistical Analysis of Reliability and Life-Testing Models: Theory and Methods, *Lee J. Bain*
25. Elementary Statistical Quality Control, *Irving W. Burr*
26. An Introduction to Probability and Statistics Using BASIC, *Richard A. Groeneveld*
27. Basic Applied Statistics, *B. L. Raktoe and J. J. Hubert*
28. A Primer in Probability, *Kathleen Subrahmaniam*
29. Random Processes: A First Look, *R. Syski*
30. Regression Methods: A Tool for Data Analysis, *Rudolf J. Freund and Paul D. Minton*
31. Randomization Tests, *Eugene S. Edgington*
32. Tables for Normal Tolerance Limits, Sampling Plans and Screening, *Robert E. Odeh and D. B. Owen*
33. Statistical Computing, *William J. Kennedy, Jr., and James E. Gentle*
34. Regression Analysis and Its Application: A Data-Oriented Approach, *Richard F. Gunst and Robert L. Mason*
35. Scientific Strategies to Save Your Life, *I. D. J. Bross*
36. Statistics in the Pharmaceutical Industry, *edited by C. Ralph Buncher and Jia-Yeong Tsay*
37. Sampling from a Finite Population, *J. Hajek*
38. Statistical Modeling Techniques, *S. S. Shapiro and A. J. Gross*
39. Statistical Theory and Inference in Research, *T. A. Bancroft and C.-P. Han*
40. Handbook of the Normal Distribution, *Jagdish K. Patel and Campbell B. Read*
41. Recent Advances in Regression Methods, *Hrishikesh D. Vinod and Aman Ullah*
42. Acceptance Sampling in Quality Control, *Edward G. Schilling*
43. The Randomized Clinical Trial and Therapeutic Decisions, *edited by Niels Tygstrup, John M Lachin, and Erik Juhl*
44. Regression Analysis of Survival Data in Cancer Chemotherapy, *Walter H. Carter, Jr., Galen L. Wampler, and Donald M. Stablein*
45. A Course in Linear Models, *Anant M. Kshirsagar*
46. Clinical Trials: Issues and Approaches, *edited by Stanley H. Shapiro and Thomas H. Louis*
47. Statistical Analysis of DNA Sequence Data, *edited by B. S. Weir*
48. Nonlinear Regression Modeling: A Unified Practical Approach, *David A. Ratkowsky*
49. Attribute Sampling Plans, Tables of Tests and Confidence Limits for Proportions, *Robert E. Odeh and D. B. Owen*

50. Experimental Design, Statistical Models, and Genetic Statistics, *edited by Klaus Hinkelmann*
51. Statistical Methods for Cancer Studies, *edited by Richard G. Cornell*
52. Practical Statistical Sampling for Auditors, *Arthur J. Wilburn*
53. Statistical Methods for Cancer Studies, *edited by Edward J. Wegman and James G. Smith*
54. Self-Organizing Methods in Modeling: GMDH Type Algorithms, *edited by Stanley J. Farlow*
55. Applied Factorial and Fractional Designs, *Robert A. McLean and Virgil L. Anderson*
56. Design of Experiments: Ranking and Selection, *edited by Thomas J. Santner and Ajit C. Tamhane*
57. Statistical Methods for Engineers and Scientists: Second Edition, Revised and Expanded, *Robert M. Bethea, Benjamin S. Duran, and Thomas L. Boullion*
58. Ensemble Modeling: Inference from Small-Scale Properties to Large-Scale Systems, *Alan E. Gelfand and Crayton C. Walker*
59. Computer Modeling for Business and Industry, *Bruce L. Bowerman and Richard T. O'Connell*
60. Bayesian Analysis of Linear Models, *Lyle D. Broemeling*
61. Methodological Issues for Health Care Surveys, *Brenda Cox and Steven Cohen*
62. Applied Regression Analysis and Experimental Design, *Richard J. Brook and Gregory C. Arnold*
63. Statpal: A Statistical Package for Microcomputers—PC-DOS Version for the IBM PC and Compatibles, *Bruce J. Chalmer and David G. Whitmore*
64. Statpal: A Statistical Package for Microcomputers—Apple Version for the II, II+, and IIe, *David G. Whitmore and Bruce J. Chalmer*
65. Nonparametric Statistical Inference: Second Edition, Revised and Expanded, *Jean Dickinson Gibbons*
66. Design and Analysis of Experiments, *Roger G. Petersen*
67. Statistical Methods for Pharmaceutical Research Planning, *Sten W. Bergman and John C. Gittins*
68. Goodness-of-Fit Techniques, *edited by Ralph B. D'Agostino and Michael A. Stephens*
69. Statistical Methods in Discrimination Litigation, *edited by D. H. Kaye and Mikel Aickin*
70. Truncated and Censored Samples from Normal Populations, *Helmut Schneider*
71. Robust Inference, *M. L. Tiku, W. Y. Tan, and N. Balakrishnan*
72. Statistical Image Processing and Graphics, *edited by Edward J. Wegman and Douglas J. DePriest*
73. Assignment Methods in Combinatorial Data Analysis, *Lawrence J. Hubert*
74. Econometrics and Structural Change, *Lyle D. Broemeling and Hiroki Tsurumi*
75. Multivariate Interpretation of Clinical Laboratory Data, *Adelin Albert and Eugene K. Harris*
76. Statistical Tools for Simulation Practitioners, *Jack P. C. Kleijnen*
77. Randomization Tests: Second Edition, *Eugene S. Edgington*

78. A Folio of Distributions: A Collection of Theoretical Quantile-Quantile Plots, *Edward B. Fowlkes*
79. Applied Categorical Data Analysis, *Daniel H. Freeman, Jr.*
80. Seemingly Unrelated Regression Equations Models: Estimation and Inference, *Virendra K. Srivastava and David E. A. Giles*
81. Response Surfaces: Designs and Analyses, *Andre I. Khuri and John A. Cornell*
82. Nonlinear Parameter Estimation: An Integrated System in BASIC, *John C. Nash and Mary Walker-Smith*
83. Cancer Modeling, *edited by James R. Thompson and Barry W. Brown*
84. Mixture Models: Inference and Applications to Clustering, *Geoffrey J. McLachlan and Kaye E. Basford*
85. Randomized Response: Theory and Techniques, *Arijit Chaudhuri and Rahul Mukerjee*
86. Biopharmaceutical Statistics for Drug Development, *edited by Karl E. Peace*
87. Parts per Million Values for Estimating Quality Levels, *Robert E. Odeh and D. B. Owen*
88. Lognormal Distributions: Theory and Applications, *edited by Edwin L. Crow and Kunio Shimizu*
89. Properties of Estimators for the Gamma Distribution, *K. O. Bowman and L. R. Shenton*
90. Spline Smoothing and Nonparametric Regression, *Randall L. Eubank*
91. Linear Least Squares Computations, *R. W. Farebrother*
92. Exploring Statistics, *Damaraju Raghavarao*
93. Applied Time Series Analysis for Business and Economic Forecasting, *Sufi M. Nazem*
94. Bayesian Analysis of Time Series and Dynamic Models, *edited by James C. Spall*
95. The Inverse Gaussian Distribution: Theory, Methodology, and Applications, *Raj S. Chhikara and J. Leroy Folks*
96. Parameter Estimation in Reliability and Life Span Models, *A. Clifford Cohen and Betty Jones Whitten*
97. Pooled Cross-Sectional and Time Series Data Analysis, *Terry E. Dielman*
98. Random Processes: A First Look, Second Edition, Revised and Expanded, *R. Syski*
99. Generalized Poisson Distributions: Properties and Applications, *P. C. Consul*
100. Nonlinear L_p-Norm Estimation, *Rene Gonin and Arthur H. Money*
101. Model Discrimination for Nonlinear Regression Models, *Dale S. Borowiak*
102. Applied Regression Analysis in Econometrics, *Howard E. Doran*
103. Continued Fractions in Statistical Applications, *K. O. Bowman and L. R. Shenton*
104. Statistical Methodology in the Pharmaceutical Sciences, *Donald A. Berry*
105. Experimental Design in Biotechnology, *Perry D. Haaland*
106. Statistical Issues in Drug Research and Development, *edited by Karl E. Peace*
107. Handbook of Nonlinear Regression Models, *David A. Ratkowsky*

108. Robust Regression: Analysis and Applications, *edited by Kenneth D. Lawrence and Jeffrey L. Arthur*
109. Statistical Design and Analysis of Industrial Experiments, *edited by Subir Ghosh*
110. *U*-Statistics: Theory and Practice, *A. J. Lee*
111. A Primer in Probability: Second Edition, Revised and Expanded, *Kathleen Subrahmaniam*
112. Data Quality Control: Theory and Pragmatics, *edited by Gunar E. Liepins and V. R. R. Uppuluri*
113. Engineering Quality by Design: Interpreting the Taguchi Approach, *Thomas B. Barker*
114. Survivorship Analysis for Clinical Studies, *Eugene K. Harris and Adelin Albert*
115. Statistical Analysis of Reliability and Life-Testing Models: Second Edition, *Lee J. Bain and Max Engelhardt*
116. Stochastic Models of Carcinogenesis, *Wai-Yuan Tan*
117. Statistics and Society: Data Collection and Interpretation, Second Edition, Revised and Expanded, *Walter T. Federer*
118. Handbook of Sequential Analysis, *B. K. Ghosh and P. K. Sen*
119. Truncated and Censored Samples: Theory and Applications, *A. Clifford Cohen*
120. Survey Sampling Principles, *E. K. Foreman*
121. Applied Engineering Statistics, *Robert M. Bethea and R. Russell Rhinehart*
122. Sample Size Choice: Charts for Experiments with Linear Models: Second Edition, *Robert E. Odeh and Martin Fox*
123. Handbook of the Logistic Distribution, *edited by N. Balakrishnan*
124. Fundamentals of Biostatistical Inference, *Chap T. Le*
125. Correspondence Analysis Handbook, *J.-P. Benzécri*
126. Quadratic Forms in Random Variables: Theory and Applications, *A. M. Mathai and Serge B. Provost*
127. Confidence Intervals on Variance Components, *Richard K. Burdick and Franklin A. Graybill*
128. Biopharmaceutical Sequential Statistical Applications, *edited by Karl E. Peace*
129. Item Response Theory: Parameter Estimation Techniques, *Frank B. Baker*
130. Survey Sampling: Theory and Methods, *Arijit Chaudhuri and Horst Stenger*
131. Nonparametric Statistical Inference: Third Edition, Revised and Expanded, *Jean Dickinson Gibbons and Subhabrata Chakraborti*
132. Bivariate Discrete Distribution, *Subrahmaniam Kocherlakota and Kathleen Kocherlakota*
133. Design and Analysis of Bioavailability and Bioequivalence Studies, *Shein-Chung Chow and Jen-pei Liu*
134. Multiple Comparisons, Selection, and Applications in Biometry, *edited by Fred M. Hoppe*
135. Cross-Over Experiments: Design, Analysis, and Application, *David A. Ratkowsky, Marc A. Evans, and J. Richard Alldredge*
136. Introduction to Probability and Statistics: Second Edition, Revised and Expanded, *Narayan C. Giri*

137. Applied Analysis of Variance in Behavioral Science, *edited by Lynne K. Edwards*
138. Drug Safety Assessment in Clinical Trials, *edited by Gene S. Gilbert*
139. Design of Experiments: A No-Name Approach, *Thomas J. Lorenzen and Virgil L. Anderson*
140. Statistics in the Pharmaceutical Industry: Second Edition, Revised and Expanded, *edited by C. Ralph Buncher and Jia-Yeong Tsay*
141. Advanced Linear Models: Theory and Applications, *Song-Gui Wang and Shein-Chung Chow*
142. Multistage Selection and Ranking Procedures: Second-Order Asymptotics, *Nitis Mukhopadhyay and Tumulesh K. S. Solanky*
143. Statistical Design and Analysis in Pharmaceutical Science: Validation, Process Controls, and Stability, *Shein-Chung Chow and Jen-pei Liu*
144. Statistical Methods for Engineers and Scientists: Third Edition, Revised and Expanded, *Robert M. Bethea, Benjamin S. Duran, and Thomas L. Boullion*
145. Growth Curves, *Anant M. Kshirsagar and William Boyce Smith*
146. Statistical Bases of Reference Values in Laboratory Medicine, *Eugene K. Harris and James C. Boyd*
147. Randomization Tests: Third Edition, Revised and Expanded, *Eugene S. Edgington*
148. Practical Sampling Techniques: Second Edition, Revised and Expanded, *Ranjan K. Som*
149. Multivariate Statistical Analysis, *Narayan C. Giri*
150. Handbook of the Normal Distribution: Second Edition, Revised and Expanded, *Jagdish K. Patel and Campbell B. Read*
151. Bayesian Biostatistics, *edited by Donald A. Berry and Dalene K. Stangl*
152. Response Surfaces: Designs and Analyses, Second Edition, Revised and Expanded, *André I. Khuri and John A. Cornell*
153. Statistics of Quality, *edited by Subir Ghosh, William R. Schucany, and William B. Smith*
154. Linear and Nonlinear Models for the Analysis of Repeated Measurements, *Edward F. Vonesh and Vernon M. Chinchilli*
155. Handbook of Applied Economic Statistics, *Aman Ullah and David E. A. Giles*
156. Improving Efficiency by Shrinkage: The James-Stein and Ridge Regression Estimators, *Marvin H. J. Gruber*
157. Nonparametric Regression and Spline Smoothing: Second Edition, *Randall L. Eubank*
158. Asymptotics, Nonparametrics, and Time Series, *edited by Subir Ghosh*
159. Multivariate Analysis, Design of Experiments, and Survey Sampling, *edited by Subir Ghosh*
160. Statistical Process Monitoring and Control, *edited by Sung H. Park and G. Geoffrey Vining*
161. Statistics for the 21st Century: Methodologies for Applications of the Future, *edited by C. R. Rao and Gábor J. Székely*
162. Probability and Statistical Inference, *Nitis Mukhopadhyay*

163. Handbook of Stochastic Analysis and Applications, *edited by D. Kannan and V. Lakshmikantham*
164. Testing for Normality, *Henry C. Thode, Jr.*
165. Handbook of Applied Econometrics and Statistical Inference, *edited by Aman Ullah, Alan T. K. Wan, and Anoop Chaturvedi*
166. Visualizing Statistical Models and Concepts, *R. W. Farebrother and Michael Schyns*
167. Financial and Actuarial Statistics, *Dale Borowiak*
168. Nonparametric Statistical Inference, Fourth Edition, Revised and Expanded, *edited by Jean Dickinson Gibbons and Subhabrata Chakraborti*
169. Computer-Aided Econometrics, *edited by David EA. Giles*
170. The EM Algorithm and Related Statistical Models, *edited by Michiko Watanabe and Kazunori Yamaguchi*
171. Multivariate Statistical Analysis, Second Edition, Revised and Expanded, *Narayan C. Giri*
172. Computational Methods in Statistics and Econometrics, *Hisashi Tanizaki*
173. Applied Sequential Methodologies: Real-World Examples with Data Analysis, *edited by Nitis Mukhopadhyay, Sujay Datta, and Saibal Chattopadhyay*
174. Handbook of Beta Distribution and Its Applications, *edited by Richard Guarino and Saralees Nadarajah*
175. Item Response Theory: Parameter Estimation Techniques, Second Edition, *edited by Frank B. Baker and Seock-Ho Kim*
176. Statistical Methods in Computer Security, *William W. S. Chen*
177. Elementary Statistical Quality Control, Second Edition, *John T. Burr*
178. Data Analysis of Asymmetric Structures, *edited by Takayuki Saito and Hiroshi Yadohisa*
179. Mathematical Statistics with Applications, *Asha Seth Kapadia, Wenyaw Chan, and Lemuel Moyé*
180. Advances on Models Character and Applications, *N. Balakrishnan and I. G. Bayramov*
181. Survey Sampling: Theory and Methods, Second Edition, *Arijit Chaudhuri and Horst Stenger*

Survey Sampling
Theory and Methods
Second Edition

Arijit Chaudhuri
Indian Statistical Institute
Calcutta, India

Horst Stenger
University of Manheim
Manheim, Germany

CRC Press
Taylor & Francis Group
Boca Raton London New York

CRC Press is an imprint of the
Taylor & Francis Group, an **informa** business
A CHAPMAN & HALL BOOK

Published in 2005 by
Chapman & Hall/CRC
Taylor & Francis Group
6000 Broken Sound Parkway NW, Suite 300
Boca Raton, FL 33487-2742

First issued in paperback 2020

ISBN 13: 978-0-367-57809-1 (pbk)
ISBN 13: 978-0-8247-5754-0 (hbk)
Library of Congress Card Number 2004058264

Library of Congress Cataloging-in-Publication Data

Chaudhuri, Arijit, 1940-
Survey sampling : theory and methods / Arijit Chaudhuri, Horst Stenger.—2nd ed.
p. cm. -- (Statistics, textbooks and monographs ; v. 181)
Includes bibliographical references and index.
ISBN 0-82475-754-8
1. Sampling (Statistics) I. Stenger, Horst, 1935- II. Title. III. Series.

QA276.6.C43 2005
519.5'2--dc22 2004058264

Visit the Taylor & Francis Web site at
http://www.taylorandfrancis.com

and the CRC Press Web site at
http://www.crcpress.com

Foreword

ARIJIT CHAUDHURI and HORST STENGER are well known in sampling theory. The present book further confirms their reputation. Here the authors have undertaken the large task of surveying the sampling literature of the past few decades to provide a reference book for researchers in the area. They have done an excellent job. Starting with the unified theory the authors very clearly explain subsequent developments. In fact, even the most modern innovations of survey sampling, both methodological and theoretical, have found a place in this concise volume. In this connection I may specially mention the authors' presentation of estimating functions. With its own distinctiveness, this book is indeed a very welcome addition to the already existing rich literature on survey sampling.

V. P. GODAMBE
University of Waterloo
Waterloo, Ontario, Canada

Preface to the Second Edition

It is gratifying that our Publishers engaged us to prepare this second edition. Since our first edition appeared in 1992, *Survey Sampling* acquired a remarkable growth to which we, too, have made a modest contribution. So, some addition seems due. Meanwhile, we have received feedback from our readers that prompts us to incorporate some modifications.

Several significant books of relevance have emerged after our write-up for the first edition went to press that we may now draw upon, by the following authors or editors: SÄRNDAL, SWENSSON and WRETMAN (1992), BOLFARINE and ZACKS (1992), S. K. THOMPSON (1992), GHOSH and MEEDEN (1986), THOMPSON and SEBER (1996), M. E. THOMPSON, (1997) GODAMBE (1991), COX (1991) and VALLIANT, DORFMAN and ROYALL (2000), among others.

Numerous path-breaking research articles have also appeared in journals keeping pace with this phenomenal progress. So, we are blessed with an opportunity to enlighten ourselves with plenty of new ideas. Yet we curb our impulse to cover the salient aspects of even a sizeable section of this current literature. This is because we are not inclined to reshape the essential structure of our original volume and we are aware of the limitations that prevent us from such a venture.

As in our earlier presentation, herein we also avoid being dogmatic—more precisely, we eschew taking sides. *Survey Sampling* is at the periphery of mainstream statistics. The speciality here is that we have a tangible collection of objects with certain features, and there is an intention to pry into them by getting hold of some of these objects and attempting an inference about those left untouched. This inference is traditionally based on a theory of probability that is used to exploit a possible link of the observed with the unobserved. This probability is not conceived as in statistics, covering other fields, to characterize the interrelation of the individual values of the variables of our interest. But this is created by a survey sampling investigator through arbitrary specification of an artifice to select the samples from the populations of objects with preassigned probabilities. This is motivated by a desire to draw a representative sample, which is a concept yet to be precisely defined. Purposive selection (earlier purported to achieve representativeness) is discarded in favor of this sampling design-based approach, which is theoretically admitted as a means of yielding a legitimate inference about an aggregate from a sampled segment and also valued for its objectivity, being free of personal bias of a sampler. NEYMAN's (1934) pioneering masterpiece, followed by survey sampling texts by YATES (1953), HANSEN, HURWITZ and MADOW (1953), DEMING (1954) and SUKHATME (1954), backed up by exquisitely executed survey findings by MAHALANOBIS (1946) in India as well as by others in England and the U.S., ensured an unstinted support of probability sampling for about 35 years.

But ROYALL (1970) and BREWER (1963) installed a rival theory dislodging the role of the selection probability as an inferential tool in survey sampling. This theory takes off postulating a probability model characterizing the possible links among the observed and the unobserved variate values associated with the survey population units. The parameter of the surveyor's inferential concern is now a random variable rather than a constant. Hence it can be predicted, not estimated.

The basis of inference here is this probability structure as modeled.

Fortunately, the virtues of some of the sampling design-supported techniques like stratification, ratio method of estimation, etc., continue to be upheld by this model-based prediction theory as well. But procedures for assessing and measuring the errors in estimation and prediction and setting up confidence intervals do not match.

The design-based approach fails to yield a best estimator for a total free of design-bias. By contrast, a model-specific best predictor is readily produced if the model is simple, correct, and plausible. If the model is in doubt one has to strike a balance over bias versus accuracy. A procedure that works well even with a wrong model and is thus robust is in demand with this approach. That requires a sample that is adequately balanced in terms of sample and population values of one or more variables related to one of the primary inferential interest. For the design-based classical approach, currently recognized performers are the estimators motivated by appropriate prediction models that are design-biased, but the biases are negligible when the sample sizes are large. So, a modern compromise survey approach called model assisted survey sampling is now popular. Thanks to the pioneering efforts by SÄRNDAL (1982) and his colleagues the generalized regression (GREG) estimators of this category are found to be very effective in practice.

Regression modeling motivated their arrival. But an alternative calibration approach cultivated since the early nineties by ZIESCHANG (1990), DEVILLE and SÄRNDAL (1992), and others renders them purely design-based as well with an assured robustness or riddance from model-dependence altogether.

A predictor for a survey population total is a sum of the sampled values plus the sum of the predictors for the unsampled ones. A design-based estimator for a population total, by contrast, is a sum of the sampled values with multiplicative weights yielded by specific sampling designs. A calibration approach adjusts these initial sampling weights, the new weights keeping close to them but satisfying certain consistency constraints or calibration equations determined by one or more auxiliary variables with known population totals.

This approach was not discussed in the first edition but is now treated at length. Adjustments here need further care to keep the new weights within certain plausible limits, for which there is considerable documentation in the literature. Here we also discuss a concern for outliers—a topic which also recommends adjustments of sampling weights. While calibration and restricted calibration estimators remain asymptotically design unbiased (ADU) and asymptotically design consistent (ADC), the other adjusted ones do not.

Earlier we discussed the QR predictors, which include (1) the best predictors, (2) projection estimators, (3) generalized regression estimators, and (4) the cosmetic predictors for which (1) and (3) match under certain conditions. Developments since 1992 modify QR predictors into restricted QR predictors (RQR) as we also recount.

SÄRNDAL (1996), DEVILLE (1999), BREWER (1999a, 1999b), and BREWER and GREGOIRE (2000) are prescribing a line of research to justify omission of the cross-product terms in the quadratic forms, giving the variance and mean square error (MSE) estimators of linear estimators of population totals, by suitable approximations. In this context SÄRNDAL (1996) makes a strong plea for the use of generalized regression estimators based either on stratified (1) simple random sampling (SRS) or (2) Bernoulli sampling (BS), which is a special case of Poisson sampling devoid of cross-product terms. This encourages us to present an appraisal of Poisson sampling and its valuable ramifications employing permanent random numbers (PRN), useful in coordination and exercise of control in rotational sampling, a topic we omitted earlier.

Among other novelties of this edition we mention the following. We give essential complements to our earlier discussion of the minimax principle. In the first edition, exact results were presented for completely symmetric situations and approximate results for large populations and samples. Now, following STENGER and GABLER (1996) an exact minimax property of the expansion estimator in connection with the LAHIRI-MIDZUNO-SEN design is presented for arbitrary sample sizes.

An exact minimax property of a Hansen-Hurwitz estimator proved by GABLER and STENGER (2000) is reviewed; in this case a rather complicated design has to be applied, as sample sizes are arbitrary.

A corrective term is added to SEN (1953) and YATES and GRUNDY's (1953) variance estimator to make it unbiased even for non-fixed-sample-size designs with an easy check for its uniform non-negativity, as introduced by CHAUDHURI and PAL (2002). Its extension to cover the generalized regression estimator analogously to HORVITZ and THOMPSON's (1952) estimator is but a simple step forward.

In multistage sampling DURBIN (1953), RAJ (1968) and J. N. K. RAO's (1975a) formulae for variance estimation need expression in general for single-stage variance formulae as quadratic forms to start with, a condition violated in RAJ (1956), MURTHY (1957) and RAO, HARTLEY and COCHRAN (1962) estimators, among others. Utilizing commutativity of expectation operators in the first and later stages of sampling, new simple formulae are derived bypassing the above constraint following CHAUDHURI, ADHIKARI and DIHIDAR (2000a, 2000b).

The concepts of borrowing strength, synthetic, and empirical Bayes estimation in the context of developing small domain statistics were introduced in the first edition. Now we clarify how in two-stage sampling an estimator for the population total may be strengthened by employing empirical Bayes estimators initiated through synthetic versions of GREG estimators for the totals of the sampling clusters, which are themselves chosen with suitable unequal probabilities. A new version of cluster sampling developed by CHAUDHURI and PAL (2003) is also recounted.

S. K. THOMPSON (1992) and THOMPSON and SEBER's (1996) adaptive and network sampling techniques have been shown by CHAUDHURI (2000a) to be generally applicable for any sampling scheme in one stage or multistages with or without stratification. It is now illustrated how adaptive sampling may help the capture of rare units with appropriate network formations; vide CHAUDHURI, BOSE and GHOSH (2003).

In the first edition as well as in the text by CHAUDHURI and MUKERJEE (1988), randomized response technique to cover qualitative features was restricted to simple random sampling with replacement (SRSWR) alone. Newly emerging extension procedures to general sampling designs are now covered.

In the first edition we failed to cover SITTER's (1992a, 1992b) mirror-match and extended BWO bootstrap procedures and discussed RAO and WU's (1985, 1988) rescaled bootstrap only cursorily; we have extended coverage on them now.

Circular systematic sampling (CSS) with probability proportional to size (PPS) is known to yield zero inclusion probabilities for paired units. But this defect may now be removed on allowing a random, rather than a predetermined, sampling interval—a recent development, which we now cover. Barring these innovations and a few stylistic repairs the second edition mimics the first.

Of course, the supplementary references are added alphabetically. We continue to remain grateful to the same persons and institutions mentioned in the first edition for their sustained support.

In addition, we wish to thank Mrs. Y. CHEN for typing and organizing typesetting of the manuscript.

ARIJIT CHAUDHURI

HORST STENGER

Preface to the First Edition

Our subject of attention is a finite population with a known number of identifiable individuals, bearing values of a characteristic under study. The main problem is to estimate the population total or mean of these values by surveying a suitably chosen sample of individuals. An elaborate literature has grown over the years around various criteria for appropriate sampling designs and estimators based on selected samples so designed. We cover this literature selectively to communicate to the reader our appreciation of the current state of development of essential aspects of theory and methods of survey sampling.

Our aim is to reach graduate and advanced level students of sampling and, at the same time, researchers in the area looking for a reference book. Practitioners will be interested in many techniques of sampling that, we believe, are not adequately covered in most textbooks. We have avoided details of foundational aspects of inference in survey sampling treated in the texts by CASSEL, SÄRNDAL and WRETMAN (1977) and CHAUDHURI and VOS (1988).

In the first four chapters we state fundamental results and provide proofs of many propositions, although often leaving some of them incomplete purposely in order to save space and invite our readers to fill in the gaps themselves. We have taken care to keep the level of discussion within reach of the average graduate-level student.

The first four chapters constitute the core of the book. Although not a prerequisite, they are nevertheless helpful in giving motivations for numerous theoretical and practical problems of survey sampling dealt with in subsequent chapters, which are rather specialized and indicate several lines of approach. We have collected widely scattered materials in order to aid researchers in pursuing further studies in areas of specific interest. The coverage is mostly review in nature, leaving wide gaps to be bridged with further reading from sources cited in the References.

In chapter 1 we first formulate the problem of getting a good point estimator for a finite population total. We suppose the number of individuals is known and each unit can be assigned an identifying label. Consequently, one may choose an appropriate sample of these labels. It is assumed that unknown values can be ascertained for the individuals sampled. First we discuss the classical design-based approach of inference and present GODAMBE (1955) and GODAMBE and JOSHI's (1965) celebrated theorems on nonexistence of the best estimator of a population total. The concepts of likelihood and sufficiency and the criteria of admissibility, minimaxity, and completeness of estimators and strategies are introduced and briefly reviewed. Uses and limitations of well-known superpopulation modeling in finding serviceable sampling strategies are also discussed. But an innovation worth mentioning is the introduction of certain preliminaries on GODAMBE's (1960b) theory of estimating equations. We illustrate its application to survey sampling, bestowing optimality properties on certain sampling strategies traditionally employed ad hoc.

The second chapter gives RAO and VIJAYAN's (1977) procedure of mean square error estimation for homogeneous linear estimators and mentions several specific strategies to which it applies.

The third chapter introduces ROYALL's (1970) linear prediction approach in sampling. Here one does not speculate about what may happen if another sample is drawn with a preassigned probability. On the contrary, the inference is based on speculation on the possible nature of the finite population vector of variate values for which one may postulate plausible models. It is also shown how and why one needs to revise appropriate predictors and optimal purposive sampling designs to guard against possible mis-specifications in models and, at the same time, seek to employ robust but nonoptimal procedures that work well even when a model is inaccurately hypothesized. This illustrates how these sampling designs may be recommended when a model is correctly but simplistically postulated. Later in the chapter, Bayes estimators for finite population totals based on simplistic priors are mentioned and requirements for their replacements by empirical Bayes methods are indicated with examples. Uses of the JAMES–STEIN technique on borrowing strength from allied sources are also emphasized, especially when one has inadequate sample data specific to a given situation.

In chapter 4 we first note that if a model is correctly postulated, a design-unbiased strategy under the model may be optimal yet poorer than a comparable optimal predictive strategy. On the other hand, the optimal predictive strategy is devoid of design-based properties and modeling is difficult. Hence the importance of relaxing design-unbiasedness for the design-based strategy and replacing the optimal predictive strategy by a nonoptimal robust alternative enriched with good design properties. The two considerations lead to inevitable asymptotics. We present, therefore, contemporary activities in exploring competitive strategies that do well under correct modeling but continue to have desirable asymptotic design-based features in case of model failures. Although achieving robustness is a guiding motive in this presentation, we do not repeat here alternative robustness preserving techniques, for example, due to GODAMBE (1982). However, the asymptotic approaches for minimax sampling strategies are duly reported to cover recently emerging developments.

In chapter 5 we address the problem of mean square error estimation covering estimators and predictors and we follow procedures that originate from twin considerations of designs and models. In judging comparative efficacies of competing procedures one needs to appeal to asymptotics and extensive empirical investigations demanding Monte Carlo simulations; we have illustrated some of the relevant findings of established experts in this regard.

Chapter 6 is intended to supplement a few recent developments of topics concerning multistage, multiphase, and repetitive sampling. The time series methods applicable for a fuller treatment are not discussed.

Chapter 7 recounts a few techniques for variance estimation involving nonlinear estimators and complex survey designs including stratification, clustering, and selection in stages.

The next chapter deals with specialized techniques needed for domain estimation, poststratification, and estimation from samples taken using inadequate frames. The chapter emphasizes the necessity for conditional inference involving speculation over only those samples having some recognizable features common with the sample at hand.

Chapter 9 introduces the topic of analytic rather than descriptive studies where the center of attention is not the survey population at hand but something that lies beyond and typifies it in some discernible respect. Aspects of various methodologies needed for regression and categorical data analyses in connection with complex sampling designs are discussed as briefly as possible.

Chapter 10 includes some accounts of methods of generating randomized data and their analyses when there is a need for protected privacy relating to sensitive issues under investigation.

Chapter 11 presents several methods of analyzing survey data when there is an appreciable discrepancy between those gathered and those desired. The material presented is culled intensively from the three-volume text on incomplete data by MADOW et al. (1983) and from KALTON's (1983a,b) texts and other sources mentioned in the references.

The concluding chapter sums up our ideas about inference problems in survey sampling.

We would like to end with the following brief remarks. In employing a good sampling strategy it is important to acquire knowledge about the background of the material under investigation. In light of the background information at one's command one may postulate models characterizing some of the essential features of the population on which an inference is to be made. While employing the model one should guard against its possible incorrectness and hence be ready to take advantage of the classical design-based approach in adjusting the inference procedures. While deriving in full the virtue of design-based arguments one should also examine if appropriate conditional inference is applicable in case some cognizable features common to the given sample are discernible. This would allow averaging over them instead of over the entire set of samples.

ARIJIT CHAUDHURI gratefully acknowledges the facilities for work provided at the Virginia Polytechnic Institute and University of Mannheim as a visiting professor and the generosity of the Indian Statistical Institute in granting him the necessary leave and opportunities for joint research with his coauthor. He is also grateful to his wife, Mrs. BINATA CHAUDHURI, for her nonacademic but silent help.

HORST STENGER gratefully acknowledges the support of the Deutsche Forschungsgemeinschaft offering the opportunity of an intensive cooperation with the coauthor. His thanks also go to the Indian Statistical Institute, where joint research could be continued. In addition, he wishes to thank Mrs. R. BENT, Mrs. H. HARYANTO, and, especially, Mrs. P. URBAN, who typed the manuscript through many versions.

Comments on inaccuracies and flaws in our presentation will be appreciated and necessary corrective measures are promised for any future editions.

ARIJIT CHAUDHURI
HORST STENGER

The Authors

ARIJIT CHAUDHURI is a CSIR (Council of Scientific and Industrial Research) Emeritus Scientist and a visiting professor at the Applied Statistics Unit, Indian Statistical Institute in Kolkata, India, where he served as a professor from 1982 to 2002. He has served as a visiting professor at the Virginia Polytechnic Institute and State University, the University of Nebraska — Lincoln, the University of Mannheim, Germany and other institutes abroad. He is the chairman of the Advanced Survey Research Centre in Kolkata and a life member of the Indian Statistical Institute, the Calcutta Statistical Association, and the Indian Society of Agricultural Statistics.

Dr. CHAUDHURI holds a Ph.D. in statistics from Calcutta University, and undertook a postdoctoral fellowship for two years at the University of Sydney. He has published more than 100 research papers in numerous journals and is the coauthor of three research monographs: the first edition of the current volume (1992), *Randomized Response: Theory and Techniques* (with Rahul Mukerjee) (1988), *Unified Theory and Strategies of Survey Sampling* (with J.W.E. Vos) (1988).

HORST STENGER, professor of statistics at the University of Mannheim, Germany, has written several journal articles and two books on survey sampling, *Stichprobentheorie* (1971) and *Stichproben* (1986). He is also the coauthor of three books on general statistics, *Grundlagen der Statistik* (1978, 1979) with A. Anderson, W. Popp, M. Schaffranek, K. Szameitat; *Bevölkerungs- und Wirtschaftsstatistik* (1983) with A. Anderson, M. Schaffranek, K. Szameitat; and *Schätzen und Testen* (1976, 1997), with A. Anderson, W. Popp, M. Schaffranek, D. Steinmetz.

Dr. STENGER is a member of the International Statistical Institute, the American Statistical Association and the Deutsche Statistische Gesellschaft. He received the Dr. rer. nat. degree (1965) in mathematical statistics and the habilitation qualification (1967) in statistics from the University of Munich, Germany. From 1967 to 1971 he was professor of statistics and econometrics at the University of Göttingen, Germany. He has been a visiting professor at the Indian Statistical Institute, Calcutta.

Contents

Chapter 1

Estimation in Finite Populations: A Unified Theory

1.1 INTRODUCTION

Suppose it is considered important to gather ideas about, for example, (1) the total quantity of food grains stocked in all the godowns managed by a state government, (2) the total number of patients admitted in all the hospitals of a country classified by varieties of their complaints, (3) the amount of income tax evaded on an average by the income earners of a city. Now, to inspect all godowns, examine all admission documents of all hospitals of a country, and make inquiries about all income earners of a city will be too expensive and time consuming. So it seems natural to select a few godowns, hospitals, and income earners, to get all relevant data for them and to be able to draw conclusions on those quantities that could be ascertained exactly only by a survey of all godowns, hospitals, and income earners. We feel it is useful to formulate mathematically as follows the essentials of the issues at hand common to the above and similar circumstances.

1.2 ELEMENTARY DEFINITIONS

Let N be a known number of units, e.g., godowns, hospitals, or income earners, each assignable identifying labels $1, 2, \ldots, N$ and bearing values, respectively, $Y_1, Y_2, \ldots, Y_N$ of a real-valued variable y, which are initially unknown to an investigator who intends to estimate the total

$$Y = \sum_1^N Y_i$$

or the mean $\overline{Y} = Y/N$.

We call the sequence $U = (1, \ldots, N)$ of labels a **population**. Selecting units leads to a sequence $s = (i_1, \ldots, i_n)$, which is called a **sample**. Here $i_1, \ldots, i_n$ are elements of U, not necessarily distinct from one another but the **order of its appearance** is maintained. We refer to $n = n(s)$ as the **size** of s, while the **effective sample size** $\nu(s) = |s|$ is the cardinality of s, i.e., the number of distinct units in s. Once a specific sample s is chosen we suppose it is possible to ascertain the values $Y_{i_1}, \ldots, Y_{i_n}$ of y associated with the respective units of s. Then

$$d = [(i_1, Y_{i_1}), \ldots, (i_n, Y_{i_n})] \quad \text{or briefly}$$
$$d = [(i, Y_i) | i \in s]$$

constitutes the **survey data**.

An **estimator** t is a real-valued function $t(d)$, which is free of Y_i for $i \notin s$ but may involve Y_i for $i \in s$. Sometimes we will express $t(d)$ alternatively by $t(s, \underline{Y})$, where $\underline{Y} = (Y_1, \ldots, Y_N)'$.

An estimator of special importance for $\overline{Y}$ is the **sample mean**

$$t(s, \underline{Y}) = \frac{1}{n(s)} \sum_{i=1}^N f_{si} Y_i = \overline{y}, \text{ say}$$

where f_{si} denotes the frequency of i in s such that

$$\sum_{i=1}^N f_{si} = n(s).$$

$N\overline{y}$ is called the **expansion estimator** for Y.

More generally, an estimator t of the form

$$t(s, \underline{Y}) = b_s + \sum_{i=1}^{N} b_{si} Y_i$$

with $b_{si} = 0$ for $i \notin s$ is called **linear** (L). Here b_s and b_{si} are free of $\underline{Y}$. Keeping $b_s = 0$ we obtain a **homogeneous linear** (HL) estimator.

We must emphasize that here $t(\underline{s}, \underline{Y})$ is linear (or homogeneous linear) in $Y_i, i \in s$. It may be a nonlinear function of two random variables, e.g., when $b_s = 0$ and $b_{si} = X / \Sigma_1^N f_{si} X_i$ so that

$$t(s, \underline{Y}) = \frac{\sum_1^N f_{si} Y_i}{\sum_1^N f_{si} X_i} X.$$

Here, X_i is the value of a variable x on $i \in \mathcal{U}$ and $X = \Sigma_1^N X_i$ (see section 2.2.)

In what follows we will assume that a sample is drawn at **random**, i.e., with each sample s is associated a selection probability $p(s)$. A **design** p may depend on related variables x, z, etc. But we assume, unless explicitly mentioned otherwise, that p is free of $\underline{Y}$. To emphasize this freedom, p is often referred to in the literature as a **noninformative design**.

If p involves any component of $\underline{Y}$ it is an **informative design**.

A design p is **without replacement** (WOR) if no repetitions occur in any s with $p(s) > 0$; otherwise, p is called **with replacement** (WR). A design p is of **fixed size** n (**fixed effective size** n) if $p(s) > 0$ implies that s is of size n (of effective size n). With respect to WOR designs there is, of course, no difference between fixed size and fixed effective size.

A design p is called **simple random sampling without replacement** (SRSWOR) if

$$p(s) = \frac{1}{\binom{N}{n} n!}$$

for s of size n without repetitions, while it is called **simple random sampling with replacement** (SRSWR) if

$$p(s) = \frac{1}{N^n}$$

for every s of size n, n fixed in advance.

The combination (p, t) denoting an estimator t based on s chosen according to a design p is called a **strategy**. Sometimes a redundant epithet **sampling** is used before design and strategy but we will avoid this usage.

Whatever $\underline{Y}$ may be, let

$$E_p(t) = \sum_s t(s, \underline{Y})p(s)$$

denote the **expectation** of t and

$$M_p(t) = E_p(t - Y)^2 = \sum_s p(s)(t(s, \underline{Y}) - Y)^2$$

the **mean square error** (MSE) of t. If $E_p(t) = Y$ for every $\underline{Y}$, then t is called a p**-unbiased estimator** (UE) of Y. In this case $M_p(t)$ becomes the **variance** of t and is written

$$V_p(t) = E_p(t - E_p(t))^2.$$

For an arbitrary design p, consider the **inclusion probabilities**

$$\pi_i = \sum_{s \ni i} p(s)\,;\ i = 1, 2, \ldots, N$$

$$\pi_{ij} = \sum_{s \ni i,j} p(s)\,;\ i \neq j = 1, 2, \ldots, N$$

and, provided $\pi_1, \pi_2, \ldots, \pi_N > 0$, the **Horvitz–Thompson** (HT) **estimator** (HTE)

$$\bar{t} = \sum_{i \in s} \frac{Y_i}{\pi_i}$$

(see HORVITZ and THOMPSON, 1952) where the sum is over $|s|$ terms while s is of length $n(s)$. It is easily seen that $\bar{t}$ is HL and p-unbiased (HLU) for Y.

REMARK 1.1 *To mention another way to write $\bar{t}$ define*

$$I_{si} = \begin{cases} 1 & \text{if} \quad i \in s \\ 0 & \text{if} \quad i \notin s \end{cases}$$

for $i = 1, 2, \ldots, N$. Then

$$\bar{t} = \bar{t}(s, \underline{Y}) = \sum_{i=1}^{N} I_{si} \frac{Y_i}{\pi_i}.$$

where the sum is over $i = 1, 2, \ldots, N$

REMARK 1.2 *Assume $i_0 \in U$ exists with $\pi_{i_0} = 0$ for a design p. Then, for an estimator t*

$$E_p t = \sum_{s \ni i_0} p(s) t(s, \underline{Y}) + \sum_{s \not\ni i_0} p(s) t(s, \underline{Y}).$$

The second term on the right of this equation is obviously free of Y_{i_0}. Since $p(s) = 0$ for all s with $i_0 \in s$, the first term is 0. Hence, $E_p t$ is free of Y_{i_0} and, especially, not equal to $Y = \Sigma_1^N Y_i$. Consequently, no p-unbiased estimator exists.

1.3 DESIGN-BASED INFERENCE

Let Σ_1 be the sum over samples for which $|t(s, \underline{Y}) - Y| \geq k > 0$ and let Σ_2 be the sum over samples for which $|t(s, \underline{Y}) - Y| < k$ for a fixed $\underline{Y}$. Then from

$$\begin{aligned} M_p(t) &= \Sigma_1 p(s)(t - Y)^2 + \Sigma_2 p(s)(t - Y)^2 \\ &\geq k^2 \text{Prob}\left[|t(s, \underline{Y}) - Y| \geq k\right] \end{aligned}$$

one derives the **Chebyshev inequality**:

$$\text{Prob}[|t(s, \underline{Y}) - Y| \geq k] \leq \frac{M_p(t)}{k^2}.$$

Hence

$$\text{Prob}[t - k \leq Y \leq t + k] \geq 1 - \frac{M_p(t)}{k^2} = 1 - \frac{1}{k^2}\left[V_p(t) + B_p^2(t)\right]$$

where $B_p(t) = E_p(t) - Y$ is the **bias** of t. Writing $\sigma_p(t) = \sqrt{V_p(t)}$ for the standard error of t and taking $k = 3\sigma_p(t)$, it follows that, whatever $\underline{Y}$ may be, the random interval $t \pm 3\sigma_p(t)$

covers the unknown Y with a probability not less than

$$\frac{8}{9} - \frac{1}{9}\frac{B_p^2(t)}{V_p(t)}.$$

So, to keep this probability high and the length of this covering interval small it is desirable that both $|B_p(t)|$ and $\sigma_p(t)$ be small, leading to a small $M_p(t)$ as well.

EXAMPLE 1.1 *Let y be a variable with values 0 and 1 only. Then, as a consequence of $Y_i^2 = Y_i$,*

$$\begin{aligned}\sigma_{yy} &= \frac{1}{N}\sum (Y_i - \overline{Y})^2 \\ &= \overline{Y}(1-\overline{Y}) \leq \frac{1}{4}.\end{aligned}$$

Therefore, with p SRSWR of size n,

$$\begin{aligned}V_p(N\overline{y}) &= N^2\tfrac{\sigma_{yy}}{n} \\ &\leq \tfrac{N^2}{4n}.\end{aligned}$$

From

$$E_p\,\overline{y} = \overline{Y}$$

we derive that the random interval

$$N\,\overline{y} \pm 3\sqrt{N^2\frac{1}{4n}} = N\left[\overline{y} \pm \frac{3}{2\sqrt{n}}\right]$$

covers the unknown $N\overline{Y}$ with a probability of at least 8/9.

It may be noted that $\underline{Y}$ is regarded as fixed (nonstochastic) and s is a random variable with a probability distribution $p(s)$ that the investigator adopts at pleasure. It is through p alone that for a fixed $\underline{Y}$ the interval $t \pm 3\sigma_p(t)$ is a random interval. In practice an upper bound of $\sigma_p(t)$ may be available, as in the above example, or $\sigma_p(t)$ is estimated from survey data d plus auxiliary information by, for example, $\hat{\sigma}_p(t)$ inducing necessary changes in the above confidence statements.

If $|B_t(t)|$ is small, then we may argue that the average value of t over repeated sampling according to p is numerically close to Y and, if $M_p(t)$ is small, then we may say that the average square error $E_p(t - Y)^2$ calculated over repeated sampling according to p is small.

Let us stress this point more fully. The parameter to be estimated may be written as $Y = \Sigma_s Y_i + \Sigma_r Y_i$, the sums being over the distinct units sampled and the remaining units of U, respectively. Its estimator is

$$t = \sum_s Y_i + \left(t - \sum_s Y_i \right).$$

Now, t is close to Y for a sample s at hand and the realized survey data $d = (i, Y_i \mid i \in s)$ if and only if $(t - \Sigma_s Y_i)$ is close to $\Sigma_r Y_i$, the first expression depending on Y_i for $i \in s$ and the second determined by Y_j for $j \notin s$. Now, so far we permit $\underline{Y}$ to be any vector of real numbers without any restrictions on the structural relationships among its coordinates. In this **fixed population setup** we have no way to claim or disclaim the required closeness of $(t - \Sigma_s Y_i)$ and $\Sigma_r Y_i$ for a given sample s. But we need a link between Y_i for $i \in s$ and Y_j for $j \notin s$ in order to provide a base on which our inference about Y from realized data d may stand. Such a link is established by the hypothesis of **repeated sampling**. The resulting **design-based** (briefly: p-based) theory following NEYMAN (1934) is developed around the faith that it is desirable and satisfactory to assess the performance of the strategy (p, t) over repeated sampling, even if in practice a sample will really be drawn once, yielding a single value for t.

This theory is unified in the sense that the performance of a strategy (p, t) is evaluated in terms of the characteristics $E_p(t)$ and $M_p(t)$, such that there is no need to refer to specific selection procedures.

1.4 SAMPLING SCHEMES

A unified theory is developed by noting that it is enough to establish results concerning (p, t) without heeding how one may actually succeed in choosing samples with preassigned probabilities. A method of choosing a sample draw by draw, assigning selection probabilities with each draw, is called a

sampling scheme. Following HANURAV (1966), we show below that starting with an arbitrary design we may construct a sampling scheme.

Suppose for each possible sample s from U the selection probability $p(s)$ is fixed. Let

$$\begin{aligned} \beta_{i1} &= p(i_1), & \beta_{i_1,i_2} &= p(i_1, i_2), \ldots, & \beta_{i_1,\ldots,i_n} &= p(i_1, \ldots, i_n) \\ \alpha_{i1} &= \Sigma_1 p(s), & \alpha_{i_1,i_2} &= \Sigma_2 p(s), \ldots, & \alpha_{i_1,\ldots,i_n} &= \Sigma_n p(s) \end{aligned}$$

where Σ_1 is the sum over all samples s with i_1 as the first entry; Σ_2 is the sum over all samples with i_1, i_2, respectively, as the first and second entries in $s, \ldots,$ and Σ_n is the sum over all samples of which the first, second, ..., nth entries are, respectively, $i_1, i_2, \ldots, i_n$.

Then, let us consider the scheme of selection such that on the first draw from U, i_1 is chosen with probability α_{i1}, a second draw from U is made with probability

$$\left(1 - \frac{\beta_{i1}}{\alpha_{i1}}\right).$$

On the second draw from U the unit i_2 is chosen with probability

$$\frac{\alpha_{i_1,i_2}}{\alpha_{i1} - \beta_{i1}}.$$

A third draw is made from U with probability

$$\left(1 - \frac{\beta_{i_1,i_2}}{\alpha_{i_1,i_2}}\right).$$

On the third draw from U the unit i_3 is chosen with probability

$$\frac{\alpha_{i_1,i_2,i_3}}{\alpha_{i_1,i_2} - \beta_{i_1,i_2}}$$

and so on. Finally, after the nth draw the sampling is terminated with a probability

$$\frac{\beta_{i_1,i_2,\ldots,i_n}}{\alpha_{i_1,\ldots,i_n}}.$$

For this scheme, then, $s = (i_1, \dots, i_n)$ is chosen with a probability

$$p(s) = \alpha_{i_1}\left(1 - \frac{\beta_{i_1}}{\alpha_{i_1}}\right)\frac{\alpha_{i_1,i_2}}{\alpha_{i_1} - \beta_{i_1}}\left(1 - \frac{\beta_{i_1,i_2}}{\alpha_{i_1,i_2}}\right)\cdots\frac{\alpha_{i_1,\dots,i_{n-1}}}{\alpha_{i_1,\dots,i_{n-2}} - \beta_{i_1,\dots,i_{n-2}}}$$
$$\times\left(1 - \frac{\beta_{i_1,\dots,i_{n-1}}}{\alpha_{i_1,\dots,i_{n-1}}}\right)\frac{\alpha_{i_1,\dots,i_n}}{\alpha_{i_1,\dots,i_{n-1}} - \beta_{i_1,\dots,i_{n-1}}}\left(\frac{\beta_{i_1,\dots,i_n}}{\alpha_{i_1,\dots,i_n}}\right)$$
$$= \beta_{i_1,\dots,i_n}$$

as it should be.

1.5 CONTROLLED SAMPLING

EXAMPLE 1.2 *Consider the population* $U = (1, 2, \dots, 9)$ *and the SRSWOR design of size* $n = 3$, p, *with the inclusion probabilities*

$$\pi_i = 1/3 \quad \textit{for} \quad i = 1, 2, \dots, 9$$
$$\pi_{ij} = 1/12 \quad \textit{for} \quad i \neq j.$$

Define

$$q(s) = 1/12$$

if s is equal to one of the following samples

(1,2,3)	(1,6,8)
(4,5,6)	(2,4,9)
(7,8,9)	(3,5,7)
(1,4,7)	(1,5,9)
(2,5,8)	(2,6,7)
(3,6,9)	(3,4,8)

and $q(s) = 0$ otherwise. Then q obviously is a design with the same inclusion probabilities as p. For the sample mean $\overline{y}$, which, as a consequence of $\pi_i = 1/3$ for all i, is identical with the HTE, we therefore have

$$E_p\overline{y} = E_q\overline{y}$$
$$V_p\overline{y} = V_q\overline{y}$$

that is, the performance characteristics of the sample mean do not change when p is replaced by q.

Now, consider an arbitrary design p of fixed size n and a linear estimator t; suppose a subset S_0 of all samples is less desirable from practical considerations like geographical location, inaccessibility, or, more generally, costliness. Then, it is advantageous to replace design p by a modified one, for example, q, which attaches minimal values $q(s)$ to the samples s in S_0 keeping

$$E_p(t) = E_q(t)$$
$$E_p(t - Y)^2 = E_q(t - Y)^2$$

and even maintaining other desirable properties of p, if any. A resulting q is called a **controlled design** and a corresponding scheme of selection is called a **controlled sampling scheme**. Quite a sizeable literature has grown around this problem of finding appropriate controlled designs. The methods of implementing such a scheme utilize theories of incomplete block designs and predominantly involve ingeneous devices of reducing the size of support of possible samples demanding trials and errors. But RAO and NIGAM (1990) have recently presented a simple solution by posing it as a linear programming problem and applying the well-known simplex algorithm to demonstrate their ability to work out suitable controlled schemes.

Taking t as the HORVITZ–THOMPSON estimator $\bar{t} = \sum_{i \in s} Y_i/\pi_i$, they minimize the objective function $F = \sum_{s \in S_0} q(s)$ subject to the linear constraints

$$\sum_{s \ni i,j} q(s) = \sum_{s \ni i,j} p(s) = \pi_{ij}$$
$$q(s) \geq 0 \quad \text{for all} \quad s$$

where π_{ij}'s are known quantities in terms of the original **uncontrolled** design p.

Chapter 2

Strategies Depending on Auxiliary Variables

Besides y there may be related variables $x, z, \ldots$, called **auxiliary variables**, with values

$$X_1, X_2, \ldots, X_N;\ Z_1, Z_2, \ldots, Z_N; \ldots$$

respectively, for the units of U. These values may be partly or fully known to the investigator; if the values of an auxiliary variable are positive, this variable may be called a **size measure** of the units of U.

In the present chapter we discuss a few strategies of interest in theory and practice. They are based on the knowledge of a size measure and are **representative**, in a sense to be explained, with respect to this measure. Unbiased estimation of the mean square error of these strategies is of special importance. A general method of estimation is presented in section 2.3. Applications to examples of representative strategies (which are less essential for later chapters) are considered in section 2.4.

2.1 REPRESENTATIVE STRATEGIES

Let p be a design. Consider a size measure x and assume that, approximately,

$$Y_i \propto X_i.$$

Then it seems natural to look for an estimator

$$t = \sum_{i=1}^{N} b_{si} Y_i$$

with $b_{si} = 0$ for $i \notin s$, such that

$$\sum_{i=1}^{N} b_{si} X_i = X$$

for all s with $p(s) > 0$. With reference to HÁJEK (1959), a strategy with this property is called **representative** with respect to $\underline{X} = (X_1, X_2, \ldots, X_N)'$.

For the mean square error (MSE) of a strategy (p, t) we have

$$\begin{aligned} M_p(t) &= E_p(t - Y)^2 \\ &= E_p\left(\sum Y_i(b_{si} - 1)\right)^2 \\ &= \sum_i \sum_j Y_i Y_j d_{ij} \end{aligned}$$

where

$$d_{ij} = E_p(b_{si} - 1)(b_{sj} - 1).$$

A strategy (p, t) is representative if and only if there exists a vector $\underline{X} = (X_1, X_2, \ldots, X_N)'$ such that $M_p(t) = 0$ for $Y_i \propto X_i$ implying

$$\sum_i \sum_j X_i X_j d_{ij} = 0.$$

It may be advisable to use strategies that are representative with respect to several auxiliary variables $x_1, x_2, \ldots, x_K$. Let

$$\underline{x}_i = (X_{i1}, X_{i2}, \ldots, X_{iK})'$$

be the vector of values of these variables for unit i and write

$$\underline{X}_1 = (X_{11}, X_{21}, \ldots, X_{N1})'$$
$$\vdots$$
$$\underline{X}_K = (X_{1K}, X_{2K}, \ldots, X_{NK})'.$$

A strategy (p, t) is representative with respect to $\underline{X}_k$; $k = 1, \ldots, K$ if $p(s) > 0$ implies

$$\sum_{i=1}^{N} b_{si} X_{ik} = \sum_{i=1}^{N} X_{ik}$$

for $k = 1, \ldots, K$, which may be written as

$$\sum_{i=1}^{N} b_{si} \underline{x}_i = \sum_{i=1}^{N} \underline{x}_i .$$

This equation is often called a **calibration equation**.

In sections 2.2, 2.3, and 2.4 we deal with representativity for $K = 1$. In section 2.5 this restriction is dropped and the concept of calibration is introduced.

2.2 EXAMPLES OF REPRESENTATIVE STRATEGIES

The **ratio estimator**

$$t_1 = X \frac{\sum_{i \in s} Y_i}{\sum_{i \in s} X_i}$$

is of special importance because of its traditional use in practice. Here, (p, t_1) is obviously representative with respect to a size measure x, more precisely to $(X_1, \ldots, X_N)$, whatever the sampling design p.

Note, however, that t_1 is usually combined with SRSWOR or SRSWR. The sampling scheme of LAHIRI–MIDZUNO–SEN (LAHIRI, 1951; MIDZUNO, 1952; SEN, 1953) (LMS) yields a design of interest to be employed in conjunction with t_1 by rendering it design unbiased.

The **Hansen–Hurwitz** (HH, 1943) **estimator** (HHE)

$$t_2 = \frac{1}{n} \sum_{i=1}^{N} f_{si} \frac{Y_i}{P_i},$$

with f_{si} as the frequency of i in s, $i \in \mathcal{U}$, combined with any design p, gives rise to a strategy representative with respect to $(P_1, \ldots, P_N)'$. For the sake of design unbiasedness, t_2 is usually based on probability proportional to size (PPS) with replacement (PPSWR) sampling, that is, a scheme that consists of n independent draws, each draw selecting unit i with probability P_i.

Another representative strategy is due to RAO, HARTLEY and COCHRAN (RHC, 1962). We first describe the sampling scheme as follows: On choosing a sample size n, the population $\mathcal{U}$ is split at random into n mutually exclusive groups of sizes suitably chosen $N_i (i = 1, \ldots, n; \sum_1^n N_i = N)$ coextensive with $\mathcal{U}$, the units bearing values P_i, the normed sizes $(0 < P_i < 1, \sum P_i = 1)$. From each of the n groups so formed independently one unit is selected with a probability proportional to its size given the units falling in the respective groups. Writing P_{ij} for the jth unit in the ith group,

$$Q_i = \sum_{i=1}^{N_i} P_{ij},$$

the selection probability of j is P_{ij}/Q_i. For simplicity, suppressing j to mean by P_i the P value for the unit chosen from the ith group, the **Rao-Hartley-Cochran estimator** (RHCE)

$$t_3 = \sum_{i=1}^{n} Y_i \frac{Q_i}{P_i},$$

writing Y_i for the y value of the unit chosen from the ith group $(i = 1, 2, \ldots, n)$. This strategy is representative with respect to $\underline{P} = (P_1, \ldots, P_N)'$ because $\Sigma_1^n Q_i = 1$.

Murthy's (1957) **estimator**

$$t_4 = \frac{1}{p(s)} \sum_{i \in s} Y_i \, p(s \mid i)$$

is based on a design p and a sampling scheme for which $p(s \mid i)$ is the conditional probability of choosing s given that i was chosen on the first draw. If P_i is the probability to select unit i

on the first draw we have

$$p(s) = \sum_{i=1}^{N} P_i \, p(s \mid i), \; \sum_{i=1}^{N} P_i = 1.$$

It is evident that the strategy so defined is representative with respect to $(P_1, P_2, \ldots, P_N)$.

2.3 ESTIMATION OF THE MEAN SQUARE ERROR

Let (p, t) be a strategy with

$$t = \sum_{i=1}^{N} b_{si} Y_i$$

where b_{si} is free of $\underline{Y} = (Y_1, \ldots, Y_N)'$ and $b_{si} = 0$ for $i \notin s$. Then, the mean square error may be written as

$$\begin{aligned} M_p(t) &= E_p\left[\sum Y_i(b_{si} - 1)\right]^2 \\ &= \sum_{i=1}^{N} \sum_{j=1}^{N} Y_i Y_j d_{ij} \end{aligned}$$

with

$$d_{ij} = E_p(b_{si} - 1)(b_{sj} - 1).$$

Let (p, t) be representative with respect to a given vector $\underline{X} = (X_1, \ldots, X_N)'$, $X_i > 0$, $i \in U$. Then, writing

$$Z_i = \frac{Y_i}{X_i}$$

we get

$$M_p(t) = \sum \sum Z_i Z_j (X_i X_j \, d_{ij})$$

such that

$$\sum_i \sum_j X_i X_j \; d_{ij} = 0.$$

Define $a_{ij} = X_i X_j \, d_{ij}$. Then

$$M_p(t) = \sum \sum Z_i Z_j \, a_{ij}$$

is a non-negative quadratic form in Z_i; $i = 1, \ldots, N$ subject to

$$\sum_i \sum_j a_{ij} = 0.$$

This implies for every $i = 1, \ldots, N$

$$\sum_j a_{ij} = 0.$$

From this $M_p(t) = \sum \sum Z_i Z_j a_{ij}$ may be written in the form

$$\begin{aligned} M_p(t) &= -\sum \sum_{i<j} (Z_i - Z_j)^2 a_{ij} \\ &= -\sum \sum_{i<j} \left(\frac{Y_i}{X_i} - \frac{Y_j}{X_j} \right)^2 X_i X_j d_{ij}. \end{aligned}$$

This property of a representative strategy leads to an unbiased quadratic estimator for $M_p(t)$, an estimator that is non-negative, uniformly in $\underline{Y}$, if such an estimator does exist. This may be shown as follows.

Let

$$m_p(t) = \sum_{i=1}^{N} \sum_{j=1}^{N} Y_i Y_j \, d_{sij}$$

be a quadratic unbiased estimator for $M_p(t)$ with d_{sij} free of $\underline{Y}$ and $d_{sij} = 0$ unless $i \in s$ and $j \in s$. Then

$$\sum_1^N \sum_1^N Y_i Y_j \, d_{ij} = \sum_s p(s) \left[\sum_1^N \sum_1^N Y_i Y_j \, d_{sij} \right]$$

or

$$\sum_1^N \sum_1^N Z_i Z_j \, X_i X_j \, d_{ij} = \sum_s p(s) \left[\sum_1^N \sum_1^N Z_i Z_j \, X_i X_j \, d_{sij} \right].$$

If $m_p(t)$ is to be uniformly non-negative, then for every s with $p(s) > 0$

$$\sum_i^N \sum_1^N X_i X_j \, d_{sij}$$

must be a uniformly non-negative quadratic form subject to

$$\sum_1^N \sum_1^N X_i X_j d_{sij} = 0$$

because $\sum_i^N \sum_1^N X_i X_j \, d_{ij} = 0$. Therefore, $m_p(t)$ is necessarily of the form

$$m_p(t) = -\sum\sum_{i<j} \left(\frac{Y_i}{X_i} - \frac{Y_j}{X_j} \right)^2 X_i X_j \, d_{sij}.$$

RESULT 2.1 *Let the strategy (p, t) be representative with respect to $\underline{X} = (X_1, X_2, \ldots, X_N)'$ and assume $\hat{M}$ is a uniformly non-negative quadratic function in Y_i, $i \in s$ such that*

$$E_p \hat{M} = M_p(t) .$$

Then, $\hat{M}$ must be of the form

$$\hat{M} = -\sum\sum_{i<j} \left(\frac{Y_i}{X_i} - \frac{Y_j}{X_j} \right)^2 X_i X_j \, d_{sij}$$

where $d_{sij} = 0$ unless $i \in s$ and $j \in s$.

REMARK 2.1 *Even if representativity does not hold for a strategy (p, t)*

$$M = \sum_i \sum_j Y_i Y_j \, d_{ij} = \sum_i Y_i^2 \, d_{ii} + \sum\sum_{i \neq j} Y_i Y_j \, d_{ij}$$

may be estimated unbiasedly, for example, by

$$m = \sum_i Y_i^2 \, d_{ii} \frac{I_{si}}{\pi_i} + \sum\sum_{i \neq j} Y_i Y_j \, d_{ij} \frac{I_{sij}}{\pi_{ij}},$$

where $I_{sij} = I_{si} I_{sj}$, provided $\pi_{ij} > 0$ for all $i \neq j$ and hence $\pi_i > 0$ for all i. But, in order that this may be uniformly non-negative, we have to ensure that d_{ij}, π_{ij}'s are so chosen as to make m a non-negative definite quadratic form, which is not easy to achieve. CHAUDHURI *and* PAL (2002) *have given the following simple solution to get over this trouble. For $X_i \neq 0$, $i \in U$ they* *define*

$$\beta_i = \sum_{j=1}^{N} d_{ij} X_j$$

and show

$$M = -\sum\sum_{1\le i< j\le N} X_i X_j d_{ij} \left(\frac{Y_i}{X_i} - \frac{Y_j}{X_j}\right)^2 + \sum_i \frac{Y_i^2}{X_i}\beta_i.$$

Consequently, they propose

$$m' = -\sum\sum_{1\le i< j\le N} X_i X_j d_{ij} \frac{I_{sij}}{\pi_{ij}} \left(\frac{Y_i}{X_i} - \frac{Y_j}{X_j}\right)^2 + \sum \frac{Y_i^2}{X_i}\beta_i \frac{I_{sij}}{\pi_i}$$

as an unbiased estimator for M above.

2.4 ESTIMATION OF $M_P(T)$ FOR SPECIFIC STRATEGIES

2.4.1 Ratio Strategy

Utilizing the theory thus developed by RAO and VIJAYAN (1977) and RAO (1979), one may write down the exact MSE of the ratio estimator t_1 about Y if t_1 is based on SRSWOR in n draws as

$$M = -\sum\sum_{1\le i<j\le N} \left[\frac{Y_i}{X_i} - \frac{Y_j}{X_j}\right]^2 \frac{X_i X_j}{\binom{N}{n}}$$
$$\times \left[X^2 \sum_{s\ni i,j} \frac{1}{(\sum_{i\in s} X_i)^2} - X \sum_{s\ni i} \frac{1}{(\sum_{i\in s} X_i)}\right.$$
$$\left. - X \sum_{s\ni j} \frac{1}{(\sum_{i\in s} X_i)} + \binom{N}{n}\right]$$

because

$$t_1 = X\left[\sum_{i\in s} Y_i\right] \Big/ \left[\sum_{i\in s} X_i\right] = \sum_1^N Y_i b_{si} I_{si} \quad \text{with} \quad b_{si} = \frac{X}{\sum_{i\in s} X_i}$$

has

$$\begin{aligned} d_{ij} &= E_p(b_{si}I_{si} - 1)(b_{sj}I_{sj} - 1) \\ &= \frac{1}{\binom{N}{n}}\left[X^2 \sum_{s\ni i,j} \frac{1}{(\sum_{i\in s} X_i)^2} - X\sum_{s\ni i}\frac{1}{(\sum_{i\in s} X_i)} \right. \\ &\quad \left. - X\sum_{s\ni j}\frac{1}{(\sum_{i\in s} X_i)} + \binom{N}{n}\right] \\ &= B_{ij}, \text{ say.} \end{aligned}$$

Writing

$$a_{ij} = X_i X_j \left[\frac{Y_i}{X_i} - \frac{Y_j}{X_j}\right]^2$$

we have

$$M = -\sum\sum_{i<j} a_{ij} B_{ij}.$$

Since for SRSWOR, $\pi_{ij} = \frac{n(n-1)}{N(N-1)}$ for every i, j $(i \neq j)$ an obvious uniformly non-negative quadratic unbiased estimator for M is

$$\hat{M} = -\frac{N(N-1)}{n(n-1)} \sum\sum_{i<j} a_{ij} B_{ij} I_{sij}.$$

It is important to observe that M and $\hat{M}$ are exact formulae, unlike the approximations

$$M' = N\frac{N-n}{N-1}\frac{1}{n}\sum_{1}^{N}(Y_i - RX_i)^2$$

$$\hat{M}' = N\frac{N(N-n)}{n(n-1)}\sum_{i\in s}(Y_i - \hat{R}X_i)^2$$

where $R = Y/X, \hat{R} = \overline{y}/\overline{x}$ and

$$\overline{y} = \frac{1}{n}\sum_{i\in s} Y_i, \overline{x} = \frac{1}{n}\sum_{i\in s} X_i$$

due to COCHRAN (1977). For the approximations n is required to be large and N much larger than n. These formulae are, however, much simpler than M and $\hat{M}$ because B_{ij} is very hard to calculate even if X_i is known for every $i = 1, \ldots, N$. To use $\hat{M}'$ it is enough to know only X_i for $i \in s$, but to use $\hat{M}$ one must know X_i for $i \notin s$ as well.

2.4.2 Hansen–Hurwitz Strategy

For the HANSEN–HURWITZ estimator t_2, which is unbiased for Y, when based on PPSWR sampling, the variance is well known to be

$$V_2 = M = \frac{1}{n}\left[\sum_1^N \frac{Y_i^2}{P_i} - Y^2\right]$$

$$= \frac{1}{n}\sum P_i \left[\frac{Y_i}{P_i} - Y\right]^2$$

$$= \frac{1}{n}\sum_{i<j}\sum P_i P_j \left[\frac{Y_i}{P_i} - \frac{Y_j}{P_j}\right]^2$$

admitting a well-known non-negative estimator

$$v_2 = \frac{1}{n^2(n-1)}\sum_{r<r'}\sum \left[\frac{y_r}{p_r} - \frac{y_{r'}}{p_{r'}}\right]^2$$

$$= \frac{1}{n(n-1)}\sum_{r=1}^n \left[\frac{y_r}{p_r} - t_2\right]^2$$

where y_r is the y value of the unit drawn in the rth place, while p_r is the probability of this unit to be drawn.

2.4.3 RHC Strategy

Again, the RHC estimator t_3 (see section 2.2) is unbiased for Y because writing E_C as the expectation operator, given the condition that the groups are already formed and E_G as the expectation operator over the formation of the groups, we have

$$E_C(t_3) = \sum_1^n \left[\sum_{j=1}^{N_i} Y_j \frac{Q_i}{P_{ij}}\frac{P_{ij}}{Q_i}\right] = \sum_1^n \sum_1^{N_i} Y_j = Y$$

and hence $E_p(t_3) = E_G[E_C(t_3)] = E_G(Y) = Y$. Also, writing V_C, V_G as operators for variance corresponding to E_C, E_G, respectively, we have

$$M = V_p(t_3) = E_G[V_C(t_3)] + V_G[E_C(t_3)]$$

$$= E_G\left[\sum_1^n \sum_{1\le j<k\le N_i} \frac{P_{ij}}{Q_i}\frac{P_{ik}}{Q_i}\left(\frac{Y_{ij}Q_i}{P_{ij}} - \frac{Y_{ik}Q_i}{P_{ik}}\right)^2\right]$$

$$= E_G\sum_1^n\left[\sum_{1\le j<k\le N_i} P_{ij}P_{ik}\left(\frac{Y_{ij}}{P_{ij}} - \frac{Y_{ik}}{P_{ik}}\right)^2\right]$$

$$= \sum_1^n\left[\frac{N_i(N_i-1)}{N(N-1)}\sum_{1\le j<k\le N} P_jP_k\left(\frac{Y_j}{P_j} - \frac{Y_k}{P_k}\right)^2\right]$$

$$= \frac{\sum_1^n N_i^2 - N}{N(N-1)}\sum_{1\le j<k\le N} P_jP_k\left(\frac{Y_j}{P_j} - \frac{Y_k}{P_k}\right)^2 = V_3.$$

By Cauchy's inequality, $n\sum_1^n N_i^2 \ge (\Sigma N_i)^2 = N^2$, hence $\sum_1^n N_i^2 \ge \frac{N^2}{n}$ and $\sum_1^n N_i^2$ is minimal if $N_i = \frac{N}{n}$ for all i provided, as assumed here, N/n is an integer. Then, t_3 has the minimal variance

$$V_p(t_3) = \frac{N-n}{(N-1)n}\sum_{1\le j<k\le N} P_iP_j\left[\frac{Y_i}{P_i} - \frac{Y_j}{P_j}\right]^2 = \frac{N-n}{N-1}V_2.$$

If $\frac{N}{n} = 1/f$ is not an integer, then to minimize $\Sigma_1^n N_i^2$ and equivalently to minimize V_3 one should take $k(<n)$ of the N_i's as equal to $[\frac{N}{n}]$ and the $(n-k)$ remaining of them equal to $[\frac{N}{n}]+1$ with k so chosen that $\sum_1^n N_i = N$. By $[x]$ we denote the largest integer not exceeding $x > 0$.

RHC have themselves given a uniformly non-negative unbiased estimator for V_3 as v_3 derived as below. Let v_3 be such that $E_p(v_3) = V_3$ and let

$$e = \sum_{i=1}^n \frac{Y_{ij}^2}{P_{ij}^2}Q_i.$$

Then, $E_p(t_3^2 - v_3) = Y^2$. Also,

$$E_p(e) = E_G\left[\sum_1^n \left(\sum_1^{N_i} \frac{Y_{ij}^2}{P_{ij}^2} Q_i \frac{P_{ij}}{Q_i}\right)\right]$$

$$= E_G\left[\sum_1^n \left(\sum_1^{N_i} \frac{Y_{ij}^2}{P_{ij}}\right)\right] = \sum_1^N \frac{Y_i^2}{P_i}.$$

Writing

$$V = \sum_1^N \frac{Y_i^2}{P_i} - Y^2, \qquad V_3 = \frac{\sum N_i^2 - N}{N(N-1)} V$$

an unbiased estimator for V is $e - (t_3^2 - v_3)$. So

$$\frac{\sum N_i^2 - N}{N(N-1)} E_p(e - t_3^2 + v_3) = V_3 = E_p(v_3)$$

or

$$\frac{\sum N_i^2 - N}{N(N-1)} E_p(e - t_3^2) = \left[1 - \frac{\sum N_i^2 - N}{N(N-1)}\right] E_p(v_3).$$

So

$$\frac{\sum N_i^2 - N}{N^2 - \sum N_i^2}(e - t_3^2)$$

is an unbiased estimator for V_3. This may be written as

$$v_3 = \left[\frac{\sum N_i^2 - N}{N^2 - \sum N_i^2}\right]\left[\sum_{i=1}^n \frac{Y_{ij}^2}{P_{ij}^2} Q_i - t_3^2\right]$$

$$= \frac{\sum N_i^2 - N}{N^2 - \sum N_i^2} \sum_1^n \left[\frac{Y_{ij}}{P_{ij}} - t_3\right]^2 Q_i$$

and taken as a uniformly non-negative unbiased estimator for V_3. These results are all given by RHC (1962).

REMARK 2.2 OHLSSON *(1989) has given the following alternative unbiased estimator for* $V_p(t_3)$

$$v_3' = \frac{\sum_1^n N_i^2 - N}{N(N-1)} \sum\sum_{i<j} \frac{Q_i}{N_i} \frac{Q_j}{N_j} \left(\frac{Y_i}{P_i} - \frac{Y_j}{P_j}\right)^2.$$

He also claimed that v'_3 possibly is better than v_3, showing their numerical illustrative comparisons based on simulated observations. But in their illustrations they allowed N_i's to deviate appreciably from

$$\left[\frac{N}{n}\right], \quad \left[\frac{N}{n}\right] + 1$$

which choice has been recommended by RHC as the optimal one for t_3. CHAUDHURI *and* MITRA (1992) *virtually nullified* OHLSSON's (1989) *claims demonstrating v_3 to remain quite competitive with v'_3 when N_i's are chosen optimally. Of course the two match completely if one may take $N_i = \frac{N}{n}$ as an integer for every $i = 1, 2, \ldots, n$, as is also noted by* OHLSSON (1989).

2.4.4 HT Estimator $\bar{t}$

Since $\bar{t}$ is unbiased for Y (see section 1.2), its MSE is the same as its variance, the following formula for which is given by HORVITZ and THOMPSON (1952)

$$V_1 = V_p(\bar{t}) = \sum \frac{Y_i^2}{\pi_i}(1 - \pi_i) + \sum\sum_{i \neq j} \frac{Y_i}{\pi_i}\frac{Y_j}{\pi_j}(\pi_{ij} - \pi_i\pi_j).$$

A formula for an unbiased estimator for V_1 is also given by HORVITZ and THOMPSON as

$$v_1 = \sum \frac{Y_i^2}{\pi_i}(1 - \pi_i)\frac{I_{si}}{\pi_i} + \sum\sum_{i \neq j} \frac{Y_i}{\pi_i}\frac{Y_j}{\pi_j}(\pi_{ij} - \pi_i\pi_j)\frac{I_{sij}}{\pi_{ij}}$$

assuming $\pi_{ij} > 0$ for $i \neq j$.

If $Y_i = c\,\pi_i$ for all $i \in U$

$$\bar{t} = \sum_{i \in s} \frac{Y_i}{\pi_i} = c\nu(s)$$

and $Y = c \sum \pi_i$. If $\nu(s) = n$ for every s with $p(s) > 0$, that is, $\bar{t}$ is based on a design p_n, then, since $\sum \pi_i = n$ as well, the strategy $(p, \bar{t})$ is representative with respect to $(\pi_1, \pi_2, \ldots, \pi_N)'$.

In this case it follows from RAO and VIJAYAN's (1977) general result of section 2.3 (noted earlier by SEN, 1953) that one may write $V_p(\bar{t})$ alternatively as

$$V_2 = \sum\sum_{i<j} (\pi_i \pi_j - \pi_{ij}) \left(\frac{Y_i}{\pi_i} - \frac{Y_j}{\pi_j} \right)^2 .$$

Hence, SEN and YATES and GRUNDY's unbiased estimator for V_2 as given by them is

$$v_2 = \sum\sum_{i<j} (\pi_i \pi_j - \pi_{ij}) \left(\frac{Y_i}{\pi_i} - \frac{Y_j}{\pi_j} \right)^2 \frac{I_{sij}}{\pi_{ij}}$$

assuming $\pi_{ij} > 0$ for all $i \neq j$. For designs satisfying $\pi_i \pi_j \geq \pi_{ij}$ for all $i \neq j$ v_2 is uniformly non-negative.

If $\nu(s)$ is not a constant for all s with $p(s) > 0$ representativity of $(p, \bar{t})$ is violated. To cover this case, CHAUDHURI (2000a) showed that writing

$$\alpha_i = 1 + \frac{1}{\pi_i} \sum_{j \neq i} \pi_{ij} - \sum \pi_j$$

for $i \in U$ one has a third formula for $V_p(\bar{t})$ as

$$V_3 = V_2 + \sum \frac{Y_i^2}{\pi_i} \alpha_i$$

and hence proposed

$$v_3 = v_2 + \sum \frac{Y_i^2}{\pi_i} \alpha_i \frac{I_{si}}{\pi_i}$$

as an unbiased estimator for $V_p(\bar{t})$. This v_3 is uniformly non-negative if

$$\pi_i \pi_j \geq \pi_{ij} \quad \text{for all } i \neq j$$
$$\alpha_i > 0 \quad \text{for all } i \in U.$$

CHAUDHURI and PAL (2002) illustrated a sampling scheme for which the above conditions simultaneously hold while representativity fails.

2.4.5 Murthy's Estimator t_4

Writing

$$a_{ij} = P_i P_j \left[\frac{Y_i}{P_i} - \frac{Y_j}{P_j}\right]^2$$

we have

$$M = V_p(t_4) = -\sum\sum_{i<j} P_i P_j \left[\frac{Y_i}{P_i} - \frac{Y_j}{P_j}\right]^2$$

$$\times E_p\left[\left(\frac{p(s\,|\,i)}{p(s)} I_{si} - 1\right)\left(\frac{p(s\,|\,j)}{p(s)} I_{sj} - 1\right)\right]$$

$$= \sum\sum_{i<j} a_{ij}\left[1 - \sum_{\substack{s\ni i,j \\ p(s)>0}} \frac{p(s\,|\,i)p(s\,|\,j)}{p(s)}\right]$$

because

$$E_p\left[\frac{p(s\,|\,i)}{p(s)} I_{si}\right] = \sum_s p(s\,|\,i) I_{si}$$

$$= \sum_{s\ni i} p(s\,|\,i) = 1 \quad \text{for} \quad i = 1, \ldots, N.$$

One obvious unbiased estimator for $V_p(t_4)$ is

$$\hat{M} = \sum\sum_{1\le i<j\le N} a_{ij} \frac{I_{sij}}{p^2(s)} [p(s\,|\,i,j)p(s) - p(s\,|\,i)p(s\,|\,j)]$$

which follows from

$$\sum_s I_{sij}\, p(s\,|\,i,j) = \sum_{s\ni i,j} p(s\,|\,i,j) = 1$$

writing $p(s\,|\,i,j)$ as the conditional probability of choosing s given that i and j are the first two units in s. It is assumed that the scheme of sampling is so adopted that it is meaningful to talk about the conditional probabilities $p(s\,|\,i)$, $p(s\,|\,i,j)$.

Consider in particular the well-known sampling scheme due to LAHIRI (1951), MIDZUNO (1952), and SEN (1953) to be referred to as LMS scheme. Then on the first draw i is chosen with probability $P_i (0 < P_i < 1,\ \Sigma_1^N P_i = 1)$, $i = 1, \ldots, N$ and subsequently $(n-1)$ distinct units are chosen from the remaining $(N-1)$ units by the SRSWOR method, leaving aside the unit chosen on the first draw. For this scheme, then

$$p(s) = \sum_{i \in s} P_i \Big/ \binom{N-1}{n-1}.$$

If based on this scheme t_4 reduces to the ratio estimator

$$t_R = \sum_{i \in s} Y_i \Big/ \sum_{i \in s} P_i.$$

Writing $C_r = \binom{N-r}{n-r}$, it follows that for this LMS scheme

$$p(s \mid i) = 1/C_1, p(s \mid i, j) = 1/C_2$$
$$E_p(t_R) = Y$$
$$M = E_p(t_R - Y)^2 = V_p(t_R)$$
$$= \sum\sum_{1 \le i < j \le N} a_{ij} \left[1 - \frac{1}{C_1} \sum_{s \ni i,j} \frac{1}{[\sum_{i \in s} P_i]} \right].$$

An unbiased estimator for M is

$$\hat{M} = \sum\sum_{1 \le i < j \le N} a_{ij} \frac{I_{sij}}{\sum_{i \in s} P_i} \left[\frac{N-1}{n-1} - \frac{1}{\sum_{i \in s} P_i} \right].$$

It may be noted that if one takes $P_i = X_i / X$, then t_R reduces to t_1, which is thus unbiased for Y if based on the LMS scheme instead of SRSWOR, which is p-biased for Y in the latter case.

2.4.6 Raj's Estimator t_5

Another popular strategy is due to RAJ (1956, 1968). The sampling scheme is called probability proportional to size without replacement (PPSWOR) with P_i's ($0 < P_i < 1, \Sigma P_i = 1$) as the normed size measures. On the first draw a unit i_1 is chosen with probability P_{i_1}, on the second draw a unit $i_2 (\neq i_1)$ is chosen with probability $P_{i_2}/(1 - P_{i_1})$ out of the units of U leaving i_1 aside, on the third draw a unit $i_3 (\neq i_1, i_2)$ is chosen with probability $P_{i_3}/(1 - P_{i_1} - P_{i_2})$ out of U leaving aside i_1, i_2, and so on. On the final nth ($n > 2$) draw a unit $i_n (\neq i_1, \ldots, i_{n-1})$ is chosen with probability

$$\frac{P_{i_n}}{1 - P_{i_1} - P_{i_2} - \ldots, -P_{i_{n-1}}}$$

out of the units of U minus $i_1, i_2, \ldots, i_{n-1}$. Then,

$$e_1 = \frac{Y_{i_1}}{P_{i_1}}$$

$$e_2 = Y_{i_1} + \frac{Y_{i_2}}{P_{i_2}}(1 - P_{i_1})$$

$$e_j = Y_{i_1} + \ldots + Y_{i_{j-1}} + \frac{Y_{i_j}}{P_{i_j}}(1 - P_{i_1} - \ldots - P_{i_{j-1}})$$

$j = 3, \ldots, n$ are all unbiased for Y because the conditional expectation

$$E_c\left[e_j \mid (i_1, Y_{i_1}), \ldots, (i_{j-1}, Y_{i_{j-1}})\right]$$

$$= (Y_{i_1} + \ldots, +Y_{i_{j-1}}) + \sum_{\substack{k=1 \\ (\neq i_1,\ldots,i_{j-1})}}^{N} Y_k = Y.$$

So, unconditionally, $E_p(e_j) = Y$ for every $j = 1, \ldots, n$, and

$$t_5 = \frac{1}{n}\sum_{j=1}^{n} e_j,$$

called **Raj's** (1956) **estimator**, is unbiased for Y.

To find an elegant formula for $M = V_p(t_5)$ is not easy, but RAJ (1956) gave a formula for an unbiased estimator for $M = V_p(t_5)$ noting e_j, e_k $(j < k)$ are pair-wise uncorrelated since

$$E_p(e_j e_k) = E\left[E_c(e_j e_k \mid (i_1, Y_{i_1}), \ldots, (i_{k-1}, Y_{i_{k-1}})\right]$$

$$= E\left[e_j E_c(e_k \mid (i_1, Y_{i_1}), \ldots, (i_{k-1}, Y_{i_{k-1}})\right]$$

$$= Y E(e_j) = Y^2 = E_p(e_j) E_p(e_k)$$

that is, $cov_p(e_j, e_k) = 0$. So,

$$V_p(t_5) = \frac{1}{n^2}\sum_{j=1}^{n} V_p(e_j)$$

and

$$v_5 = \frac{1}{n(n-1)}\sum_{j=1}^{n}(e_j - t_5)^2$$

is a non-negative unbiased estimator for $V_p(t_5)$.

Incidentally, it can be shown that $V_p(t_5)$ is smaller than the variance of t_2 with respect to PPSWR:

$$V_p(e_1) = \sum_1^N \frac{Y_i^2}{P_i} - Y^2 = \sum P_i \left[\frac{Y_i}{P_i} - Y\right]^2$$
$$= \sum_{1 \le i < j \le N}\sum P_i P_j \left[\frac{Y_i}{P_i} - \frac{Y_j}{P_j}\right]^2$$
$$= V.$$

And

$$V_p(e_2) = E_p[V_p(e_2 \mid (i_1, Y_{i_1}))] + V_p[E_p(e_2 \mid (i_1, Y_{i_1}))]$$
$$= E\left[\sum_{\substack{1 \le i < j \le N \\ (i,j \ne i_1)}}\sum Q_i Q_j \left[\frac{Y_i}{Q_i} - \frac{Y_j}{Q_j}\right]^2\right], \text{ writing } Q_i = \frac{P_i}{1 - P_{i_1}}$$
$$= E\left[\sum_{\substack{1 \le i < j \le N \\ (i,j \ne i_1)}}\sum P_i P_j \left[\frac{Y_i}{P_i} - \frac{Y_j}{P_j}\right]^2\right]$$
$$= \sum_{1 \le i < j \le N}\sum (1 - P_i - P_j)\, P_i P_j \left[\frac{Y_i}{P_i} - \frac{Y_j}{P_j}\right]^2 < V$$

$$V_p(e_3) = E\left[\sum_{\substack{1 \le i < j \le N \\ (i,j \ne i_1)}}\sum R_i R_j \left[\frac{Y_i}{P_i} - \frac{Y_j}{P_j}\right]^2\right]$$

$$\left(\text{writing } R_k = \frac{P_k}{1 - P_{i_1} - P_{i_2}} = \frac{P_k/(1 - P_{i_1})}{1 - \frac{P_{i_2}}{1 - P_{i_1}}} = \frac{Q_k}{1 - Q_{i_2}}\right)$$

$$= E\sum_{\substack{1 \le i < j \le N \\ (i,j \ne i_1, i_2)}}\sum Q_i Q_j \left[\frac{Y_i}{Q_i} - \frac{Y_j}{Q_j}\right]^2$$
$$= E\sum_{\substack{1 \le i < j \le N \\ (i,j \ne i_1, i_2)}}\sum (1 - Q_i - Q_j) Q_i Q_j \left[\frac{Y_i}{Q_i} - \frac{Y_j}{Q_j}\right]^2 < V_p(e_2).$$

Similarly, $V_p(e_k) < V_p(e_j)$ for every $j < k = 2, \ldots, n$. So,

$$V_p(t_5) = \frac{1}{n^2}\sum_{j=1}^{n} V_p(e_j) < \frac{V_p(e_1)}{n} = \frac{V}{n}$$

which is the variance of t_2 with respect to PPSWR.

Clearly, t_5 depends on the order in which the units are drawn in the sample s. So, one may apply **Murthy's** (1957) **unordering** on t_5 to get the estimator

$$t_6 = \sum_{s'\sim s} p(s')t_5(s', \underline{Y}) \Big/ \sum_{s'\sim s} p(s')$$

for which $V_p(t_6) < V_p(t_5) < V_p(t_2)$. Here $s = (i_1, \ldots, i_n)$ is a sample drawn by PPSWOR scheme and $\sum_{s'\sim s}$ denotes the sum over all samples obtained by permuting the coordinates of s. This estimator t_6 is called **Murthy's** (1957) **symmetrized Des Raj estimator** (SDE) based on PPSWOR sampling.

2.4.7 Hartley–Ross Estimator t_7

Another estimator based on SRSWOR due to HARTLEY and ROSS (1954), called **Hartley-Ross estimator** (HRE) is defined as follows.

Let

$$R_i = \frac{Y_i}{X_i}, i = 1, 2, \ldots, N.$$

$$\overline{R} = \frac{1}{N}\sum \frac{Y_i}{X_i}, \overline{r} = \frac{1}{n}\sum_{i\in s} R_i$$

Define

$$C = \frac{1}{N}\sum_{i=1}^{N}\left[\frac{Y_i}{X_i} - \frac{1}{N}\sum_{j=1}^{N}\frac{Y_j}{X_j}\right]\left[X_i - \frac{1}{N}\sum_{j=1}^{N}X_j\right]$$

$$= \frac{1}{N}\sum_{1}^{N} Y_i - \frac{X}{N}\frac{1}{N}\sum_{1}^{N}\frac{Y_i}{X_i} = \overline{Y} - \overline{X}\,\overline{R}.$$

Then $\overline{r}$ and

$$\hat{C} = \frac{N-1}{N}\frac{1}{n-1}\sum_{i\in s}(R_i - \overline{r})(X_i - \overline{x}) = \frac{(N-1)n}{N(n-1)}(\overline{y} - \overline{r}\,\overline{x})$$

based on SRSWOR in n draws are unbiased estimators of $\overline{R}$ and C, respectively. So,

$$\overline{X}\overline{r} + \frac{(N-1)n}{N(n-1)}(\overline{y} - \overline{r}\,\overline{x})$$

is an unbiased estimator of $\overline{Y}$ and the HRE

$$t_7 = X\overline{r} + \frac{(N-1)n}{N(n-1)}(\overline{y} - \overline{r}\,\overline{x})$$

is an unbiased estimator of Y. t_7 is regarded as a ratio-type estimator that is exactly unbiased for Y. Other strategies will be mentioned in subsequent chapters.

2.5 CALIBRATION

Consider a design p and the corresponding HT estimator $\bar{t}$. Such a strategy may not be representative with respect to a relevant size measure x with values $X_1, X_2, \ldots, X_N$. Then, it is important to look for an estimator

$$\sum b_{si} Y_i$$

which, in combination with p, is representative with respect to $(X_1, X_2, \ldots, X_N)'$ and, at the same time, is closer to $\bar{t}$ in an appropriate topology than all other estimators yielding representative strategies.

The relevant ideas of DEVILLE (1988) and DEVILLE and SÄRNDAL (1992) are presented below in a general framework, with auxiliary variables $x_1, x_2, \ldots, x_k$. Define (see section 2.1)

$$\underline{x}_i = (X_{i1}, X_{i2}, \ldots, X_{ik})'$$

$$\underline{x} = \sum_{i=1}^{N} \underline{x}_i$$

and consider an estimator

$$t = t(s, \underline{Y}) = \sum_{i=1}^{N} a_{si} Y_i$$

with **weights** a_{si} not satisfying the calibration equation

$$\sum_{i=1}^{N} a_{si}\,\underline{x}_i = \underline{x}$$

(see section 2.1). Then we may look for new weights b_{si} satisfying the calibration equation but kept close to the original weights a_{si}. Let a measure of the distance between the new and the original weights be a function

$$\sum_{i \in s}(b_{si} - a_{si})^2 / Q_i \tag{2.1}$$

with $Q_i > 0; i = 1, 2, \dots, N$ to be determined; note that $a_{si} = b_{si} = 0$ for $i \notin s$.

RESULT 2.2 *Minimizing Eq. (2.1) subject to the calibration equation*

$$\sum b_{si}\underline{x}_i = \underline{x}$$

leads to

$$\begin{aligned}\tilde{t} &= \sum_{i=1}^{N} b_{si} Y_i \\ &= \sum_{i=1}^{N} a_{si} Y_i + \left[\underline{x} - \sum_{i=1}^{N} a_{si}\underline{x}_i\right]' \left[\sum_{i=1}^{N} Q_i \underline{x}_i \underline{x}_i'\right]^{-1} \sum_{i=1}^{N} Q_i \underline{x}_i Y_i.\end{aligned} \tag{2.2}$$

PROOF: *Consider the Lagrange function*

$$\sum_{i=1}^{N}(b_{si} - a_{si})^2 / Q_i - 2 \cdot \underline{\lambda}' \left(\sum_{i=1}^{N} b_{si}\underline{x}_i - \underline{x}\right)$$

with partial derivative $\partial / \partial b_{si}$

$$2(b_{si} - a_{si})/Q_i - 2\underline{\lambda}'\underline{x}_i$$

where $\underline{\lambda} = (\lambda_1, \dots, \lambda_k)'$ *is a vector of Lagrange factors. Equating the partial derivative to* 0 *yields*

$$b_{si} = Q_i \underline{\lambda}' \underline{x}_i + a_{si}$$

leading to

$$\sum_{i=1}^{N} (Q_i \underline{\lambda}' \underline{x}_i + a_{si}) \underline{x}_i' = \underline{x}'$$

$$\underline{\lambda}' = \left[\underline{x} - \sum_{i=1}^{N} a_{si} \underline{x}_i \right]' \left[\sum_{i=1}^{N} Q_i \underline{x}_i \underline{x}_i' \right]^{-1} .$$

and the estimator $\tilde{t}$ *stated in Eq. (2.2).*

EXAMPLE 2.1 *Let*

$$a_{si} = \frac{1}{\pi_i} \quad \textit{for} \quad i \in s$$

(and 0 *otherwise) for which the calibrated estimator takes the form*

$$\tilde{t}_\pi = \sum_{i \in s} Y_i/\pi_i + \left[\underline{x} - \sum_{i \in s} \underline{x}_i/\pi_i \right]' \left[\sum_{i \in s} Q_i \underline{x}_i \underline{x}_i' \right]^{-1} \sum_{i \in s} Q_i \underline{x}_i Y_i$$

$\tilde{t}_\pi$ *coincides with the* ***generalized regression*** *(GREG)* ***estimator*** *which was introduced by* CASSEL, SÄRNDAL *and* WRETMAN *(1976) with a totally different approach, which we will discuss in section 6.1.*

Chapter 3

Choosing Good Sampling Strategies

3.1 FIXED POPULATION APPROACH

3.1.1 Nonexistence Results

Let a design p be given and consider a p-unbiased estimator t, that is, $B_p(t) = E_p(t - Y) = 0$ uniformly in $\underline{Y}$. The performance of such an estimator is assessed by $V_p(t) = E_p(t - Y)^2$ and we would like to minimize $V_p(t)$ uniformly in $\underline{Y}$. Assume t^* is such a **uniformly minimum variance** (UMV) **unbiased estimator** (UMVUE), that is, for every unbiased t (other than t^*) one has $V_p(t^*) \leq V_p(t)$ for every $\underline{Y}$ and $V_p(t^*) < V_p(t)$ at least for one $\underline{Y}$.

Let Ω be the range (usually known) of $\underline{Y}$; for example, $\Omega = \{\underline{Y} : a_i < Y_i < b_i, i = 1, \ldots, N\}$ with $a_i, b_i (i = 1, \ldots, N)$ as known real numbers. If $a_i = -\infty$ and $b_i = +\infty$, then Ω coincides with the N-dimensional Euclidean space $\mathbb{R}^N$; otherwise Ω is a subset of $\mathbb{R}^N$. Let us choose a point $\underline{A} = (A_1, \ldots, A_i, \ldots, A_N)'$ in Ω and consider as an estimator for Y

$$\begin{aligned} t_A &= t_A(s, \underline{Y}) \\ &= t^*(s, \underline{Y}) - t^*(s, \underline{A}) + A \end{aligned}$$

where $A = \Sigma A_i$. Then,

$$E_p(t_A) = E_p t^*(s, \underline{Y}) - E_p t^*(s, \underline{A}) + A = Y - A + A = Y$$

that is, t_A is unbiased for Y. Now the value of

$$V_p(t_A) = E_p[t^*(s, \underline{Y}) - t^*(s, \underline{A}) + A - Y]^2$$

equals zero at the point $\underline{Y} = \underline{A}$. Since t^* is supposed to be the UMVUE, $V_p(t^*)$ must also be zero when $\underline{Y} = \underline{A}$. Now $\underline{A}$ is arbitrary. So, in order to qualify as the UMVUE for Y, the t^* must have its variance identically equal to zero. This is possible only if one has a census, that is, every unit of U is in s rendering t^* coincident with Y. So, for no design except a **census design**, for which the entire population is surveyed, there may exist a UMV estimator among all UE's for Y. The same is true if, instead of Y, one takes $\overline{Y}$ as the estimand. This important non-existence result is due to GODAMBE and JOSHI (1965) while the proof presented above was given by BASU (1971).

Let us now seek a UMV estimator for Y within the restricted class of HLU estimators of the form

$$t = t_b = t(s, \underline{Y}) = \sum_{i \in s} b_{si} Y_i.$$

Because of the unbiasedness of the estimator we need, uniformly in $\underline{Y}$, Y equal to

$$E(t_b) = \sum_s p(s) \left[\sum_{i \in s} b_{si} Y_i \right] = \sum_{i=1}^{N} Y_i \left[\sum_{s \ni i} b_{si} p(s) \right].$$

Allowing Y_j to be zero for every $j = 1, \ldots, N$ we derive for all i

$$\sum_{s \ni i} b_{si} p(s) = 1.$$

To find the UMV estimators among such estimators based on a fixed design p, we have to minimize

$$E_p(t_b^2) = \sum_s p(s) \left[\sum_{i \in s} b_{si} Y_i \right]^2$$

subject to

$$\sum_{s \ni i} b_{si} p(s) = 1 \quad \text{for} \quad i = 1, \ldots, N.$$

Hence, we need to solve

$$0 = \frac{\partial}{\partial b_{si}} \left[\sum_s p(s) \left(\sum_{i \in s} b_{si} Y_i \right)^2 - \sum_1^N \lambda_i \left(\sum_{s \ni i} b_{si} p(s) - 1 \right) \right]$$

$$= \left[2Y_i \sum_{i \in s} b_{si} Y_i - \lambda_i \right] p(s)$$

introducing Lagrangian undetermined multipliers λ_i. Therefore, for s with $p(s) > 0$ and $s \ni i$

$$\sum_{j \in s} b_{sj} Y_j = \frac{\lambda_i}{2Y_i}$$

for all $\underline{Y}$ with $Y_i \neq 0$. Letting $Y_i \neq 0, Y_j = 0$ for every $j \neq i$ this leads to a possible solution

$$b_{si} = \frac{\lambda_i}{2Y_i^2} = b_i, \text{ say}$$

free of s, leading to $b_i = 1/\pi_i$.

From the above it follows that the UMV estimator, if available, is identical with the HT estimator and, in addition, satisfies

$$\sum_{j \in s} \frac{Y_j}{\pi_j} = \frac{\lambda_i}{2Y_i}$$

for every $s \ni i$ with $p(s) > 0$, provided $Y_i \neq 0$. For example, if

$$s_1 \ni i, s_2 \ni i, p(s_1) > 0, p(s_2) > 0, Y_i \neq 0$$

then we need

$$\sum_{s_1} \frac{Y_i}{\pi_i} = \sum_{s_2} \frac{Y_i}{\pi_i} \quad \text{for all} \quad \underline{Y}$$

for the existence of a UMV estimator in the class of homogeneous linear unbiased estimators (HLUE). This cannot be realized unless the design p satisfies the conditions that for s_1, s_2 with $p(s_1) > 0$, $p(s_2) > 0$, either $s_1 \cap s_2$ is empty or $s_1 \sim s_2$, meaning that s_1 and s_2 are **equivalent** in the sense of both containing an identical set of distinct units of U.

Such a design, for example, one corresponding to a systematic sample, is called a **unicluster design** (UCD). Any design that does not meet these stringent conditions is called a **non-unicluster design** (NUCD). For a UCD it is possible to realize

$$\sum_{s_1} \frac{Y_i}{\pi_i} = \sum_{s_2} \frac{Y_i}{\pi_i}$$

uniformly in $\underline{Y}$, but not for an NUCD. So, for any NUCD, a UMV estimator does not exist among the HLUE's.

This celebrated nonexistence result really opened up the modern problem of finite population inference. It is due to GODAMBE (1955); the exceptional character of uni-cluster designs was pointed out by HEGE (1965) and HANURAV (1966).

If the class of estimators is extended to that of **linear unbiased estimators** (LUE) of the form

$$t_L = b_s + \sum_{i \in s} b_{si} Y_i$$

with b_s free of $\underline{Y}$ such that

$$E_p(b_s) = 0, E_p(t_L) = Y$$

uniformly in $\underline{Y}$, then it is easy to apply BASU's (1971) approach to show that, again, a UMV estimator does not exist. However, if $b_s = 0$, then BASU's proof does not apply and GODAMBE's (1955) result retains its importance covering the HLUE subclass.

3.1.2 Rao-Blackwellization

An estimator $t = t(s, \underline{Y})$ may depend on the order in which the units appear in s and may depend on the multiplicities of the appearances of the units in s.

EXAMPLE 3.1 *Let* P_i $(0 < P_i < 1, \Sigma_1^N P_i = 1)$ *be known numbers associated with the units* i *of* U. *Suppose on the first draw a unit* i *is chosen from* U *with probability* P_i *and on the second draw a unit* $j (\neq i)$ *is chosen with probability* $\frac{P_j}{1-P_i}$.

Consider RAJ's *(1956) estimator (see section 2.4.6)*

$$t_D = t(i,j) = \frac{1}{2}\left[\frac{Y_i}{P_i} + \left(Y_i + \frac{Y_j}{P_j}(1-P_i)\right)\right] = \frac{1}{2}(e_1 + e_2), \quad say.$$

Now,

$$E_p(e_1) = E_p\left[\frac{Y_i}{P_i}\right] = \sum_1^N \frac{Y_i}{P_i}P_i = Y$$

and

$$e_2 = Y_i + \frac{Y_j}{P_j}(1-P_j)$$

has the ***conditional expectation****, given that* (i, Y_i) *is observed on the first draw,*

$$E_C(e_2) = Y_i + \sum_{j\neq i}\left[\frac{Y_j}{P_j}(1-P_i)\right]\frac{P_j}{1-P_i} = Y_i + \sum_{j\neq i} Y_j = Y$$

and hence the unconditional expectation $E_p(e_2) = Y$. *So* t_D *is unbiased for* Y*, but depends on the order in which the units appear in the sample* $s = (i, j)$ *that is, in general*

$$t_D(i,j) \neq t_D(j,i).$$

EXAMPLE 3.2 *Let* n *draws be independently made choosing the unit* i *on every draw with the probability* P_i *and let* t *be an estimator for* Y *given by*

$$t = \frac{1}{n}\sum_{r=1}^{n}\frac{y_r}{p_r}$$

where y_r *is the value of* y *for the unit selected on the* r*th draw* $(r = 1, \ldots, n)$ *and* p_r *the value* P_i *if the* r*th draw produces the unit* i*. This* t*, usually attributed to* HANSEN *and* HURWITZ *(1943), may also be written as*

$$t_{HH} = \frac{1}{n}\sum_{i=1}^{N}\frac{Y_i}{P_i}f_{si}$$

and, therefore, depends on the multiplicity f_{si} *of* i *in* s *(see section 2.2).*

With an arbitrary sample $s = (i_1, i_2, \ldots, i_n)$, let us associate the sample

$$\hat{s} = \{j_1, j_2, , \ldots, j_k\}$$

which consists of all distinct units in s, with their order and/or multiplicity in s ignored; this $\hat{s}$ thus is equivalent to s $(s \sim \hat{s})$.

By Ω let us denote the **parameter space**, that is, the set of all vectors $\underline{Y}$ relevant in a situation, say, the cases

$$\Omega = \mathbb{R}^N$$
$$\Omega = \{\underline{Y} : 0 \leq Y_i \text{ for } i = 1, 2, \ldots, N\}$$
$$\Omega = \{\underline{Y} : Y_i = 0, 1 \text{ for } i = 1, 2, \ldots, N\}$$
$$\Omega = \{\underline{Y} : 0 \leq Y_i \leq X_i \text{ for } i = 1, 2, \ldots, N\}$$

with $X_1, X_2, \ldots, X_N > 0$, being of special importance.

Now consider any design p, yielding the survey data

$$d = (i, Y_i | i \in s) = ((i_1, Y_{i_1}), \ldots, (i_n, Y_{i_n}))$$

compatible with the subset

$$\Omega_d = \{\underline{Y} \in \Omega : Y_i \quad \text{as observed for} \quad i \in s\}$$

of the parameter space. The **likelihood** of $\underline{Y}$ given d is

$$L_d(\underline{Y}) = p(s) I_d(\underline{Y}) = P_{\underline{Y}}(d)$$

which is the probability of observing d when $\underline{Y}$ is the underlying parametric point, writing

$$I_d(\underline{Y}) = 1(0) \quad \text{if} \quad \underline{Y} \in \Omega_d (\notin \Omega_d).$$

Define the **reduced data**

$$\hat{d} = (i, Y_i | i \in \hat{s}).$$

Then, for all d

$$I_d(\underline{Y}) = I_{\hat{d}}(\underline{Y})$$

and

$$L_{\hat{d}}(\underline{Y}) = p(\hat{s}) I_{\hat{d}}(\underline{Y}) = P_{\underline{Y}}(\hat{d}).$$

For simplicity we will suppress $\underline{Y}$ in $P_{\underline{Y}}(d)$ and write $P(d | \hat{d})$ to denote the conditional probability of observing d when $\hat{d}$ is given. Since

$$P(d) = P(d \cap \hat{d}) = P(\hat{d})P(d|\hat{d}) \quad \text{or}$$
$$p(s)I_d(\underline{Y}) = p(\hat{s})I_{\hat{d}}(\underline{Y})P(d|\hat{d})$$

it follows that for $p(\hat{s}) > 0$, $P(d|\hat{d}) = p(s)/p(\hat{s})$ implying that $\hat{d}$ is a **sufficient statistic**, assuming throughout that p is a noninformative design. Let $t = t(d)$ be any function of d that is also a sufficient statistic. If for any two samples s_1, s_2 with $p(s_1), p(s_2) > 0$ and corresponding entities $\hat{s_1}, \hat{s_2}, d_1, d_2, \hat{d}_1, \hat{d}_2$ it is true that $t(d_1) = t(d_2)$, then it follows that

$$\begin{aligned} P(d_1) = P(d_1 \cap t(d_1)) &= P(t(d_1))P(d_1|t(d_1)) \\ &= P(t(d_2))P(d_1|t(d_1)) \\ &= \frac{P(d_2)}{P(d_2|t(d_2))}P(d_1|t(d_1)) \end{aligned}$$

and hence

$$p(\hat{s_1})I_{\hat{d}_1}(\underline{Y}) \propto p(\hat{s_2})I_{\hat{d}_2}(\underline{Y})$$

implying that $\hat{d}_1 = \hat{d}_2$ and hence that $\hat{d}$ is the **minimal sufficient statistic** derived from d. Thus a maximal reduction of data d sacrificing no relevant information on $\underline{Y}$ yields $\hat{d}$.

Starting with any estimator $t = t(s, \underline{Y})$ for Y depending on the order and/or multiplicity of the units in s chosen with probability $p(s)$, let us construct a new estimator as the conditional expectation

$$t^* = E_p(t|\hat{d})$$

that is,

$$t^*(s, \underline{Y}) = \sum_{s' \sim s} t(s', \underline{Y})p(s') \Big/ \sum_{s' \sim s} p(s').$$

Here $\sum_{s' \sim s}$ refers to summation over all samples s' equivalent to s.

Then

$$E_p(t^*) = E_p(t)$$
$$E_p(tt^*) = E_p[E_p(tt^*|\hat{d})] = E_p[t^* E_p(t|\hat{d})] = E_p(t^{*2})$$

and

$$E_p(t - t^*)^2 = E_p(t^2) + E_p(t^{*2}) - 2E_p(tt^*) = E_p(t^2) - E_p(t^{*2})$$

giving $E_p(t^2) \geq E_p(t^{*2})$; hence

$$V_p(t) \geq V_p(t^*)$$

equality holding if and only if for every s with $p(s) > 0$, $t(s, \underline{Y}) = t^*(s, \underline{Y})$. The **Rao-Blackwellization** of t is t^*. We may state this as:

RESULT 3.1 *Given any design p and an unbiased estimator t for Y depending on order and/or multiplicity of units in s, define the Rao-Blackwellization t^* of t by*

$$t^*(s, \underline{Y}) = \sum_{s': s' \sim s} t(s', \underline{Y}) p(s') \Big/ \sum_{s': s' \sim s} p(s')$$

where the summation is over all s' consisting of the units of s, possibly in other orders and/or using their various multiplicities.

Then, t^* is unbiased for Y and is independent of order and/or multiplicity of units in s with

$$V_p(t^*) \leq V_p(t)$$

equality holding uniformly in $\underline{Y}$ if and only if $t^* = t$ for all s with $p(s) > 0$, that is, if t itself shares the property of t^* in being free of order and/or multiplicity of units in s.

So, within the class of all unbiased estimators for Y based on a given design p, the subclass of unbiased estimators independent of the order and/or multiplicity of the units in s is a **complete class**, C, in the sense that given any estimator in the class UE but outside C there exists one inside C that is better, that is, has a uniformly smaller variance. This result is essentially due to MURTHY (1957) but in fact is a straightforward application of the Rao-Blackwellization technique in the finite population context.

EXAMPLE 3.3 *Reconsider Example 3.3.1. For $i \neq j$ and $s = (i, j)$*

$$s' = (j, i)$$

is the only sample with $p(s') > 0$ *and* $s' \sim s$. *From*

$$p(i, j) = \frac{P_i P_j}{1 - P_i}$$

$$\frac{p(i,j)}{p(i,j) + p(j,i)} = \frac{\dfrac{1}{1-P_i}}{\dfrac{1}{1-P_i} + \dfrac{1}{1-P_j}} = \frac{\alpha_i}{\alpha_i + \alpha_j}, \text{ say}$$

we derive

$$\begin{aligned} t^*(s, \underline{Y}) &= t((i,j), \underline{Y}) \frac{\alpha_i}{\alpha_i + \alpha_j} + t((j,i), \underline{Y}) \frac{\alpha_j}{\alpha_i + \alpha_j} \\ &= \frac{\alpha_i}{\alpha_i + \alpha_j} \frac{Y_i}{P_i} + \frac{\alpha_j}{\alpha_i + \alpha_j} \frac{Y_j}{P_j} \end{aligned}$$

which is symmetric in i *and* j*, that is, independent of the order in which the units are drawn.*

To consider an application of Result 3.1 suppose p is a UCD and $t_b = \Sigma_{i \in s} b_{si} Y_i$ with $\Sigma_{s \ni i} b_{si} p(s) = 1$ for every i is an HLUE for Y. If a particular $t_b^* = \Sigma b_{si}^* Y_i$ is to be the UMVHLUE for Y, then it must belong to the complete subclass C_H of the HLUE class. Let s_0 be a typical sample containing i; then for every other sample $s \ni i$, which is equivalent to s_0 because p is UCD, we must have $b_{si}^* = b_{s_0 i}^*$ as a consequence of $t_b^* \in C_H$. So, $1 = b_{s_0 i}^* \Sigma_{s \ni i} p(s) = b_{s_0 i}^* \pi_i$ giving $b_{s_0 i}^* = b_{si}^* = \frac{1}{\pi_i}$ for every $s \ni i$, that is, t_b^* must equal the HT estimator $\bar{t}$, which is the unique member of C_H. Consequently, $\bar{t}$ is the unique UMVHLUE for a UCD. This result is due to HEGE (1965) and HANURAV (1966) with the proof later refined by LANKE (1975).

3.1.3 Admissibility

Next we consider a requirement of admissibility of an estimator in the absence of UMVUEs for useful designs in a meaningful sense.

An unbiased estimator t_1 for Y is **better** than another unbiased estimator t_2 for Y if $V_p(t_1) \leq V_p(t_2)$ for every $\underline{Y} \in \Omega$ and $V_p(t_1) < V_p(t_2)$ at least for one $\underline{Y} \in \Omega$. Subsequently, the four cases mentioned in section 3.1.2 are considered for Ω without explicit reference.

If there does not exist any unbiased estimator for Y better than t_1, then t_1 is called an **admissible estimator** for Y within the UE class. If this definition is restricted throughout within the HLUE class, then we have admissibility within HLUE.

RESULT 3.2 *The HTE*

$$\bar{t} = \sum_{i \in s} \frac{Y_i}{\pi_i}$$

is admissible within the HLUE class.

PROOF: *For t_b in the HLUE class and for the HTE $\bar{t}$ we have*

$$V_p(t_b) = \sum_i Y_i^2 \left[\sum_{s \ni i} b_{si}^2 p(s) \right] + \sum\sum_{i \neq j} Y_i Y_j \left[\sum_{s \ni i,j} b_{si} b_{sj} p(s) \right] - Y^2$$

$$V_p(\bar{t}) = \sum_i Y_i^2 / \pi_i + \sum\sum_{i \neq j} Y_i Y_j \frac{\pi_{ij}}{\pi_i \pi_j} - Y^2.$$

Evaluated at a point $\underline{Y}_0^{(i)} = (0, \ldots, Y_i \neq 0, \ldots, 0)$, $[V_p(t_b) - V_p(\bar{t})]$ equals

$$Y_i^2 \left[\sum_{s \ni i} b_{si}^2 p(s) - \frac{1}{\pi_i} \right] \geq 0 \tag{3.1}$$

on applying Cauchy's inequality. This degenerates into an equality if and only if $b_{si} = b_i$, for every $s \ni i$, rendering t_b equal to the HTE $\bar{t}$. So, for t_b other than $\bar{t}$,

$$[V_p(t_b) - V_p(\bar{t})]_{\underline{Y} = \underline{Y}_0^{(i)}} > 0.$$

This result is due to GODAMBE *(1960a). Following* GODAMBE *and* JOSHI *(1965) we have:*

RESULT 3.3 *The HTE $\bar{t}$ is admissible in the wider UE class.*

PROOF: *Let, if possible, t be an unbiased estimator for Y better than the HTE $\bar{t}$. Then, we may write*

$$t = t(s, \underline{Y}) = \bar{t}(s, \underline{Y}) + h(s, \underline{Y}) = \bar{t} + h$$

with $h = h(s, \underline{Y}) = t - \bar{t}$ as an unbiased estimator of zero. Thus,

$$0 = E_p(h) = \sum_s h(s, \underline{Y}) p(s). \tag{3.2}$$

For t to be better than $\bar{t}$, we need $V_p(t) \leq V_p(\bar{t})$

$$\text{or} \quad \sum_s h^2(s, \underline{Y})p(s) \leq -2\sum_s \bar{t}(s, \underline{Y})h(s, \underline{Y})p(s). \quad (3.3)$$

Let $\underline{X}_i (i = 0, 1, \ldots, N)$ consist of all vectors $\underline{Y} = (Y_1, \ldots, Y_N)'$ such that exactly i of the coordinates in them are non-zero. Now, if $\underline{Y} \in \underline{X}_0$, then $\bar{t}(s, \underline{Y}) = 0$, giving $h^2(s, \underline{Y})p(s) = 0$ implying $h(s, \underline{Y})p(s) = 0$ for every s and for $\underline{Y} \in \underline{X}_0$.

Let us suppose that $r = 0, 1, \ldots, N - 1$ exists with $h(s, \underline{Y})p(s) = 0$ for every s and every

$$\underline{Y} \in \underline{X}_r. \quad (3.4)$$

Then, it will follow that $h(s, \underline{Y})p(s) = 0$ for every s and every $\underline{Y}$ in $\underline{X}_{r+1}$. To see this, let $\underline{Z}$ be a point in $\underline{X}_{r+1}$. Then, by Eq. (3.2) and Eq. (3.3), we have

$$0 = \sum_s p(s)h(s, \underline{Z})$$

$$\sum_s p(s)h^2(s, \underline{Z}) \leq -2\sum_s p(s)\bar{t}(s, \underline{Z})h(s, \underline{Z}).$$

Let S denote the totality of all possible samples s with $p(s) > 0$ and S_i the collection of samples s in S such that exactly i of the coordinates Z_j of $\underline{Z}$ with j in s are non-zero. Then, each S_i is disjoint with each S_k for $i \neq k$ and S is the union of $S_i, i = 0, 1, \ldots, r + 1$. So we may write

$$0 = \sum_0^{r+1} \sum_{s \in S_i} p(s)h(s, \underline{Z})$$

$$\sum_0^{r+1} \sum_{s \in S_i} p(s)h^2(s, \underline{Z}) \leq -2\sum_0^{r+1} \sum_{s \in S_i} p(s)\bar{t}(s, \underline{Z})h(s, \underline{Z}).$$

Now, by Eq. (3.4),

$$p(s)h(s, \underline{Z}) = 0 \quad \text{for every } s \text{ in } S_i, i = 0, 1, \ldots, r. \quad (3.5)$$

So it follows that

$$0 = \sum_{s \in S_{r+1}} p(s)h(s, \underline{Z})$$

$$\sum_{s \in S_{r+1}} p(s)h^2(s, \underline{Z}) \leq -2 \sum_{s \in S_{r+1}} p(s)\bar{t}(s, \underline{Z})h(s, \underline{Z}). \quad (3.6)$$

But, for every s in S_{r+1}

$$\bar{t}(s, \underline{Z}) = \sum_{i \in s} \frac{Z_i}{\pi_i} \text{ equals } \sum_{i=1}^{N} \frac{Z_i}{\pi_i}.$$

Since the latter is a constant (for every s) we may write by Eq. (3.6),

$$\sum_{s \in S_{r+1}} p(s) h^2(s, \underline{Z}) \le -2 \left[\sum_{i}^{N} \frac{Z_i}{\pi_i} \right] \sum_{s \in S_{r+1}} p(s) h(s, \underline{Z}) = 0,$$

leading to $p(s)h^2(s, \underline{Z}) = 0$ *for every s in* S_{r+1} *or* $p(s)h(s, \underline{Z}) = 0$ *for every s in* $S_i, i = 0, 1, \ldots, r + 1$ *using Eq. (3.5), that is,* $h(s, \underline{Z})p(s) = 0$ *for every s in S, that is,* $h(s, \underline{Y})p(s) = 0$ *for every s and every* $\underline{Y}$ *in* $\underline{X}_{r+1}$*. But* $h(s, \underline{Y})p(s) = 0$ *for every s and every* $\underline{Y}$ *in* $\underline{X}_0$ *as already shown. So, it follows that* $h(s, \underline{Y})p(s) = 0$ *for every s and every* $\underline{Y}$ *in* Ω *if t is to be better than* $\bar{t}$*. So, for every sample s with* $p(s) > 0$*, t must coincide with* $\bar{t}$ *itself.*

Admissibility, however, is hardly a very selective criterion. There may be infinitely many admissible estimators for Y among UEs. For example, if we fix any point $\underline{A} = (A_1, \ldots, A_N)'$ in Ω, then with $A = \sum_1^N A_i$ we can take an estimator for Y as

$$t_A = \sum_{i \in s} \frac{Y_i - A_i}{\pi_i} + A$$

Obviously, t_A is unbiased for Y. Writing $W_i = Y_i - A_i$ and considering the space or totality of points $\underline{W} = (W_1, \ldots, W_N)'$ and assuming it is feasible to assign zero values to any number of its coordinates, it is easy to show that t_A is also admissible for Y within UE class. The estimator t_A is called a **generalized difference estimator** (GDE). If the parameter space of $\underline{Y}$ is restricted to be a close neighborhood $N(\underline{A})$ of the fixed point $\underline{A}$, then it is easy to see that $E_p(\bar{t}) = Y = E_p(t_A)$ but $V_p(t_A) < V_p(\bar{t})$ for every $\underline{Y}$ in $N(\underline{A})$ showing inadmissibility of $\bar{t}$ when the parametric space is thus restricted. In practice, the parametric spaces are in fact restricted. A curious reader may consult GHOSH (1987) for further details.

3.2 SUPERPOPULATION APPROACH

3.2.1 Concept

With the fixed population approach considered so far it is difficult, as we have just seen, to hit upon an appropriately optimal strategy or an estimator for Y or $\overline{Y}$ based on a fixed sampling design. So, one approach is to regard $\underline{Y} = (Y_1, \ldots, Y_N)'$ as a particular realization of an N-dimensional random vector $\underline{\eta} = (\eta_1, \ldots, \eta_N)'$, say, with real-valued coordinates. The probability distribution of $\underline{\eta}$ defines a population, called a **superpopulation**. A class of such distributions is called a **superpopulation model** or just a **model**, in brief. Our central objective remains to estimate the total (or mean) for the particular realization $\underline{Y}$ of $\underline{\eta}$. But the criteria for the choice of strategies (p, t) may now be changed suitably.

We assume that the superpopulation model is such that the expectations, variances of η_i, and covariances of η_i, η_j exist. To simplify notations we write E_m, V_m, C_m as operators for expectations, variances, and covariances with respect to a model and write Y_i for η_i pretending that $\underline{Y}$ is itself a random vector.

Let (p_1, t_1) and (p_2, t_2) be two unbiased strategies for estimating Y, that is, $E_{p_1}t_1 = E_{p_2}t_2 = Y$. Assume that p_1, p_2 are suitably comparable in the sense of admitting samples of comparable sizes with positive selection probabilities. We might have, for example, the same average effective sample sizes; that is,

$$\sum |s| p_1(s) = \sum |s| p_2(s)$$

where $\sum$ extends over all samples and $|s|$ is the cardinality of s.

Then, (p_1, t_1) will be preferred to (p_2, t_2) if

$$E_m V_{p_1}(t_1) \leq E_m V_{p_2}(t_2)$$

REMARK 3.1 *We assume that the expectation operators E_p and E_m commute. This assumption is automatically fulfilled in most situations. But to illustrate a case where E_p and E_m may* *not commute, let*

$$p(s) = \frac{1}{\binom{N-1}{n-1}} \sum_s X_i/X \quad \text{and} \quad t = X \sum_s Y_i \Big/ \sum_s X_i$$

where $X = \sum_1^N X_i$ *and* X_i*'s,* $i = 1, \dots, N$ *are independent realizations on a positive valued random variable* x*. Define* $\underline{X} = (X_1, \dots, X_N)'$ *and let* E_C*,* E_x *denote, respectively, operators of expectation conditional on a given realization* $\underline{X}$ *and the expectation over the distribution of* x*. Then, we may meaningfully evaluate the expectation*

$$E_m E_p(t) = E_x E_C E_p(t)$$

where again we may interchange E_C *and* E_p *to get*

$$E_C E_p(t) = E_p E_C(t) = X E_p \left(\frac{\sum_s E_C(Y_i|\underline{X})}{\sum_s X_i} \right).$$

But here we cannot meaningfully evaluate $E_p E_m(t) = E_p E_x E_C(t)$ *because* $p(s)$ *involves* X_i*'s that occur in* t *on which* $E_m = E_x E_C$ *operates. Such a pathological case, however, may not arise in case* X_i*'s are nonstochastic. To avoid complications we assume commutativity of* E_p *and* E_m*.*

3.2.2 Model $\mathcal{M}_1$

Let us consider a particular model, $\mathcal{M}_1$, such that for $i = 1, 2, \dots, N$

$$Y_i = \mu_i + \sigma_i \varepsilon_i$$

with

$$\begin{aligned} \mu_i \in \mathbb{R}, \sigma_i &> 0 \\ E_m \varepsilon_i &= 0 \\ V_m \varepsilon_i &= 1 \\ C_m(\varepsilon_i, \varepsilon_j) &= 0 \quad \text{for} \quad i \neq j \end{aligned}$$

that is,

$$\begin{aligned} E_m(Y_i) &= \mu_i \\ V_m(Y_i) &= \sigma_i^2 \\ C_m(Y_i, Y_j) &= 0 \quad \text{for} \quad i \neq j. \end{aligned}$$

Then, we derive for any UE t

$$\begin{aligned}
E_m V_p(t) &= E_m E_p (t - Y)^2 = E_p E_m (t - Y)^2 \\
&= E_p E_m \left[(t - E_m(t)) + (E_m(t) - E_m(Y)) \right. \\
&\quad \left. - (Y - E_m Y)\right]^2 \\
&= E_p V_m(t) + E_p \Delta_m^2(t) - V_m(Y) \qquad (3.7)
\end{aligned}$$

writing $\Delta_m(t) = E_m(t - Y)$. The same is true for $\bar{t}$ and any other HLUE t_b. Thus,

$$\begin{aligned}
&E_m V_p(t_b) - E_m V_p(\bar{t}) \\
&= E_p \left[\sum_{i \in s} \sigma_i^2 b_{si}^2 - \sum_{i \in s} \sigma_i^2 / \pi_i^2\right] + E_p \left[\Delta_m^2(t_b) - \Delta_m^2(\bar{t})\right] \\
&= \sum \sigma_i^2 \left[\sum_{i \in s} b_{si}^2 p(s) - \frac{1}{\pi_i}\right] \\
&\quad + E_p \left[(E_m t_b - \mu)^2 - \left[\sum_{i \in s} \frac{\mu_i}{\pi_i} - \mu\right]^2\right] \\
&\geq E_p \left[(E_m t_b - \mu)^2 - \left[\sum_{i \in s} \frac{\mu_i}{\pi_i} - \mu\right]^2\right] \qquad (3.8)
\end{aligned}$$

by Cauchy's inequality (writing $\mu = \Sigma \mu_i$).

To derive a meaningful inequality we will now impose conditions on the designs. By p_n we shall denote a design for which $p_n(s) > 0$ implies that the effective size of s is equal to n. If, in addition, $\pi_i = n\mu_i/\mu$ for every $i = 1, 2, \ldots, N$, we write p_n as $p_{n\mu}$.

Then, from Eq. (3.8) we get

$$E_m V_{p_{n\mu}}(t_b) - E_m V_{p_{n\mu}}(\bar{t}) \geq E_{p_{n\mu}} [E_m(t_b) - \mu]^2 \geq 0$$

because, for $p_{n\mu}$,

$$\sum_{i \in s} \frac{\mu_i}{\pi_i} = \mu.$$

Thus, we may state:

RESULT 3.4 *Let $p_{n\mu}$ be a design of fixed size n with inclusion probabilities*

$$\pi_i = n\frac{\mu_i}{\mu} \ ; \ i = 1, 2, \ldots, N.$$

Then, for model $\mathcal{M}_1$, we have

$$E_m V_{p_{n\mu}}(t_b) \geq E_m V_{p_{n\mu}}(\bar{t})$$

where t_b is an arbitrary HLUE and

$$\bar{t} = \sum_{i \in s} \frac{Y_i}{\pi_i} = \frac{\mu}{n} \sum \frac{Y_i}{\mu_i}.$$

Thus, among the competitors $(p_{n\mu}, t_b)$ the strategy $(p_{n\mu}, \bar{t})$ is optimal.

However, this optimality result due to GODAMBE (1955) is not very attractive. This is because $p_{n\mu}$ is well suited to $\bar{t}$ since $V_p(\bar{t}) = E_p[\sum_{i \in s} \frac{Y_i}{\pi_i} - Y]^2$ equals zero if $\pi_i = nY_i/Y$ and although such a π_i cannot be implemented, it may be approximated by $\pi_i = nX_i/X$ if Y_i is closely proportional to X_i; or, if $E_m(Y_i) \propto X_i$, $V_p(\bar{t})$ based on $p_{n\mu}$ should be under control. But this does not justify forcing this design on every competing estimator t_b, each of which may have $V_p(t_b)$ suitably controlled when combined with an appropriate design p_n.

3.2.3 Model $\mathcal{M}_2$

To derive optimal strategies among all (p, t) with t unbiased for Y let us postulate that $Y_1, Y_2, \ldots, Y_N$ are not only uncorrelated, but even independent. We write $\mathcal{M}_2$ for $\mathcal{M}_1$ together with this independence assumption.

Thus, the model $\mathcal{M}_2$ may be specified as follows:

Assume for $\underline{Y} = (Y_1, Y_2, \ldots, Y_N)'$

$$Y_i = \mu_i + \sigma_i \varepsilon_i$$

with μ_i, σ_i as constants and ε_i $(i = 1, 2, \ldots, N)$ as independent random variables subject to

$$E_m \varepsilon_i = 0$$
$$V_m \varepsilon_i = 1.$$

Consider a design p and an estimator

$$t = t(s, \underline{Y}) = \bar{t} + h$$

with

$$\bar{t} = \sum_{i \in s} \frac{Y_i}{\pi_i}$$

and

$$h = h(s, \underline{Y})$$

subject to

$$E_p(h) = \sum h(s, \underline{Y}) p(s) = 0$$

implying that

$$\sum_{s: i \in s} h(s, \underline{Y}) p(s) = - \sum_{s: i \notin s} h(s, \underline{Y}) p(s)$$

for all $i = 1, 2, \ldots, N$. Then, for $m = \mathcal{M}_2$,

$$\begin{aligned} E_p C_m(\bar{t}, h) &= E_p E_m \left[\sum_{i \in s} \frac{Y_i - \mu_i}{\pi_i} \right] h(s, \underline{Y}) \\ &= E_m \sum_1^N \left[\frac{Y_i - \mu_i}{\pi_i} \right] \sum_{s \ni i} h(s, \underline{Y}) p(s) \\ &= -E_m \sum_1^N \left[\frac{Y_i - \mu_i}{\pi_i} \right] \sum_{s \not\ni i} h(s, \underline{Y}) p(s) \\ &= 0. \end{aligned}$$

where the last equality holds by the independence assumption.

By Eq. (3.7) we derive for $t = \bar{t} + h$

$$E_m V_p(t) = E_p V_m(\bar{t}) + E_p V_m(h) + E_p \Delta_m^2(t) - V_m(Y). \quad (3.9)$$

Writing

$$t_\mu = t_\mu(s, \underline{Y}) = \sum_{i \in s} \left[\frac{Y_i - \mu_i}{\pi_i} \right] + \mu = \bar{t} + h_\mu$$

with

$$h_\mu = -\sum_{i\in s}\frac{\mu_i}{\pi_i} + \mu$$

we note that $V_m(h_\mu) = 0$, $\Delta_m(t_\mu) = 0$ and so,

$$\begin{aligned} E_m V_p(t_\mu) &= E_p V_m(\bar{t}) - V_m(Y) \\ &= \sum \sigma_i^2 \left(\frac{1}{\pi_i} - 1\right). \end{aligned} \tag{3.10}$$

From Eq. (3.9) and Eq. (3.10) we obtain

$$E_m V_p(t) - E_m V_p(t_\mu) = E_p V_m(h) + E_p \Delta_m^2(t) \geq 0 \tag{3.11}$$

and therefore

$$\begin{aligned} E_m V_p(t) &\geq E_m V_p(t_\mu) \\ &= \sum \sigma_i^2 \left(\frac{1}{\pi_i} - 1\right). \end{aligned}$$

RESULT 3.5 *Let p be an arbitrary design with inclusion probabilities $\pi_i > 0$ and*

$$t_\mu = \sum_{i\in s}\frac{Y_i - \mu_i}{\pi_i} + \mu \tag{3.12}$$

($\mu = \sum \mu_i$). Then, under model $\mathcal{M}_2$

$$\begin{aligned} E_m V_p(t) &\geq E_m V_p(t_\mu) \\ &= \sum \sigma_i^2 \left(\frac{1}{\pi_i} - 1\right) \end{aligned}$$

for any UE t.

In order to specify designs for which $\Sigma\sigma_i^2[\frac{1}{\pi_i} - 1]$ may attain its minimal value, let us restrict to designs p_n. Then Cauchy's inequality applied to

$$\sum_1^N \pi_i \sum_1^N \frac{\sigma_i^2}{\pi_i}$$

gives

$$\sum_{i=1}^N \frac{\sigma_i^2}{\pi_i} \geq \frac{\left(\sum \sigma_i\right)^2}{n}.$$

Writing $p_{n\sigma}$ for a design p_n with

$$\pi_i = \frac{n\sigma_i}{\sum \sigma_i} \tag{3.13}$$

we have

$$E_m V_{p_n}(t) \geq E_m V_{p_n}(t_\mu) = \sum \sigma_i^2 \left[\frac{1}{\pi_i} - 1\right]$$
$$\geq \frac{\left(\sum \sigma_i\right)^2}{n} - \sum \sigma_i^2 = E_m V_{p_{n\sigma}}(t_\mu).$$

RESULT 3.6 *Let p_n and $p_{n\sigma}$ be fixed size n designs, $p_{n\sigma}$ satisfying Eq. (3.13). Then, under $\mathcal{M}_2$,*

$$E_m V_{p_n}(t) \geq E_m V_{p_{n\sigma}}(t_\mu)$$
$$= \frac{\left(\sum \sigma_i\right)^2}{n} - \sum \sigma_i^2$$

for any UE t; here μ_i, σ_i^2 are defined in $\mathcal{M}_2$ and

$$t_\mu = \sum_{i \in s} \frac{Y_i - \mu_i}{\pi_i} + \mu.$$

REMARK 3.2 *Obviously,*

$$t_\mu = \sum_{i \in s} \frac{Y_i}{\pi_i} - \left(\frac{\sum_1^N \sigma_i}{n}\right) \sum_{i \in s} \frac{\mu_i}{\sigma_i} + \mu. \tag{3.14}$$

If we have, in particular, $\mu_i > 0$ and

$$\sigma_i \propto \mu_i$$

for $i = 1, 2, \ldots, N$, then t_μ reduces to the HTE

$$\bar{t} = \sum \frac{Y_i}{\pi_i} = \frac{\sum_1^N \sigma_i}{n} \sum_{i \in s} \frac{Y_i}{\sigma_i} \tag{3.15}$$

because of

$$\pi_i = n\sigma_i \Big/ \sum_i^N \sigma_i.$$

3.2.4 Model $\mathcal{M}_{2\gamma}$

Now, $p_{n\sigma}$ and t_μ are practicable only if $\sigma_1, \sigma_2, \ldots, \sigma_N$ and $\mu_1, \mu_2, \ldots, \mu_N$, respectively, are known up to proportionality factors. A useful case is

$$\sigma_i^2 \propto X_i^\gamma$$
$$\mu_i \propto X_i$$

where $X_1, X_2, \ldots, X_N > 0$ are given size measures and $\gamma \geq 0$ is known. The superpopulation model defined by $\mathcal{M}_2$ with these proportionality conditions is denoted by $\mathcal{M}_{2\gamma}$.

Consider, for example, $\mathcal{M}_{22}$. This model postulates independence of $\varepsilon_1, \varepsilon_2, \ldots, \varepsilon_N$ and for $i = 1, \ldots, N$

$$Y_i = X_i\beta + \sigma X_i \varepsilon_i$$

with

$$E_m \varepsilon_i = 0$$
$$V_m \varepsilon_i = 1.$$

Assume $\mathcal{M}_{22}$ and Eq. (3.13). Then $\pi_i \propto X_i$ and t_μ reduces to

$$\bar{t} = \frac{X}{n} \sum_{i \in s} \frac{Y_i}{X_i}.$$

Then, according to Result 3.6

$$E_m V_{p_n}(t) \geq E_m V_{p_{nx}}(\bar{t}) = \sigma^2 \left[\frac{X^2}{n} - \sum X_i^2 \right]$$

if $\sigma_i^2 = \sigma^2 X_i^2$ for $i = 1, 2, \ldots, N$.

RESULT 3.7 *Let $m = \mathcal{M}_{22}$, i.e., $\mathcal{M}_2$ with*

$$\mu_i \propto X_i$$
$$\sigma_i^2 \propto X_i^2.$$

Let t be a UE with respect to the fixed size n design p_n while p_{nx} is a fixed size n design with inclusion probabilities $\pi_i = n\frac{X_i}{X}$. Then

$$E_m V_{p_n}(t) \geq E_m V_{p_{nx}}(\bar{t})$$
$$= \sigma^2 \left[\frac{X^2}{n} - \sum X_i^2 \right]$$

if $\sigma_i^2 = \sigma^2 X_i^2$ for $i = 1, 2, \ldots, N$.

This optimality property of the HTE follows from the works of GODAMBE and JOSHI (1965), GODAMBE and THOMPSON (1977), and HO (1980).

3.2.5 Comparison of RHCE and HTE under Model $\mathcal{M}_{2\gamma}$

Incidentally, we have already noted that if a fixed sample-size design is employed with $\pi_i \propto Y_i$, then $V_p(\bar{t}) = 0$. But $\underline{Y}$ is unknown. So, if $\underline{X} = (X_1, \ldots, X_i, \ldots, X_N)'$ is available such that Y_i is approximately proportional to X_i, for example, $Y_i = \beta X_i + \varepsilon_i$, with β an unknown constant, ε_i's small and unknown but X_i's known and positive, then taking $\pi_i \propto X_i$, one may expect to have $V_p(\bar{t})$ under control. Any sampling design p with $\pi_i \propto X_i$ is called an IPPS or πPS design—more fully, an **inclusion probability proportional to size design**. Numerous schemes are available that satisfy or approximate this πPS criterion for $n \geq 2$. One may consult BREWER and HANIF (1983) and CHAUDHURI and VOS (1988) for a description of many of them along with a discussion of their properties and limitations. We need not repeat them here.

Supposing n as the common fixed sample size and $N/n = 1/f$ as an integer let us compare $\bar{t}$ based on a πPS scheme with t_3 based on the RHC scheme with N/n as the common group size and $P_i = X_i/X$ as the normed size measures. For this we postulate a superpopulation model $\mathcal{M}_{2\gamma}$:

$$Y_i = \beta X_i + \varepsilon_i,\ E_m(\varepsilon_i) = 0,\ \ V_m(\varepsilon_i) = \sigma^2 X_i^\gamma$$

where σ, γ are non-negative unknown constants and Y_i's are supposed to be independently distributed. Then, with $\pi_i = nP_i = nX_i/X$

$$E_m[V_p(t_3) - V_p(\bar{t})]$$

$$= E_m\left[\frac{N-n}{N-1}\frac{1}{n}\sum\sum_{i<j} X_i X_j \left(\frac{Y_i}{X_i} - \frac{Y_j}{X_j}\right)^2 - \sum\sum_{i<j}(\pi_i\pi_j - \pi_{ij})\left(\frac{Y_i}{\pi_i} - \frac{Y_j}{\pi_j}\right)^2\right]$$

$$= \sigma^2 \left[\frac{N-n}{N-1} \frac{1}{n} \sum\sum_{i<j} X_i X_j \left(X_i^{\gamma-2} + X_j^{\gamma-2} \right) \right.$$

$$\left. - \sum\sum_{i<j} \left(X_i X_j - \frac{X^2 \pi_{ij}}{n^2} \right) \left(X_i^{\gamma-2} + X_j^{\gamma-2} \right) \right]$$

$$= \sigma^2 \left[\frac{N-n}{N-1} \frac{1}{n} \left(X \sum X_i^{\gamma-1} - \sum X_i^{\gamma} \right) \right.$$

$$\left. - \left(X \sum X_i^{\gamma-1} - \sum X_i^{\gamma} \right) + \frac{n-1}{n} X \sum X_i^{\gamma-1} \right]$$

$$= \sigma^2 \frac{(n-1)}{n(N-1)} \left[N \sum X_i^{\gamma} - \left(\sum X_i \right) \left(\sum X_i^{\gamma-1} \right) \right]$$

$$= \frac{\sigma^2 N^2 (n-1)}{(N-1)n} \, cov \left(X_i^{\gamma-1}, X_i \right).$$

Writing $\gamma - 1 = a$ and noting that $X_i > 0$ for all $i = 1, \ldots, N$, it follows that $X_i \geq X_j \Rightarrow X_i^a \geq X_j^a$ if $a \geq 0$ and $X_i \geq X_j \Rightarrow X_i^a \leq X_j^a$ if $a \leq 0$, implying that for $\gamma \leq 1, cov(X_i^{\gamma-1}, X_i) \leq 0$ and for $\gamma \geq 1, cov(X_i^{\gamma-1}, X_i) \geq 0$ and, of course, for $\gamma = 1$, $cov(X_i^{\gamma-1}, X_i) = 0$. So,

for $\gamma < 1, E_m V_p(RHCE) < E_m V_p(HTE)$,

for $\gamma > 1, E_m V_p(RHCE) > E_m V_p(HTE)$,

for $\gamma = 1, E_m V_p(RHCE) = E_m V_p(HTE)$.

Thus, when $\gamma < 1$, HTE is not optimal when based on any πPS design relative to other available strategies. So, it is necessary to have more elaborate comparisons among available strategies under superpopulation models coupled with empirical and simulated studies. Many such exercises are known to have been carried out. Relevant references are RAO and BAYLESS (1969) and BAYLESS and RAO (1970), and for a review, CHAUDHURI and VOS (1988).

Under the same model $\mathcal{M}_{2\gamma}$ above, CHAUDHURI and ARNAB (1979) compared these two strategies with the strategy involving t_R based on LMS scheme (see section 2.4.5) taking the same n, X_i, and $P_i = X_i/X$ as above for all the three strategies. Their finding is stated below, omitting the complicated proof.

$$\text{for } \gamma < 1, E_m V_p(t_R) < E_m V_p(RHCE) < E_m V_p(HTE),$$
$$\text{for } \gamma > 1, E_m V_p(t_R) > E_m V_p(RHCE) > E_m V_p(HTE),$$
$$\text{for } \gamma = 1, E_m V_p(t_R) = E_m V_p(RHCE) = E_m V_p(HTE).$$

3.2.6 Equicorrelation Model

Following CSW (1976, 1977), consider the model of **equicorrelated** Y_i's for which

$$E_m(Y_i) = \alpha_i + \beta X_i$$

α_i known with mean $\overline{\alpha}$, β unknown, $0 < X_i$ known with $\Sigma X_i = N$,

$$V_m(Y_i) = \sigma^2 X_i^2$$
$$C_m(Y_i, Y_j) = \rho\sigma^2 X_i X_j, -\frac{1}{N-1} < \rho < 1.$$

Linear unbiased estimators (LUE) for $\overline{Y}$ are of the form

$$t = t(s, \underline{Y}) = a_s + \sum_{i \in s} b_{si} Y_i$$

with a_s, b_{si} free of $\underline{Y}$ such that for a fixed design p

$$E_p(a_s) = 0, \sum_{s \ni i} b_{si}\, p(s) = \frac{1}{N} \text{ for } i = 1, \ldots, N.$$

To find an optimal strategy (p, t) let us proceed as follows. First note that writing $c_{si} = b_{si} X_i$,

$$1 = \frac{X}{N} = \frac{1}{N} \sum_1^N X_i = \sum_1^N \sum_{s \ni i} c_{si}\, p(s) = \sum_s p(s) \left[\sum_{i \in s} c_{si} \right]. \tag{3.16}$$

Again we have

$$E_m V_p(t) = E_p V_m(t) + E_p [E_m(t) - E_m(\overline{Y})]^2 - V_m(\overline{Y})$$

$$= E_p \left[\sigma^2 \sum b_{si}^2 X_i^2 + \rho\sigma^2 \sum\sum_{i \neq j \in s} b_{si} b_{sj} X_i X_j \right]$$

$$+ E_p \left[a_s + \sum_{i \in s} b_{si}(\alpha_i + \beta X_i) - \overline{\alpha} - \beta \right]^2$$

$$- \frac{1}{N^2} \left[\sigma^2 \sum X_i^2 + \rho\sigma^2 \sum\sum_{i \neq j} X_i X_j \right]$$

$$= \sigma^2 \sum_s p(s) \left[\sum c_{si}^2 + \rho \sum\sum_{i \neq j \in s} c_{si} c_{sj} \right]$$

$$+ E_p \left[a_s - \overline{\alpha} + \sum_{i \in s} \alpha_i b_{si} + \beta \sum_{i \in s} c_{si} - \beta \right]^2$$

$$- \frac{\sigma^2}{N^2} \left[\sum X_i^2 + \rho \left\{ \left(\sum X_i \right)^2 - \sum X_i^2 \right\} \right].$$

Note that

$$\sum_s p(s) \left[\sum_{i \in s} c_{si}^2 + \rho \sum\sum_{i \neq j \in s} c_{si} c_{sj} \right]$$

$$= \sum p(s) \left[\{1 - (1 - \rho)\} \left(\sum_{i \in s} c_{si} \right)^2 + (1 - \rho) \sum_{i \in s} c_{si}^2 \right]$$

$$= \sum p(s) \left(\sum_{i \in s} c_{si} \right)^2 - (1 - \rho) \left[\sum p(s) \left\{ \left(\sum_{i \in s} c_{si} \right)^2 - \sum_{i \in s} c_{si}^2 \right\} \right]$$

$$\geq 1 - (1 - \rho) \left[\sum_s p(s) \left\{ \left(\sum_{i \in s} c_{si} \right)^2 - \sum_{i \in s} c_{si}^2 \right\} \right] \qquad (3.17)$$

by Cauchy's inequality and Eq. (3.16).

To maximize the second term in Eq. (3.17) subject to Eq. (3.16) we need to solve the following equation:

$$0 = \frac{\partial}{\partial c_{si}}\left[\sum_s p(s)\left(\sum_{i\in s} c_{si}\right)^2 - \sum_s p(s)\left(\sum_{i\in s} c_{si}^2\right)\right.$$
$$\left. - \lambda\left(\sum_s p(s)\sum_{i\in s} c_{si} - 1\right)\right]$$
$$= 2p(s)\left(\sum_{i\in s} c_{si}\right) - 2c_{si}\,p(s) - \lambda p(s)$$

where a Lagrangian multiplier λ has been introduced. Then, for $p(s) > 0$,

$$\sum_{i\in s} c_{si} - c_{si} = \frac{\lambda}{2}.$$

Assuming a design p_n, we get by summing up over $i \in s$

$$\sum_{i\in s} c_{si} = \frac{n\lambda}{2(n-1)}$$

giving

$$1 = \sum_s p(s)\sum_{i\in s} c_{si} = \frac{n\lambda}{2(n-1)}$$

hence

$$\sum_{i\in s} c_{si} = 1 \quad \text{and} \quad c_{si} = \frac{1}{n}.$$

Note that equality holds in Eq. (3.17) for $c_{si} = \frac{1}{n}$. Since

$$b_{si} = \frac{c_{si}}{X_i} = \frac{1}{nX_i}$$

we derive, following CSW (1976, 1977),

$$E_p\left[a_s - \overline{\alpha} + \sum_{i\in s} \alpha_i b_{si} + \beta\sum_{i\in s} c_{si} - \beta\right]^2 = 0,$$

choosing

$$a_s = \overline{\alpha} - \frac{1}{n}\sum_{i\in s}\frac{\alpha_i}{X_i}.$$

This leads to the optimal estimator

$$t_\alpha = \overline{\alpha} + \frac{1}{N}\sum_{i\in s}\frac{Y_i - \alpha_i}{\pi_i}, \qquad \pi_i = \frac{nX_i}{X} = \frac{nX_i}{N}.$$

It follows that

$$\begin{aligned} E_m V_{p_n}(t) &\geq E_m V_{p_{nx}}(t_\alpha) \\ &= \sigma^2\left[1 - (1-\rho)\left(1 - \frac{1}{n}\right)\right] \\ &\quad - \frac{\sigma^2}{N^2}\left[\sum X_i^2 + \rho\left(N^2 - \sum X_i^2\right)\right] \\ &= \sigma^2\frac{(1-\rho)}{n}\left[1 - f\,\frac{\sum X_i^2}{N}\right] \end{aligned}$$

where we have written $f = \frac{n}{N}$ as will be done throughout.

RESULT 3.8 *Consider the equicorrelation model*

$$Y_i = \alpha_i + \beta X_i + X_i\varepsilon_i$$

with $E_m\varepsilon_i = 0$ *and*

$$\begin{aligned} V_m(\varepsilon_i) &= \sigma^2 \\ C_m(\varepsilon_i, \varepsilon_j) &= \rho\sigma^2, i \neq j. \end{aligned}$$

Define $\overline{\alpha} = \Sigma\alpha_i/N$ *and*

$$t_\alpha = \overline{\alpha} + \frac{1}{n}\sum_{i\in s}\frac{Y_i - \alpha_i}{X_i}.$$

Then, for any linear estimator t that is unbiased for $\overline{Y}$,

$$\begin{aligned} E_m V_{p_n}(t) &\geq E_m V_{p_{nx}}(t_\alpha) \\ &= \sigma^2\frac{1-\rho}{n}\left[1 - f\,\frac{\sum X_i^2}{N}\right]. \end{aligned}$$

3.2.7 Further Model-Based Optimality Results and Robustness

Avoiding details, we may briefly mention a few recently available optimality results of interest under certain superpopulation models related to the models considered so far.

Postulating independence of Y_i's subject to

(a) $E_m(Y_i) = \alpha_i + \beta X_i$

with $X_i(> 0), \underline{\alpha} = (\alpha_1, \ldots, \alpha_N)'$, β known

and

(b) $V_m(Y_i) = \sigma^2 f_i^2$

$\sigma(> 0)$ unknown, $f_i(> 0)$ known, $i = 1, \ldots, N$

GODAMBE (1982) showed that a strategy (p_n^*, e^*) is optimal among all strategies (p_n, e) with $E_{p_n}(e) = Y$ in the sense that

$$E_m V_{p_n}(e) \geq \sigma^2 \left[\left(\sum f_i \right)^2 / n - \sum f_i^2 \right] = E_m V_{p_n^*}(e^*)$$

for all $\underline{Y}$. Here p_n^* is a p_n for which π_i equals

$$\pi_i^* = n f_i \Big/ \sum_{j=1}^{N} f_j$$

and

$$e^* = \sum_{i \in s} (Y_i - \alpha_i - \beta X_i)/\pi_i^* + \sum_1^N (\alpha_i + \beta X_i)$$
$$= t(\underline{\alpha}, \beta), \text{ say}$$

which is the generalized difference estimator (GDE) in this case.

TAM (1984) revised the above model, relaxing independence and postulating the covariance structure specified by

$$C_m(Y_i, Y_j) = \rho \sigma^2 f_i f_j$$

with $\rho(0 \leq \rho \leq 1)$ unknown, but considered only LUEs

$$e = a_s + \sum_{i \in s} b_{si} Y_i = e_L, \text{ say.}$$

With this setup he showed that

$$E_m V_{p_n}(e_L) = E_m E_{p_n}(e_L - Y)^2 \geq \sigma^2(1-\rho)\left[\frac{(\sum f_i)^2}{n} - \sum f_i^2\right]$$
$$= E_m E_{p_n^*}(e^* - Y)^2$$
$$= E_m V_{p_n^*}(e^*).$$

It is important to observe here that the same strategy (p_n^*, e^*) is optimal under both GODAMBE's (1982) and TAM's (1984) models provided one admits only linear design-unbiased estimators based on fixed sample-size designs.

If in (a), β is unknown but $\underline{\alpha}$ is known, then adopting a design p_{nx} for which

$$\pi_i = \frac{nX_i}{X}, i = 1, \ldots, N$$

one may employ the estimator

$$\frac{X}{n}\sum_{i\in s}\left[\frac{Y_i - \alpha_i}{X_i}\right] + \sum_1^N \alpha_i = t(\underline{\alpha}), \text{ say,}$$

to get rid of β in $(\underline{\alpha}, \beta)$. But $E_m V_{p_{nx}}[t(\underline{\alpha})]$ will differ from $E_m V_{p_n^*}(e^*)$ under GODAMBE's (1982) and TAM's (1984) models and the extent of the deviation will depend on the variation among the $X_i/f_i, i = 1, \ldots, N$. So, $t(\underline{\alpha})$ is optimal if $X_i \propto f_i$ and remains nearly so if X_i/f_i's vary within a narrow range.

If both $\underline{\alpha}$ and β are unknown, then a course to follow is to try the HORVITZ–THOMPSON (1952) estimator

$$\bar{t} = \sum_{i\in s}\frac{Y_i}{\pi_i}$$

instead of the optimal estimator $t(\underline{\alpha}, \beta)$. Then, since

$$E_m V_p(\bar{t}) = E_p V_m(\bar{t}) + E_p \Delta_m^2(\bar{t}) - V_m(Y)$$

where $\Delta_m(e) = E_m(e - Y)$, for any p-unbiased estimator e of Y, GODAMBE (1982) suggests employing a p_n design p_{n0}, say, such that each of

(a) $E_{p_{n0}}\Delta_m^2(\bar{t})$

(b) $E_{p_{n0}}(\bar{t} - t(\underline{\alpha}, \beta))^2$

(c) $E_{p_{n0}}\Delta_m^2(\bar{t}) - E_{p_{n0}}\Delta_m^2(t(\underline{\alpha}, \beta))$

is small so that $E_m V_{p_{n0}}(\bar{t})$ may not appreciably exceed $E_m V_{p_n^*}(t(\underline{\alpha}, \beta))$. If these conditions can be realized then it will follow that $\bar{t}$, which is optimal in the special case when $\alpha_i = 0$, $i = 1, \ldots, N$ and $f_i \propto X_i$, approximately remains so even otherwise. Such a property of a strategy is called **robustness**. A reader may consult GODAMBE (1982) for further discussions and also for reviews IACHAN (1984) and CHAUDHURI and VOS (1988).

MUKERJEE and SENGUPTA(1989) considered e_L as above, but a more general model stipulating

$$E_m(Y_i) = \mu_i, C_m(Y_i, Y_j) = v_{ij}$$

and obtained the optimality result

$$\begin{aligned} E_m V_{p_n}(e_L) &= E_m E_{p_n}(e_L - Y)^2 \geq \underline{1}'\Phi^{-1}\underline{1} - \underline{1}'V\,\underline{1} \\ &= E_m E_{\overline{p}_n}(\overline{e}_L - Y)^2 \\ &= E_m V_{\overline{p}_n}(\overline{e}_L) \end{aligned}$$

Here $V = (v_{ij})$, $\underline{1}$ is the $N \times 1$ vector with each entry as unity, $\Phi = (\Phi_{ij})$, $\Phi_{ij} = \sum_{s \ni i,j} v_s^{ij} p_n(s)$, $v_s^{ij} = ij$th element of the inverse of the matrix V_s, which is an $n \times n$ submatrix of V containing only the entries for $i \in s$. Further,

$$\underline{\lambda} = \Phi^{-1}\underline{1}.$$

$\underline{\lambda}_s$ is an $n \times 1$ subvector of $\underline{\lambda}$ with only entries for $i \in s$, $\underline{b}_s$ is an $n \times 1$ vector with entries b_{si} for $i \in s$, and

$$\underline{\bar{b}}_s = V_s^{-1}\underline{\lambda}_s$$

$$\bar{a}_s = \sum_1^N \mu_i - \sum_{i \in s} \bar{b}_{si}\mu_i.$$

$\overline{e}_L$ is e_L evaluated at $a_s = \bar{a}_s$ and $\underline{b}_s = \underline{\bar{b}}_s$ and $\overline{p}_n$ is a p_n design for which $\underline{1}'\Phi^{-1}\underline{1}$ is the least.

An important point noted by these authors with due illustrations and emphasis in this case is that the optimal estimator $\overline{e}_L$ here need not be the GDE.

A common limitation of each of these three optimality results above is the dependence, except in special cases, of both the design and the estimator components of the optimal strategies on model parameters, which in practice should be unknown. One way to circumvent this is to use a simpler strategy that is free of unknown parameters but optimal when a special case of a model obtains and identify circumstances when it continues to be so at least closely under more comprehensive modeling, which we have just illustrated. A second course may be to substitute unknown parameters in the optimal strategies by their suitable estimators. How to ensure good properties for the resulting strategies thus revised is a crucial issue in survey sampling, which we will discuss further in chapter 6.

3.3 ESTIMATING EQUATION APPROACH

Following the pioneering work of GODAMBE (1960b) and later developments by GODAMBE and THOMPSON (1986a, 1986b) we shall discuss an alternative approach of deriving suitable sampling strategies.

3.3.1 Estimating Functions and Equations

Suppose $\underline{Y} = (Y_1, \ldots, Y_N)'$ is a random vector and $\underline{X} = (X_1, \ldots, X_N)'$ is a vector of known numbers $X_i (> 0)$, $i = 1, \ldots, N$. Let the Y_i's be independent and normally distributed with means and variances, respectively

$$\theta X_i \text{ and } \sigma_i^2, \quad i = 1, \ldots, N.$$

If all the Y_i's $i = 1, \ldots, N$ are available for observation, then from the joint probability density function (pdf) of $\underline{Y}$

$$p(\underline{Y}, \theta) = \prod_{i=1}^{N} \frac{1}{\sigma_i \sqrt{2\pi}} e^{-\frac{1}{2\sigma_i^2}(Y_i - \theta X_i)^2}$$

one gets the well-known maximum likelihood estimator (MLE) θ_0, based on $\underline{Y}$, for θ, given by the solution of the **likelihood equation**

$$\frac{\partial}{\partial \theta} \log p(\underline{Y}, \theta) = 0$$

as

$$\theta_0 = \left[\sum_1^N Y_i X_i / \sigma_i^2\right] \Big/ \left[\sum_1^N X_i^2 / \sigma_i^2\right].$$

On the other hand, let the normality assumption above be dropped, everything else remaining unchanged, that is, consider the linear model

$$Y_i = \theta X_i + \varepsilon_i$$

with ε_i's distributed independently and

$$E_m(\varepsilon_i) = 0, V_m(\varepsilon_i) = \sigma_i^2, \quad i = 1, \ldots, N.$$

Then, if (Y_i, X_i), $i = 1, \ldots, N$ are observed, one may derive the same θ_0 above as the least squares estimator (LSE) or as the best linear unbiased estimator (BLUE) for θ.

Such a θ_0, based on the entire finite population vector $\underline{Y} = (Y_1, \ldots, Y_N)'$, is really a parameter of this population itself and will be regarded as a **census estimator**.

If $X_i = 1, \sigma_i = \sigma$ for all i above, then θ_0 reduces to $Y/N = \overline{Y}$.

We shall next briefly consider the theory of estimating functions and estimating equations as a generalization that unifies (see GHOSH, 1989) both of these two principal methods of point estimation and, in the next section, illustrate how the theory may be extended to yield estimators in the usual sense of the term based on a sample of Y_i values rather than on the entire $\underline{Y}$ itself.

We start with the supposition that $\underline{Y}$ is a random vector with a probability distribution belonging to a class C of distributions each identified with a real-valued parameter θ. Let

$$g = g(\underline{Y}, \theta)$$

be a function involving both $\underline{Y}$ and θ such that

(a) $\frac{\partial g}{\partial \theta}(\underline{Y}, \theta)$ exists for every $\underline{Y}$

(b) $E_m g(\underline{Y}, \theta) = 0$, called the **unbiasedness** condition

(c) $E_m \frac{\partial g}{\partial \theta}(\underline{Y}, \theta) \neq 0$

(d) the equation $g(\underline{Y}, \theta) = 0$ admits a unique solution $\theta_0 = \theta_0(\underline{Y})$

Such a function $g = g(\underline{Y}, \theta)$ is called an **unbiased estimating function** and the equation

$$g(\underline{Y}, \theta) = 0$$

is called an **unbiased estimating equation**.

Let G be a class of such unbiased estimating functions for a given C. Furthermore, let g be any estimating function and θ the true parameter. If $\underline{Y}$ happens to be such that $|g(\underline{Y},\theta)|$ is small while $|\frac{\partial g}{\partial\theta}(\underline{Y},\theta)|$ is large, then θ_0 with $g(\underline{Y},\theta_0)=0$ should be close to θ; note that using TAYLOR's expansion this is quite obvious if $g(\underline{Y},\theta)$ is linear in θ.

Since $g(\underline{Y},\theta)$ and $\frac{\partial g}{\partial\theta}(\underline{Y},\theta)$ are random variables, this observation motivated GODAMBE (1960b) to call a function g_0 in G as well as the corresponding estimating equation $g_0=0$ **optimal** if for all $g\in G$

$$\frac{E_m(g_0^2(\underline{Y},\theta))}{\left[E_m\frac{\partial g_0}{\partial\theta}(\underline{Y},\theta)\right]^2}\le\frac{E_m(g^2(\underline{Y},\theta))}{\left[E_m\frac{\partial g}{\partial\theta}(\underline{Y},\theta)\right]^2}. \tag{3.18}$$

If in a particular case $\underline{Y}$ has the density function $p(\underline{Y},\theta)$, not necessarily normal but satisfying certain regularity conditions (cf. GODAMBE, 1960b) usually required for MLEs to have their well-known properties (cf. CRAMÉR, 1966), then this optimal g_0 turns out to be the function

$$\frac{\partial}{\partial\theta}\log\,p(\underline{Y},\theta).$$

Consequently, the likelihood equation

$$\frac{\partial}{\partial\theta}\log\,p(\underline{Y},\theta)=0$$

is the optimal unbiased estimating equation, implying that the MLE is a desired good estimator θ_0 for θ.

Without requiring a knowledge of the density function of $\underline{Y}$ and thus intending to cover more general situations, let it be possible to find unbiased estimating functions

$$\phi_i(Y_i,\theta),\ i=1,\ldots,N$$

that is,

(a) $E_m\phi_i(Y_i,\theta)=0$
(b) $\frac{\partial}{\partial\theta}\phi_i(Y_i,\theta)$ exists for all $\underline{Y}$
(c) $E_m\frac{\partial}{\partial\theta}\phi_i(Y_i,\theta)\neq 0$.

Then,

$$g = g(\underline{Y}, \theta) = \sum_{1}^{N} \phi_i(Y_i, \theta) a_i(\theta) = \sum_{1}^{N} \phi_i a_i, \text{ say,}$$

with differentiable functions $a_i(\theta)$ is an unbiased estimating function, which is called **linear** in $\phi_i(Y_i, \theta)$; $i = 1, 2, \ldots, N$. If we restrict to such a class $L(\phi)$, then a function $g_0 \in L(\phi)$, satisfying Eq. (3.18) for all $g \in L(\phi)$, is called **linearly optimal**.

If, in particular, the Y_i's are assumed to be independently distributed, then a sufficient condition for linear optimality of

$$g_0 = g_0(\underline{Y}, \theta) = \sum \phi_i(Y_i, \theta)$$

is that

$$E_m \frac{\partial \phi_i}{\partial \theta}(Y_i, \theta) = k(\theta) E_m \phi_i^2(Y_i, \theta), \tag{3.19}$$

for $i = 1, 2, \ldots, N$, where $k(\theta)$ is a non-zero constant free of $\underline{Y}$.

The condition Eq. (3.18), taking $g = \Sigma \phi_i a_i$ and $g_0 = \Sigma \phi_i$ in $L(\phi)$, may be checked on noting that for

$$u = \frac{\sum \phi_i a_i}{E_m \frac{\partial}{\partial \theta}\left(\sum \phi_i a_i\right)}, v = \frac{\sum \phi_i}{E_m \frac{\partial}{\partial \theta} \sum \phi_i}$$

one has $E_m(uv) = E_m(v^2)$, giving $E_m(u^2) - E_m(v^2) = E_m(u - v)^2 \geq 0$.

EXAMPLE 3.4 *Let the Y_i's be independently distributed with $E_m(Y_i) = \theta X_i$, X_i known, $V_m(Y_i) = \sigma_i^2$. Taking*

$$\phi_i(Y_i, \theta) = \frac{X_i(Y_i - \theta X_i)}{\sigma_i^2}$$

and checking Eq. (3.19) one gets

$$g_0 = \sum_{i}^{N} \frac{X_i(Y_i - \theta X_i)}{\sigma_i^2}$$

and as a solution of $g_0 = 0$:

$$\theta_0 = \frac{\sum_1^N Y_i X_i / \sigma_i^2}{\sum_1^N X_i^2 / \sigma_i^2}.$$

This is the same MLE and LSE derived under stipulations considered earlier.

3.3.2 Applications to Survey Sampling

A further line of approach is now required because θ_0 itself needs to be estimated from survey data

$$d = (i, Y_i | i \in s)$$

available only for the Y_i's with $i \in s$, s a sample supposed to be selected with probability $p(s)$ according to a design p for which we assume

$$\pi_i = \sum_{s \ni i} p(s) > 0 \quad \text{for all} \quad i = 1, 2, \ldots, N.$$

With the setup of the preceding section, let the Y_i's be independent and consider unbiased estimating functions $\phi_i(Y_i, \theta)$; $i = 1, 2, \ldots, N$. Let

$$\theta_0 = \theta_0(\underline{Y})$$

be the solution of $g(\underline{Y}, \theta) = 0$ where

$$g(\underline{Y}, \theta) = \sum_{1}^{N} \phi_i(Y_i, \theta)$$

and consider estimating this θ_0 using survey data $d = (i, Y_i | i \in s)$. For this it seems natural to start with an **unbiased sampling function**

$$h = h(s, \underline{Y}, \theta)$$

which is free of Y_j for $j \notin s$ and satisfies

(a) $\frac{\partial h}{\partial \theta}(s, \underline{Y}, \theta)$ exists for all $\underline{Y}$
(b) $E_m \frac{\partial h}{\partial \theta}(s, \underline{Y}, \theta) \neq 0$
(c) $E_p h(s, \underline{Y}, \theta) = g(\underline{Y}, \theta)$ for all $\underline{Y}$, the unbiasedness condition.

Let H be a class of such unbiased sampling functions. Following the extension of the approach in section 3.3.1 by GODAMBE and THOMPSON (1986a), we may call a member

$$h_0 = h_0(s, \underline{Y}, \theta)$$

of H and the corresponding equation $h_0 = 0$, **optimal** if

$$\frac{E_m E_p h^2(s, \underline{Y}, \theta)}{\left[E_m E_p \frac{\partial h}{\partial \theta}(s, \underline{Y}, \theta)\right]^2} \tag{3.20}$$

as a function of $h \in H$ is minimal for $h = h_0$.

Because of the unbiasedness condition (c) above, one may check that

$$E_m E_p \left[\frac{\partial h}{\partial \theta}\right] = E_m \left[\frac{\partial g}{\partial \theta}\right]$$
$$E_p(h - g)^2 = E_p h^2 - g^2.$$

So, to minimize Eq. (3.20) it is enough to minimize

$$E_m E_p (h - E_p h)^2.$$

This is in line with the criterion considered in section 3.2.

It follows that the optimal h_0 is given by

$$h_0 = h_0(s, \underline{Y}, \theta) = \sum_{i \in s} \frac{\phi_i(Y_i, \theta)}{\pi_i}$$

To see this, let

$$\alpha = \alpha(s, \underline{Y}, \theta) = h(s, \underline{Y}, \theta) - h_0(s, \underline{Y}, \theta).$$

Then, noting $0 = E_p \alpha(s, \underline{Y}, \theta)$, and checking, with the arguments as in section 3.1.3 that $E_p \alpha h_0 = 0$, one may conclude that

$$E_m E_p h^2 = E_m E_p (h_0 + \alpha)^2 = E_m E_p h_0^2 + E_m E_p (h - h_0)^2$$
$$\geq E_m E_p h_0^2$$

thereby deriving the required optimality of h_0.

On solving the equation

$$h_0(s, \underline{Y}, \theta) = 0$$

for θ one derives an estimator $\hat{\theta}_0$, based on d, which may be regarded as the **optimal sample estimator** for θ_0, the census estimator for θ based on $\underline{Y}$ derived on solving the equation

$$g(\underline{Y}, \theta) = 0.$$

EXAMPLE 3.5 *Consider the model*

$$Y_i = \theta + \varepsilon_i$$

where the ε_i*'s are independent with* $E_m\varepsilon_i = 0,\ V_m\varepsilon_i = \sigma_i^2$. *Then the estimating function*

$$\sum_i^N \phi_i(Y_i, \theta) = \sum_i^N \frac{(Y_i - \theta)}{\sigma_i^2}$$

is linearly optimal, but does not define the survey population parameter $\overline{Y}$, *which is usually of interest. Therefore, we may consider the estimating equation* $g_0 = 0$ *where*

$$g_0 = \sum \phi_i(Y_i, \theta) = \sum (Y_i - \theta)$$

is unbiased and, while not linearly optimal, defines

$$\theta_0 = \overline{Y}$$

and the optimal sample estimator

$$\hat{\theta}_0 = \frac{\sum_s Y_i/\pi_i}{\sum_s 1/\pi_i}$$

for θ_0. *Incidentally, this estimator was proposed earlier by* HÁJEK *(1971).*

In general, the solution θ_0 of

$$g = \sum \phi_i(Y_i, \theta) = 0$$

where $\phi_i(Y_i, \theta)$, $i = 1, 2, \ldots, N$ are unbiased estimating functions is an estimator of the parameter θ of the superpopulation model, provided all $Y_1, Y_2, \ldots, Y_N$ are known. In any case, it may be of interest in itself, that is, an interesting parameter of the population. The solution $\hat{\theta}_0$ of the optimal unbiased sampling equation $h_0 = 0$ is used as an estimator for the population parameter θ_0.

If g is linearly optimal, then the population parameter θ_0 is especially well-motivated by the superpopulation model.

EXAMPLE 3.6 *Consider, for example, the model*

$$Y_i = \theta X_i + \varepsilon_i$$

with $X_1, X_2, \ldots, X_N > 0$, $\varepsilon_1, \varepsilon_2, \ldots, \varepsilon_N$ *independent and*

$$E_m\varepsilon_i = 0, V_m\varepsilon_i = \sigma^2 X_i^{\gamma}, \gamma \geq 0.$$

Define

$$\phi_i(Y_i, \theta) = \frac{X_i(Y_i - \theta X_i)}{X_i^\gamma}.$$

It is easily seen that

$$\sum \phi_i(Y_i, \theta) = 0$$

is linearly optimal. So the solution

$$\theta_0 = \frac{\sum X_i Y_i / X_i^\gamma}{\sum X_i^2 / X_i^\gamma}$$

should be estimated by the solution of

$$\sum_{i \in s} \frac{\phi_i(Y_i, \theta)}{\pi_i} = 0$$

that is, by

$$\hat{\theta}_0 = \frac{\sum_{i \in s} Y_i X_i^{1-\gamma} / \pi_i}{\sum_{i \in s} X_i^{2-\gamma} / \pi_i}.$$

Two cases of special importance are

(a) $\gamma = 1$. *Then*

$$\theta_0 = \frac{\sum_1^N Y_i}{\sum_1^N X_i} = \frac{\overline{Y}}{\overline{X}} \qquad \hat{\theta}_0 = \frac{\sum_{i \in s} Y_i / \pi_i}{\sum_{i \in s} X_i / \pi_i}.$$

(b) $\gamma = 2$. *Then*

$$\theta_0 = \frac{1}{N} \sum \frac{Y_i}{X_i} \qquad \hat{\theta}_0 = \frac{\sum_{i \in s} Y_i / X_i \pi_i}{\sum_{i \in s} 1/\pi_i}.$$

Finally, it is worth noting that among designs p_n with $p_n(s) > 0$ only for samples s containing a fixed number n of units, each distinct, the subclass $p_{n\phi}$ for which

$$\pi_i = n \left[E_m \phi_i^2 \Big/ \sum_1^N E_m \phi_i^2 \right]^{1/2}, i = 1, 2, \ldots, N$$

is optimal because for each of them the value of

$$E_m E_p \left[\sum_{i \in s} \frac{\phi_i(Y_i, \theta)}{\pi_i} \right]^2 = \sum_i^N \frac{E_m(\phi_i^2)}{\pi_i}$$

is minimized.

Thus, among all strategies

$$(p_n, t(d))$$

the optimal class of strategies is

$$(p_{n\phi}, \hat{\theta}(d))$$

where $\hat{\theta} = \hat{\theta}(d)$ is derived on solving

$$\sum_{i \in s} \frac{\phi_i(Y_i, \theta)}{\pi_i} = 0 \text{ in } \theta.$$

3.4 MINIMAX APPROACH

3.4.1 The Minimax Criterion

So far, the performance of a strategy (p, t) has been described by its MSE $M_p(t)$, which is a function defined as the parameter space Ω, the set of all vectors $\underline{Y}$ relevant in a given situation.

Now, Ω may be such that

$$\sup_{\underline{Y} \in \Omega} M_p(t) = R_p(t), \text{ say,}$$

is finite for some strategies (p, t) of a class Δ fixed in advance, especially by budget restrictions. Then it may be of interest to look for a strategy minimizing $R_p(t)$, with respect to the pair (p, t).

Let Δ be the class of all available strategies and $R_p(t)$ be finite for at least some elements of Δ. Then

$$r^* = \inf_{(p,t) \in \Delta} R_p(t) = \inf_{(p,t) \in \Delta} \sup_{\underline{Y} \in \Omega} M_p(t) < \infty$$

and r^* is called **minimax value** with respect to Ω and Δ; a strategy $(p^*, t^*) \in \Delta$ is called a **minimax strategy** if

$$R_{p^*}(t^*) = r^*.$$

For given size measures x and z with

$$\begin{aligned} 0 < X_i; \quad & i = 1, 2, \ldots, N \\ 0 < Z_i \leq Z/2; \quad & i = 1, 2, \ldots, N \end{aligned}$$

where $Z = \sum_1^N Z_i$ let us define the parameter space

$$\Omega_{xz} = \left\{ \underline{Y} \in \mathbb{R}^N \ : \sum \frac{X_i}{X} \left(\frac{Y_i}{Z_i} - \frac{Y}{Z} \right)^2 \leq 1 \right\}.$$

Of special importance is the class of strategies

$\Delta_n = \{(p,t) : p$ of fixed effective size n, t homogeneously linear$\}$.

3.4.2 Minimax Strategies of Sample Size 1

We first consider the special case Δ_1, consisting of all pairs (p,t) such that

$$p(s) > 0 \text{ implies } |s| = 1$$
$$t(s, \underline{Y}) = t(i, \underline{Y}) = Y_i / q_i, \ q_i \neq 0.$$

Writing $p_i = p(i)$ each strategy in Δ_1 may be identified with a pair $(\underline{p}, \underline{q})$; $\underline{p}, \underline{q} \in \mathbb{R}^N$, and its MSE is

$$\sum p_i \left[\frac{Y_i}{q_i} - Y \right]^2.$$

Now, following STENGER (1986), we show that

$$\sup_{\underline{Y} \in \Omega_{xz}} \sum p_i \left[\frac{Y_i}{q_i} - Y \right]^2$$

is minimum for

$$p_i = \frac{X_i}{X} = p_i^*, \text{ say,}$$

$$q_i = \frac{Z_i}{Z} = q_i^*, \text{ say,}$$

$(i = 1, 2, \ldots, N)$ such that $(\underline{p}^*, \underline{q}^*)$ is a minimax strategy. $\underline{Y} \in \Omega_{xz}$ implies $\underline{Y} + \lambda \underline{Z} \in \Omega_{xz}$ for every real λ and the MSE of a strategy $(\underline{p}, \underline{q})$ evaluated for $\underline{Y} + \lambda \underline{Z}$ is

$$\sum p_i \left[\frac{Y_i + \lambda Z_i}{q_i} - Y - \lambda Z \right]^2.$$

This quadratic function of λ is bounded if and only if

$$\frac{Z_i}{q_i} - Z = 0$$

which is equivalent to $q_i = q_i^*$. So $R_p(t) < \infty$ for $(\underline{p}, \underline{q}) = (p, t) \in \Delta_1$ if and only if $\underline{q} = \underline{q}^*$. Now, for

$$A(\underline{p}) = \sup_{\underline{Y} \in \Omega_{xz}} \sum p_i \left[\frac{Y_i}{q_i^*} - Y\right]^2$$

we have

$$A(p^*) = \sup_{\underline{Y} \in \Omega_{xz}} \sum p_i^* \left[\frac{Y_i}{q_i^*} - Y\right]^2 = Z^2.$$

For $p \neq p^*$ there exists j with $p_j = p_j^* + \varepsilon, \varepsilon > 0$.

It is easily seen that

$$p_j^* - 2p_j^* q_j^* + q_j^{*2} > 0.$$

So we may define

$$Y_i^{(j)} = q_j^* \Big/ \sqrt{p_j^* - 2p_j^* q_j^* + q_j^{*2}} \quad \text{for} \quad i = j$$
$$= 0 \quad \text{for} \quad i \neq j.$$

The total $Y^{(j)}$ of $\underline{Y}^{(j)}$ is equal to $Y_j^{(j)}$ and

$$\sum p_i \left[\frac{Y_i^{(j)}}{q_i^*} - Y^{(j)}\right]^2 = Z^2 \frac{p_j - 2p_j q_j^* + q_j^{*2}}{p_j^* - 2p_j^* q_j^* + q_j^{*2}}$$
$$= Z^2 \left[1 + \frac{\varepsilon(1 - 2q_j^*)}{p_j^* - 2p_j^* q_j^* + q_j^{*2}}\right]$$
$$\geq Z^2$$

because $Z_j \leq Z/2$ implies $1 - 2q_j^* \geq 0$.

Obviously, $\underline{Y}^{(j)} \in \Omega_{xz}$ and

$$A(\underline{p}) \geq Z^2 = A(\underline{p}^*)$$

for all $\underline{p}$.

RESULT 3.9 *Consider the class of strategies (p, t) where p is a fixed size 1 design, and t is homogeneously linear (HL).*

In this class the minimax strategy with respect to Ω_{xz} is as follows: Select unit i with probability

$$p_i^* = \frac{X_i}{X}$$

and use the estimator

$$\frac{Y_i}{q_i^*}$$

where $q_i^* = \frac{Z_i}{Z}$ *and* $Z_i \leq Z/2$ *for all* i.

Note that the minimax strategy is unbiased if and only if $\underline{X}$ and $\underline{Z}$ are proportionate.

Consider the special case $X_i = Z_i$ for $i = 1, 2, \ldots, N$. The minimax strategy for Ω_{xx} and Δ_1 obviously consists in selecting a unit with x-proportionate probabilities and using the estimator

$$\frac{Y_i}{X_i} X$$

if the unit i is selected.

REMARK 3.3 *The same strategy has been shown to be minimax in another context by* SCOTT *and* SMITH (*1975*). *Their parameter space is*

$$\Omega_x = \{\underline{Y} \in \mathbb{R}^N \ : \ 0 \leq Y_i \leq X_i \ \ for \ \ i = 1, 2, \ldots, N\}$$

where it is assumed that a subset U_0 *of* $U = \{1, 2, \ldots, N\}$ *exists with*

$$\sum_{i \in U_0} X_i = X/2.$$

They prove that the above strategy is minimax within the set Δ_1^-, say, of all strategies (p, t), p an arbitrary design of fixed sample size 1 and

$$t(i, \underline{Y}) = XY_i/X_i.$$

This result may also be stated as follows: The design of fixed sample size 1 with x-proportionate selection probabilities is minimax if Ω_x is relevant and $t(i, \underline{Y}) = XY_i/X_i$ is prescribed. An exact generalization for arbitrary sample sizes n is not available, but an asymptotic result will be presented in chapter 6.

3.4.3 Minimax Strategies of Sample Size $n \geq 1$

In the special case $X_i = Z_i = 1$ we have the parameter space

$$\Omega_{11} = \left\{ \underline{Y} \in \mathbb{R}^N \ : \ \frac{1}{N} \sum (Y_i - \overline{Y})^2 \leq 1 \right\}$$

and, according to the above result, the minimax strategy within Δ_1 consists of choosing every unit with a probability $1/N$ and employing the estimator NY_i for Y if the unit i is selected.

A much stronger result has been proved by AGGARWAL (1959) and BICKEL and LEHMANN (1981). They consider Ω_{11} and the class Δ_n^+ of all strategies (p_n, t), p_n a design of fixed effective size n and t arbitrary, and show that the expansion estimator $N\overline{y}$ based on SRSWOR of size n is minimax.

Unfortunately, it seems impossible to find analogously general results for other choices of $\underline{X}$ and $\underline{Z}$; however, in chapter 6 we report some results valid at least for large samples.

In the present section we give two results for $n \geq 1$ postulating additional conditions on n in relation to N and X_1, $X_2, \ldots, X_N$.

Assume for $i = 1, 2, \ldots, N$

$$Z_i = 1$$

and

$$\frac{X_i}{X} > \frac{n-1}{n} \frac{1}{N-2}. \tag{3.21}$$

According to the last condition, the variance of the values $X_1, X_2, \ldots, X_N$ must be small. This condition implies that

$$P_i = n \frac{N-2}{N-2n} \frac{X_i}{X} - \frac{n-1}{N-2n} \tag{3.22}$$

$(i = 1, 2, \ldots, N)$ are positive with sum 1. Denote by p_{LMS} the LAHIRI-MIDZUNO-SEN design based on the probabilities P_1, $P_2, \ldots, P_N$, that is, in the first draw unit i is selected with probability P_i ; $i = 1, 2, \ldots, N$ and subsequently $n-1$ distinct units are selected by SRSWOR from the $N-1$ units left after the first draw. STENGER and GABLER (1996) have shown:

RESULT 3.10 *Let $\tilde{t}$ be the expansion estimator for Y and p_{LMS} the* LAHIRI-MIDZUNO-SEN *design based on $P_1, P_2, \ldots, P_N$*

defined in Eq. (3.22). Then

$$(p_{LMS}, \widetilde{t})$$

is minimax in Δ_n with respect to the parameter space

$$\Omega_{x1} = \left\{\underline{Y} \in \mathbb{R}^N \ : \ \sum \frac{X_i}{X}(Y_i - \overline{Y})^2 \le 1\right\}$$

provided Eq. (3.21) is true. The minimax value is

$$\frac{N}{n}\frac{N-n}{N-1}.$$

Another example of a very general nature seems to be important. GABLER and STENGER (2000) assume

$$N - 2n \ge \sum \sqrt{1 - X_i/X_o}$$

where $X_o = \max\{X_1, X_2, \ldots, X_N\}$. By this inequality, situations are eliminated in which the x values of one or a few units add up to 1 or nearly so, such that random sampling is not suggestive. The inequality ensures that

$$(N - 2n)z = \sum_1^N \sqrt{z^2 - X_i}$$

admits a unique solution z_o. We define for $i = 1, 2, \ldots, N$

$$d_i = \frac{z_o + \sqrt{z_o^2 - X_i}}{X_i}$$

and obtain the estimator

$$t^*(s, \underline{Y}) = \sum_{i \in s} a_{si}^* Y_i = \frac{\sum_{i \in s} d_i Y_i}{\sum_{i \in s} d_i X_i}$$

which is of fundamental importance. Defining $\alpha_i = d_i X_i$ for $i = 1, 2, \ldots, N$, $t^*(s, \underline{Y})$ can be written as a HANSEN–HURWITZ type estimator

$$t^*(s, \underline{Y}) = \frac{\sum_{i \in s} \alpha_i \frac{Y_i}{X_i}}{\sum_{i \in s} \alpha_i}$$

The parameter space is assumed to be defined as

$$\Omega = \{\underline{Y} \in \mathbb{R}^N \ : \ \underline{Y}' U \underline{Y} \le 1\}$$

where U is a $N \times N$ non-negative definite matrix with

$$U\underline{X} = \underline{0}.$$

The α_i's do not depend on U. For

$$D = diag(d_1, d_2, \ldots, d_N)$$

$$V^* = D^{-1}\left(I - \frac{\underline{1}\,\underline{1}'}{n}\right)D^{-1} + \underline{X}\,\underline{X}'$$

GABLER and STENGER (1999) show that

$$\sup_{\underline{Y}\in\Omega} MSE(\underline{Y};\ p,t) \geq \frac{1}{tr(UV^*)}$$

for all strategies $(p,t) \in \Delta_n$.

Under the assumption that the variance of $X_1, X_2, \ldots, X_N$ is not too large a design, p^* is constructed such that (p^*, t^*) is minimax.

REMARK 3.4 GABLER (*1990*) *assumes that designs p with $\Sigma|s|p(s) = n$, n fixed, are prescribed while all LEs*

$$t(s, \underline{Y}) = b_s + \sum_{i\in s} b_{si} Y_i$$

are admitted. He considers Ω_x and derives the minimax value

$$r^* = \frac{1}{4n}\left[\overline{X}^2\left(1 - \frac{n}{N}\right) - \frac{n}{N}\sigma_{xx}\right]$$

where

$$\sigma_{xx} = \frac{1}{N}\sum(X_i - \overline{X})^2.$$

We will not discuss GABLER's class of strategies. His result is mentioned especially because the same minimax value r^* will play an important role in our asymptotic discussion of Ω_x and Δ_n in chapter 6.

Chapter 4

Predictors

Writing a finite population total Y as $Y = \Sigma_i Y_i = \Sigma_s Y_i + \Sigma_r Y_i$ an estimator $t = t(s, \underline{Y})$ for it may be written as $t = \Sigma_s Y_i + (t - \Sigma_s Y_i)$, where $\Sigma_s(\Sigma_r)$ is the sum over the distinct units sampled (unsampled). Here a sample s is supposed to be chosen yielding the survey data $d = (i, Y_i | i \in s)$. To find a value $t(d)$ close to Y is equivalent to deriving from $Y_i, i \in s$ a quantity, $t(d) - \Sigma_s Y_i$, which is close to $\Sigma_r Y_i$. In order to achieve this we need a link between $Y_i, i \notin s$ and $Y_i, i \in s$. So far, a link established by a design p has been exploited. Even where a superpopulation model entered the scene, we did not use it to bridge the "gap" between $Y_i, i \in s$ and $Y_i, i \notin s$. We only took advantage of the model when deciding for a specific strategy (p, t) and then based our conclusions on p alone.

In section 4.1 we follow ROYALL (1970, 1971, 1988), considering an approach for estimation founded on a superpopulation from which $\underline{Y}$ at hand is just a realization.

In section 4.2 we assume that a suitable prior density function of $\underline{Y}$ is given and derive Bayes estimators.

4.1 MODEL-DEPENDENT ESTIMATION

We assume that the values Y_i; $i = 1, \ldots, N$ may be considered to be realizations of random variables, also denoted as Y_i; $i = 1, \ldots, N$ and satisfying the conditions of a linear model (regression model). In sections 4.1.1–4.1.4 models with only one explanatory variable are considered, sections 4.1.5–4.1.7 deal with the linear model in its general form.

4.1.1 Linear Models and BLU Predictors

Let a superpopulation be modeled as follows:

$$Y_i = \beta X_i + \varepsilon_i, \ i = 1, \ldots, N$$

where X_i's are the known positive values of a nonstochastic real variable x; ε_i's are random variables with

$$E_m(\varepsilon_i) = 0, \ V_m(\varepsilon_i) = \sigma_i^2, \ C_m(\varepsilon_i, \varepsilon_j) = \rho_{ij}\sigma_i\sigma_j,$$

writing E_m, V_m, C_m as operators for expectation, variance and covariance with respect to the modeled distribution.

To estimate $Y = \Sigma_s Y_i + \Sigma_r Y_i$, where $\Sigma_r Y_i$ is the value of a random variable, is actually to predict this value, add that predicted value to the observed quantity $\Sigma_s Y_i$, and hence obtain a predicted value of Y, which also is a random variable in the present formulation of the problem.

Since

$$\sum_r Y_i = \beta \sum_r X_i + \sum_r \varepsilon_i$$

with $E_m \Sigma_r \varepsilon_i = 0$, a predictor for $\Sigma_r Y_i$ may be $\hat{\beta}\Sigma_r X_i$. Here $\hat{\beta}$ is a function of d (and $\underline{X}$) and for simplicity we will take it as linear in $\underline{Y}$,

$$\hat{\beta} = \sum_s B_i Y_i, \text{ say.}$$

The resulting predictor for Y

$$t = \sum_s Y_i + \hat{\beta} \sum_r X_i$$

will then be **model-unbiased** (m-unbiased) if

$$\begin{aligned}
0 &= E_m(t - Y) \\
&= E_m\left(\sum_s Y_i + \hat{\beta}\sum_r X_i - \sum_s Y_i - \sum_r Y_i\right) \\
&= E_m\left(\hat{\beta}\sum_r X_i - \beta\sum_r X_i - \sum_r \varepsilon_i\right) \\
&= [E_m(\hat{\beta}) - \beta]\sum_r X_i
\end{aligned}$$

that is, if

$$\begin{aligned}
\beta &= E_m\hat{\beta} \\
&= E_m\sum_{i\in s} B_i(\beta X_i + \varepsilon_i) \\
&= \beta\sum_{i\in s} B_i X_i
\end{aligned}$$

which is equivalent to

$$\sum_{i\in s} B_i X_i = 1.$$

Note that the predictor for Y then takes the form

$$\begin{aligned}
t &= \sum_{i\in s}\left(1 + B_i\sum_r X_j\right)Y_i \\
&= \sum_{i\in s} a_{si}Y_i, \text{ say,}
\end{aligned}$$

and

$$\begin{aligned}
\sum a_{si}X_i &= \sum_{i\in s} X_i\left(1 + B_i\sum_r X_j\right) \\
&= \sum_s X_i + \sum_s X_i B_i \cdot \sum_r X_j \\
&= X.
\end{aligned}$$

This is the equation known from representativity and calibration.

For a linear m-unbiased predictor a measure of error is

$$V_m(t-Y) = E_m\,[(t-Y) - E_m(t-Y)]^2$$
$$= E_m\left[\hat{\beta}\sum_r X_i - \sum_r Y_i\right]^2$$
$$= E_m\left[\left(\sum_r X_i\right)(\hat{\beta}-\beta) - \sum_r (Y_i - \beta X_i)\right]^2$$
$$= M, \text{ say.}$$

M is a function of the coefficients B_i, $i \in s$ and may be minimized under the restriction $\Sigma_s B_i X_i = 1$. Let B_{oi}, $i \in s$ be the minimizing coefficients. The corresponding predictor

$$t_o = \sum_s Y_i + \sum_r X_i \sum_s B_{oi} Y_i$$

is naturally called the **best linear unbiased** (BLU) **predictor** (BLUP) for Y.

EXAMPLE 4.1 *For illustration purposes, let us simplify the above model by assuming* $\sigma_i = \sigma X_i (\sigma > 0$*, unknown*) *and* $\rho_{ij} = \rho[-\frac{1}{N-1} < \rho < 1$*, unknown*]. *Then,*

$$M = \left(\sum_r X_i\right)^2 E_m\left[\sum_s B_i(Y_i - \beta X_i)\right]^2 + E_m\left[\sum_r (Y_i - \beta X_i)\right]^2$$
$$-2\sum_r X_i E_m\left[\sum_s B_i(Y_i - \beta X_i)\sum_r (Y_i - \beta X_i)\right]$$
$$= \sigma^2\left[\left(\sum_r X_i\right)^2\left\{\sum_s B_i^2 X_i^2 + \rho\sum\sum_{i\neq j\in s} B_i B_j X_i X_j\right\}\right.$$
$$\left. + \sum_r X_i^2 + \rho\sum\sum_{i\neq j\in r} X_i X_j - 2\left[\sum_r X_i\right]\rho\sum\sum_{i\in s, j\notin s} B_i X_i X_j\right]$$
$$= \sigma^2\left[\left(\sum_r X_i\right)^2\left\{\rho + (1-\rho)\sum_s B_i^2 X_i^2\right\} + (1-\rho)\sum_r X_i^2\right.$$
$$\left. + \rho\left(\sum_r X_i\right)^2 - 2\rho\left(\sum_r X_i\right)^2\right].$$

A choice of B_i *that minimizes* M *subject to* $\Sigma_{i\in s} B_i X_i = 1$ *is* $B_i = 1/nX_i$ *for* $i \in s$, *assuming* n *as the size of* s. *The resulting* *minimal value of M, M_0 is*

$$M_0 = \sigma^2(1-\rho)\left[\sum_r X_i^2 + \left(\sum_r X_i\right)^2 \Big/ n\right]$$
$$= V_m(t_0 - Y) = E_m(t_0 - Y)^2$$

writing t_0 for the linear m-unbiased predictor with the above B_i's called BLUP, that is,

$$t_0 = \sum_s Y_i + \frac{1}{n}\left[\sum_s \frac{Y_i}{X_i}\right]\left[\sum_r X_i\right] = \sum_s Y_i + \hat{\beta}\sum_r X_i.$$

It is easy to see that

$$\hat{\beta} = \frac{1}{n}\sum_s \frac{Y_i}{X_i}$$

occurring in t_0, is the BLU estimator of β.

EXAMPLE 4.2 *Now, we assume, $\rho_{ij} = 0$ for all $i \neq j$. Hence $E_m(Y_i) = \beta X_i$, $V_m(Y_i) = \sigma_i^2$ but $C_m(Y_i, Y_j) = 0, i \neq j$, that is, we have (cf. section 3.2.2) $\mathcal{M}_1$ with $\mu_i = \beta X_i$. Then the BLUP for Y comes out as*

$$t_{BLU} = \sum_s Y_i + \left[\frac{\sum_s Y_i X_i/\sigma_i^2}{\sum_s X_i^2/\sigma_i^2}\right]\left[\sum_r X_i\right]$$

which reduces to the well-known ratio estimator, now to be called the ***ratio predictor****,*

$$t_R = \sum_s Y_i + \left[\frac{\sum_s Y_i}{\sum_s X_i}\right]\left[\sum_r X_i\right] = X\left[\sum_s Y_i\right]\Big/\left[\sum_s X_i\right] = X\overline{y}/\overline{x},$$

if in particular, $\sigma_i^2 = \sigma^2 X_i, i = 1, \ldots, N$, writing $\overline{y}$ ($\overline{x}$) as the sample mean of y (x). It follows, under this model, that

$$M_0 = V_m(t_R - Y) = E_m(t_R - Y)^2$$
$$= E_m\left[\frac{\sum_s Y_i}{\sum_s X_i}\sum_r X_i - \sum_r Y_i\right]^2$$
$$= E_m\left[\frac{\sum_r X_i}{\sum_s X_i}\sum_s (Y_i - \beta X_i) - \sum_r (Y_i - \beta X_i)\right]^2$$
$$= \frac{N^2}{n}(1-f)\frac{\overline{X}\overline{x}_r}{\overline{x}}\sigma^2,$$

writing $\overline{x}_r$ for the mean of the $(N-n)$ unsampled units.

4.1.2 Purposive Selection

We introduce some notations for easy reference to several models.

Arbitrary random variables $Y_1, Y_2, \ldots, Y_N$ may be written as

$$Y_i = \mu_i + \varepsilon_i$$

where $\varepsilon_1, \varepsilon_2, \ldots, \varepsilon_N$ are random variables with

$$E_m(\varepsilon_i) = 0, \;\; V_m(\varepsilon_i) = \sigma_i^2, \;\; C_m(\varepsilon_i, \varepsilon_j) = \rho_{ij}\sigma_i\sigma_j$$

for $i, j = 1, 2, \ldots, N$ and $i \neq j$.

A superpopulation model of special importance is defined by the restrictions

$$\mu_i = \beta X_i$$
$$\sigma_i^2 = \sigma^2 X_i^\gamma$$

with known positive values X_i of a nonstochastic variable x. This model is denoted by

$$\mathcal{M}_{0\gamma} \quad \text{if } \rho_{ij} = \rho \text{ for all } i \neq j$$
$$\mathcal{M}_{1\gamma} \quad \text{if } \rho_{ij} = 0 \text{ for all } i \neq j$$
$$\mathcal{M}_{2\gamma} \quad \text{if } \varepsilon_1, \varepsilon_2, \ldots, \varepsilon_N \text{ are independent}$$

(cf. section 3.2.4). If the assumption $\mu_i = \beta X_i$ is replaced by

$$\mu_i = \alpha + \beta X_i$$

we write $\mathcal{M}'_{j\gamma}$ instead of $\mathcal{M}_{j\gamma}$ for $j = 0, 1, 2$.

In the previous section we have shown that the ratio predictor t_R is BLU under $\mathcal{M}_{11}$ and has the MSE

$$M_0 = \frac{N^2}{n}(1 - f)\frac{\overline{X}\overline{x}_r}{\overline{x}}\sigma^2.$$

It follows from the last formula that if the n units with the largest X_i's are chosen as to constitute the sample on which to base the BLUP t_R, then the value of M_0 will be minimal. So, an optimal sampling design is a purposive one that prescribes to select with probability one a sample of n units with the largest X_i values.

Let the optimal purposive design be denoted as p_{no}. It follows that

$$E_{p_{no}} V_m(t_R - Y) = E_{p_{no}} E_m(t_R - Y)^2 \leq E_{p_n} E_m(t_R - Y)^2$$

for any other design of fixed sample size n.

Consider the model $\mathcal{M}'_{10}$, that is,

$$Y_i = \alpha + \beta X_i + \varepsilon_i$$

with uncorrelated $\varepsilon_1, \varepsilon_2, \ldots, \varepsilon_N$ of equal variance σ^2. Let

$$t = t(s, \underline{Y}) = \sum_s Y_i + \sum_s g_i Y_i$$

be an m-unbiased linear predictor for $Y = \sum_s Y_i + \sum_r Y_i$, that is,

$$E_m\left(t - \sum_s Y_i\right) = E_m\left(\sum_s g_i Y_i\right) = \sum_r (\alpha + \beta X_i).$$

This implies

$$\text{(a)} \quad \sum_s g_i = N - n$$

$$\text{(b)} \quad \sum_s g_i X_i = \sum_r X_i.$$

Note that (a) and (b) may be written as

$$\sum_s g_i X_i^k = \sum_r X_i^k;\ k = 0, 1.$$

Obviously,

$$\begin{aligned} M = V_m(t - Y) &= E_m(t - Y)^2 \\ &= E_m\left[\sum_s g_i Y_i - \sum_r (\alpha + \beta X_i) - \sum_r (Y_i - \alpha - \beta X_i)\right]^2 \\ &= E_m\left[\sum_s g_i Y_i - E_m\left(\sum_s g_i Y_i\right) - \sum_r \varepsilon_j\right]^2 \\ &= E_m\left(\sum_s g_i \varepsilon_i - \sum_r \varepsilon_i\right)^2 \\ &= \left(\sum g_i^2 + N - n\right)\sigma^2. \end{aligned}$$

To minimize this, subject to (a), (b), we are to solve

$$0 = \frac{\partial}{\partial g_i}\left[M - \lambda\left(\sum_s g_i - N + n\right) - \mu\left(\sum_s g_i X_i - \sum_r X_i\right)\right]$$

taking λ, μ as Lagrangian multipliers and derive

$$g_i = \left(\frac{N}{n} - 1\right) + \frac{N(\overline{X} - \overline{x})}{\sum_s (X_i - \overline{x})^2}(X_i - \overline{x}) = g_{io}, \text{ say.}$$

The resulting BLU predictor

$$t_0 = \sum_s Y_i + \sum_s g_{i0} Y_i = N\left[\overline{y} + b(\overline{X} - \overline{x})\right]$$

with

$$b = \sum_s (Y_i - \overline{y})(X_i - \overline{x}) \Big/ \sum_s (X_i - \overline{x})^2$$

is usually called a **regression predictor**. The model variance of t_0 is

$$M_0 = V_m(t_0 - Y) = \left[(N - n) + \sum_s g_{i0}^2\right]\sigma^2$$

$$= N^2\left[\frac{1}{n}(1 - f) + \frac{(\overline{x} - \overline{X})^2}{\sum_s (X_i - \overline{x})^2}\right]\sigma^2.$$

M_0 achieves a minimum if $\overline{x}$ equals $\overline{X}$. So, the optimal design is again a purposive one that prescribes choosing one of the samples of size n that has $\overline{x}$ closest to $\overline{X}$. Note that for $\overline{x} = \overline{X}$ the predictor t_0 is identical with the **expansion predictor** $N\overline{y}$. Analogous optimal purposive designs may also be derived for more general models.

RESULT 4.1 *Let $\mathcal{M}_{10}'$ be given. Then, the* ***regression predictor***

$$t_0 = t_0(s, \underline{Y})$$

$$= N\left[\overline{y} - \frac{\sum_s (Y_i - \overline{y})(X_i - \overline{x})}{\sum_s (X_i - \overline{x})^2}(\overline{x} - \overline{X})\right]$$

is BLU for Y. Its MSE is minimum if

$$\overline{x} = \overline{X}$$

in which case

$$t_0(s, \underline{y}) = N\overline{y}.$$

REMARK 4.1 *Consider the model $\mathcal{M}_{02}$ with the BLUP t_0 given in Example 4.1.*

$V_m(t_0 - Y)$ is minimized for the purposive design p_{n0}. If, in addition, the ϵ_i's are supposed independent, that is, $\mathcal{M}_{22}$ is assumed, then $V_m(t_0 - Y)$ reduces to

$$\sigma^2\left[\sum_r X_i^2 + \frac{\left(\sum_r X_i\right)^2}{n}\right].$$

For this same model an optimal p-unbiased strategy was found in section 3.2.4 as $(p_{nx}, \bar{t})$ among all competitors (p_n, t) with

$$E_{p_n}(t) = Y \quad \text{for every } \underline{Y}$$

in terms of the criterion $E_m E_{p_n}(t - Y)^2$. We may note that for p_{nx}

$$\bar{t} = \sum_s \frac{Y_i}{\pi_i} = \frac{X}{n}\sum_s \frac{Y_i}{X_i}$$

has $E_m(\bar{t}) = \beta X$, that is, like $t_0 = \sum_s Y_i + \frac{1}{n}(\sum_s \frac{Y_i}{X_i})\sum_r X_i$ the HTE $\bar{t}$ is m-unbiased. So, it follows that

$$\begin{aligned} E_m E_{p_{no}}(t_o - Y)^2 &= E_{p_{no}} E_m(t_o - Y)^2 \\ &\leq E_{p_{nx}} E_m(t_o - Y)^2 \\ &\leq E_{p_{nx}} E_m(\bar{t} - Y)^2 = E_m E_{p_{nx}}(\bar{t} - Y)^2 \end{aligned}$$

Thus, the strategy (p_{no}, t_o) is superior to the strategy $(p_{nx}, \bar{t})$, which is optimal in the class of all (p_n, t), t p_n-unbiased.

For any p-unbiased estimator for Y that is also m-unbiased under any specific model, a similar conclusion will follow. So, if a model is acceptable and mathematically tractable, there is obviously an advantage in adopting an optimal model-based strategy involving an optimal purposive design and the pertinent BLUP rather that a p-unbiased estimator.

4.1.3 Balancing and Robustness for $\mathcal{M}_{11}$

In practice, we never will be sure as to which particular model is appropriate in a given situation. Let us suppose that the model $\mathcal{M}_{11}$ is considered adequate and one contemplates adopting the optimal strategy (p_{no}, t_R) for which

$$V_m(t_R - Y) = M_0 = \frac{N^2(1-f)}{n}\frac{\overline{X}\overline{x}_r}{\overline{x}}\sigma^2$$

as noted in section 4.1.1. We intend to examine what happens to the performance of this strategy if the correct model is $\mathcal{M}'_{11}$.

Under $\mathcal{M}'_{11}$,

$$E_m(t_R) = N\alpha\frac{\overline{X}}{\overline{x}} + \beta X$$

and thus t_R has the bias

$$B_m(t_R) = E_m(t_R - Y) = N\alpha\left(\frac{\overline{X}}{\overline{x}} - 1\right)$$

which vanishes if and only if $\overline{x}$ equals $\overline{X}$. So, if instead of the design p_{no}, which is optimal under $\mathcal{M}_{11}$, one adopts a design for which $\overline{x}$ equals $\overline{X}$, then t_R, which is m-unbiased under $\mathcal{M}_{11}$, continues to be m-unbiased under $\mathcal{M}'_{11}$ as well.

A sample for which $\overline{x}$ equals $\overline{X}$ is called a **balanced sample** and a design that prescribes choosing a balanced sample with probability one is called a **balanced design**. Hence, based on a balanced sample, t_R is **robust** in respect of model failure.

It is important to note that t_R based on a balanced sample is identical to the expansion predictor $N\overline{y}$.

REMARK 4.2 *Of course, a balanced design may not be available, for example, if there exists no sample of a given size admitting $\overline{x}$ equal to $\overline{X}$. In that case, an approximately balanced design suggests itself, namely the one that chooses with probability one a sample of a given size for which $\overline{x}$ is the closest to $\overline{X}$. If the sample size n is large, then simple random sampling (SRS) without replacement (WOR) leads with high probability to a sample, which is approximately balanced. This is so because by* CHEBYSHEV*'s inequality, under SRSWOR,*

$$Prob[|\overline{x} - \overline{X}| \leq \varepsilon] \geq 1 - \frac{N-n}{Nn}\frac{S^2}{\varepsilon^2}, \quad \text{for any } \varepsilon > 0,$$

writing $S^2 = \frac{1}{N-1}\sum_1^N (X_i - \overline{X})^2$.

An obvious way to achieve a balance in samples is to stratify a population in terms of the values of x, keeping each stratum internally as homogeneous as possible.

Let the sizes $N_1, N_2, \dots, N_H$ of the H strata be sufficiently large (with $\sum_1^H N_h = N$) and assume that samples are drawn from the H strata independently, by SRSWOR of sufficiently large sizes $n_1, n_2, \dots, n_H$ ($\sum_1^H n_h = n$) with n_h/N_h small relative to 1. Then, the stratum sample mean $\overline{x}_h$ will be quite close to the stratum mean $\overline{X}_h$ of x for $h = 1, 2, \dots, H$. ROYALL *and* HERSON *(1973) is a reference for this approach.*

4.1.4 Balancing for Polynomial Models

We return to the model $\mathcal{M}'_{10}$ of 4.1.2 and consider an extension $\mathcal{M}_k$ defined as follows:

$$Y_i = \sum_{j=0}^{k} \beta_j X_i^j + \varepsilon_i$$

$$E_m(\varepsilon_i) = 0, V_m(\varepsilon_i) = \sigma^2, C_m(\varepsilon_i, \varepsilon_j) = 0, \text{ for } i \neq j$$

where $i, j = 1, 2, \dots, N$. By generalizing the developments of section 4.1.2, we derive.

RESULT 4.2 *Let $\mathcal{M}_k$ be given. Then, the MSE of the BLU predictor t_o for Y is minimum for a sample s of size n if*

$$\frac{1}{n}\sum_s X_i^j = \frac{1}{N}\sum_1^N X_i^j \text{ for } j = 0, 1, \dots, k.$$

If these equalities hold we have

$$t_o(s, \underline{Y}) = N\overline{y}.$$

A sample satisfying the equalities in Result 4.2 is said to be **balanced up to order k**.

Now, assume the true model $\mathcal{M}_{k'}$ agrees with a statistician's working model $\mathcal{M}_k$ in all respects except that

$$E_m(Y_i) = \sum_0^{k'} \beta_j X_i^j$$

with $k' > k$. The statistician will use t_o instead of t'_o, the BLU predictor for Y on the base of $\mathcal{M}_{k'}$. However, if he selects a sample that is balanced up to order k'

$$t_o'(s, \underline{Y}) = t_o(s, \underline{Y}) = N\overline{y}$$

and his error does not cause losses.

It is, of course, too ambitious to realize exactly the balancing conditions even if k' is of moderate size, for example, $k' = 4$ or 5. But if n is large the considerations outlined in Result 4.1 apply again for SRSWOR or SRSWOR independently from within strata after internally homogeneous strata are priorly constructed.

But how it fares in respect to its model mean square error under incorrect modeling is more difficult to examine. Since a model cannot be postulated in a manner that is correct and acceptable without any dispute and a classical design-based but model-free alternative is available, it is considered important to examine how a specific model-based predictor, for example, t_m, fares in respect to design characteristics if it is based on a sample s chosen according to some design p. On such a sample may also be based a design-based estimator t_d, and one may be inclined to compare the magnitudes of the design mean square errors $M_p(t_m) = E_p(t_m - Y)^2$ and $M_p(t_d) = E_p(t_d - Y)^2$. Since $M_p(t_m) = V_p(t_m) + B_p^2(t_m)$ and $M_p(t_d) = V_p(t_d) + B_p^2(t_d)$ it may be argued that if the sample size is sufficiently large, as is the case in large scale sample surveys, in practice both $V_p(t_m)$ and $V_p(t_d)$ may be considered to be small in magnitudes. But $|B_p(t_m)|$ is usually large and appreciably dominates both $|B_p(t_d)|$ and $V_p(t_m)$ and, consequently, for large samples $M_p(t_m)$ often explodes relative to $M_p(t_d)$, especially if t_m is based on an incorrect model.

The estimator t_d itself may or may not be model-based, but even if it is suggested by considerations of an underlying model, its model-based properties need not be invoked; it may be judged only in terms of the design, and, if it has good design properties, it may be considered robust because its performance is evaluated without appeal to a model and hence there is no question of model failures. However, if the sample size is small and the model is not grossly inaccurate, then in terms of model- and design-based mean square error criteria m-based procedures may do better than t_d, as we have seen already.

These discussions suggest the possibility of considering estimators that may be appropriately based on both model and design characteristics so that they may perform well in terms of model-based bias and mean square error when the model is correct, but will also do well in terms of design-based bias and mean square error irrespective of the truth or falsity of the postulated model. To examine such possibilities, in view of what has been discussed above it is necessary to relax the condition of design unbiasedness and to avoid small sample sizes. In the next section we examine the prospects of exploration in some other directions, but we will pursue this problem in chapters 5 and 6.

4.1.5 Linear Models in Matrix Notation

Suppose $x_1, x_2, \ldots, x_k$ are real variables, called **auxiliary** or **explanatory variables**, each closely related to the variable of interest y. Let

$$\underline{x}_i = (X_{i1}, X_{i2}, \ldots, X_{ik})'$$

be the vector of explanatory variables for unit i and assume the linear model

$$Y_i = \underline{x}_i' \underline{\beta} + \varepsilon_i$$

for $i = 1, 2, \ldots, N$. Here

$$\underline{\beta} = (\beta_1, \beta_2, \ldots, \beta_k)'$$

is the vector of (unknown) **regression parameters**; $\varepsilon_1, \varepsilon_2, \ldots, \varepsilon_N$ are random variables satisfying

$$E_m \varepsilon_i = 0$$

$$V_m \varepsilon_i = v_{ii}$$

$$C_m(\varepsilon_i, \varepsilon_j) = v_{ij}, i \neq j$$

where E_m, V_m, C_m are operators for expectation, variance, and covariance with respect to the model distribution; and the matrix $V = (v_{ij})$ is assumed to be known up to a constant σ^2.

To have a more compact notation define

$$\underline{Y} = (Y_1, Y_2, \ldots, Y_N)'$$

$$\underline{X} = (\underline{x}_1, \underline{x}_2, \ldots, \underline{x}_N)' = (X_{ij})$$

$$\underline{\varepsilon} = (\varepsilon_1, \varepsilon_2, \ldots, \varepsilon_N)'$$

and write the linear model as

$$\underline{Y} = \underline{X}\underline{\beta} + \underline{\varepsilon}$$

where

$$E_m\underline{\varepsilon} = \underline{0}$$
$$V_m(\underline{\varepsilon}) = V$$

Assume that n components of $\underline{Y}$ may be observed with the objective to estimate $\underline{\beta}$ or to predict the sum of all $N - n$ components of $\underline{Y}$ that are not observed. It is not restrictive to assume that

$$\underline{Y}_s = (Y_1, Y_2, \ldots, Y_n)'$$

is observed; define

$$\underline{Y}_r = (Y_{n+1}, \ldots, Y_N)'$$

and partition $\underline{X}$ and V correspondingly such that

$$\underline{X} = \begin{pmatrix} \underline{X}_s \\ \underline{X}_r \end{pmatrix}$$

$$V = \begin{pmatrix} V_{ss} & V_{sr} \\ V_{rs} & V_{rr} \end{pmatrix}$$

Assume

$$\sum_1^N \gamma_i Y_i = \underline{\gamma}'\underline{Y}$$

is to be predicted. Modifying slightly the approach of section 4.1.1 (to predict $\underline{1}'\underline{Y}$) we use $\underline{g}_s'\underline{Y}_s$ as a predictor of $\underline{\gamma}_r' Y_r$ and add the predicted value to the known quantity

$$\underline{\gamma}_s'\underline{Y}_s$$

to get as a predictor for $\underline{\gamma}'\underline{Y}$

$$(\underline{\gamma}_s + \underline{g}_s)'\underline{Y}_s$$

where $\underline{\gamma}_s = (\gamma_1, \gamma_2, \ldots, \gamma_n)'$ and $\underline{g}_s = (g_1, g_2, \ldots, g_n)'$.

$\underline{g}_s$ will be chosen such that

$$E_m[(\underline{\gamma_s} + g_s)'\underline{Y}_s - \underline{\gamma}'\underline{Y}] = 0$$

and

$$V_m[(\underline{\gamma_s} + g_s)'\underline{Y}_s - \underline{\gamma}'\underline{Y}]^2$$

is minimized. The linear predictor defined by these two properties is called the **best linear unbiased** (BLU) **predictor** (BLUP) of $\underline{\gamma}'\underline{Y}$. Assuming that the inverses of the occurring matrices exist it may be shown:

RESULT 4.3 *The BLU predictor of $\underline{\gamma}'\underline{Y}$ is*

$$t_0 = \underline{\gamma}_s'\underline{Y}_s + \underline{\gamma}_r'\left[\underline{X}_r\underline{\hat{\beta}} + V_{rs}V_{ss}^{-1}(\underline{Y}_s - \underline{X}_s\underline{\hat{\beta}})\right]$$

where

$$\underline{\hat{\beta}} = (\underline{X}_s'V_{ss}^{-1}\underline{X}_s)^{-1}\underline{X}_s'V_{ss}^{-1}\underline{Y}_s$$

is the BLU estimator of β. Further,

$$\begin{aligned} V_m(t_0) = {} & \gamma_r'(V_{rr} - V_{rs}V_{ss}^{-1}V_{sr})\gamma_r \\ & + \gamma_r'(\underline{X}_r - V_{rs}V_{ss}^{-1}V_{sr})(\underline{X}_s'V_{ss}^{-1}\underline{X}_s)^{-1} \\ & \times (\underline{X}_r - V_{rs}V_{ss}^{-1}V_{sr})'\underline{\gamma}_r . \end{aligned}$$

For a proof we refer to VALLIANT, DORFMAN, and ROYALL (2000).

4.1.6 Robustness Against Model Failures

Consider the general linear model described in section 4.1.4. TAM (1986) has shown that a necessary and sufficient condition for

$$\underline{T}'\underline{Y}_s = \sum_s T_i Y_i$$

to be BLU for $Y = \underline{1}'\underline{Y}$ is that

(a) $\underline{T}\,\underline{X}_s = \underline{1}'\underline{X}$
(b) $V_{ss}\underline{T} - K\underline{1} \in M(\underline{X}_s)$

where

$$K = (V_{ss}, V_{sr}),$$

and $M(\underline{X}_s)$ is the column space of $\underline{X}_s$.

In case $V_{rs} = 0$ these conditions reduce to (q) and

$$(b)' \quad V_{ss}(T - \underline{1}_s) \in M(\underline{X}_s)$$

as given earlier by PEREIRA and RODRIGUES (1983).

By TAM's (1986) results one may deduce the following.

If the true model is as above, $\mathcal{M}$, but one employs the best predictor postulating a wrong model, say $\mathcal{M}^*$, using $\underline{X}^*$ instead of $\underline{X}$ throughout where

$$\underline{X} = (\underline{X}^*, \underline{\tilde{X}}),$$

then the best predictor under $\mathcal{M}^*$ is still best under $\mathcal{M}$ if and only if

$$\underline{T}'\underline{\tilde{X}}_s = \underline{1}'\underline{\tilde{X}}$$

using obvious notations. This evidently is a condition that the predictor should remain model-unbiased under the correct model $\mathcal{M}$. Thus, choosing a right sample meeting this stipulation, one may achieve robustness. But, in practice, $\underline{\tilde{X}}$ will be unknown and one cannot realize this robustness condition at will, although for large samples this condition may hold approximately. In this situation, it is advisable to adopt suitable unequal probability sampling designs that assign higher selection probabilities to samples for which this condition should hold approximately, provided one may guess effectively the nature for variables omitted but influential in explaining variabilities in y values. If a sample is thus rightly chosen one may preserve optimality even under modeling deficient as above. On the other hand, if one employs the best predictor using $\underline{W}^*$ instead of $\underline{X}$ when $\underline{W}^* = (\underline{X}, \underline{W})$, then this predictor continues to remain best if and only if the condition (b) above still holds. But this condition is too restrictive, demanding correct specification of the nature of V, which should be too elusive in practice. ROYALL and HERSON (1973), TALLIS (1978), SCOTT, BREWER and HO (1978), PEREIRA and RODRIGUES (1983), RODRIGUES (1984), ROYALL and PFEFFERMANN (1982), and PFEFFERMANN (1984) have derived results relevant to this context of robust prediction.

4.2 PRIOR DISTRIBUTION–BASED APPROACH

4.2.1 Bayes Estimation

Fruitful inference through the likelihood based $\hat{d}$ cannot be obtained without postulating suitable structures on $\underline{Y}$. If $\underline{Y}$ is given a suitable prior density function $q(\underline{Y})$, then a posterior given d is

$$q_d^*(\underline{Y}) = q(\underline{Y})\, I_d(\underline{Y})\, c(d)$$

where $c(d)$ is a function of d required for normalization. This form is simplistic if $q(\underline{Y})$ is so. If a square error loss function is assumed, then the BAYES estimator (BE) for Y is

$$t_B = E_{q*}(Y|d) = \sum_s Y_i + \sum_r E_{q*}(Y_i|d)$$

writing E_{q*} for an operator for expectation with respect to the posterior pdf $q*$. If q is suitably postulated in a mathematically tractable and realistically acceptable manner, then it is easy to find Bayes estimators for Y. Let us illustrate as follows.

Suppose $Y_i \sim N(\theta, \sigma^2)$ and $\theta \sim N(\mu, \phi^2)$, meaning that Y_i's are independently, identically distributed (iid) normally with a mean θ and variance σ^2 and θ itself is distributed normally with a mean μ and variance ϕ^2. As a consequence, θ is distributed independently of $\varepsilon_i = Y_i - \theta,\ i = 1, \ldots, N$. Then, writing $\psi = \frac{\sigma^2}{\phi^2}$, $W = 1 - [1 - \frac{n}{N}]\frac{\psi}{\psi+n}$, for a sample s of size n with sample mean $\overline{y}$, the BAYES estimator of Y is

$$t_B = N\,[W\,\overline{y} + (1 - W)\mu].$$

Of course it cannot be implemented unless μ, σ, and ϕ, or at least μ and ψ, are known.

Leaving this issue aside for the time being, it is important to observe that an optimal sampling design to choose a sample on which a t_B is to be based is again purposive, as in the case of using m-based predictors. For optimality one must assign a selection probability 1 to a sample that yields the minimal value for the posterior mean square error of t_B to be called the **posterior risk,** in this case with a square error loss, viz $E_{q*}(t_B - Y)^2$. This is a function of s plus other parameters involved in q. Because of the appearance of unknown parameters here, to implement a Bayesian strategy in large-scale surveys is practically impossible. However, there is a way out in situations where one may have enough survey data that may be utilized to obtain plausible estimates of the parameters involved in the BAYES estimator. Substituting these estimates for the nuisance parameters in the Bayes estimator (BE) one gets what is called an **empirical Bayes estimator** (EBE), which is often quite useful. Let us illustrate a situation where an EBE may be available.

4.2.2 James–Stein and Empirical Bayes Estimators

Suppose $\theta_1, \ldots, \theta_k$ are $k \geq 3$ finite population parameters, that is, totals of a variable for mutually exclusive population groups required to be estimated. Let independent estimators $t_1, \ldots t_k$, respectively, be available for them and suppose it is reasonable to postulate that $t_i \sim N(\theta_i, \sigma^2)$ with σ^2 known.

Then, writing $S = \sum_1^k t_i^2$ it can be shown, following JAMES and STEIN (1961), that

$$\underline{\delta} = (\delta_1, \ldots, \delta_k)' \quad \text{where} \quad \delta_i = \left[1 - \frac{k-2}{S}\sigma^2\right] t_i$$

is a better estimator for $\underline{\theta} = (\theta_1, \ldots, \theta_k)'$ than $\underline{t} = (t_1, \ldots, t_k)'$ in the sense that

$$\sum_1^k E_{\theta i}(\delta_i - \theta_i)^2 \leq \sum_1^k E_{\theta i}(t_i - \theta_i)^2 = k\,\sigma^2.$$

This **shrinkage estimator** $\underline{\delta}$ is usually called the **James–Stein estimator** (JSE). But a limitation of its applicability is that all t_i must have a common variance σ^2, which must be known.

Assume further that it is plausible to postulate, in view of the assumed closeness among θ_i's, that $\theta_i \sim N(0, \phi^2)$, with ϕ as a known positive number. Then the BEs for θ_i are

$$t_{Bi} = \left[1 - \frac{\sigma^2}{\sigma^2 + \phi^2}\right] t_i,\ i = 1, \ldots, k.$$

Now $S/(\sigma^2 + \phi^2)$ follows a χ^2 distribution with k degrees of freedom and, therefore,

$$E\left[\frac{k-2}{S}\sigma^2\right] = \frac{\sigma^2}{\sigma^2 + \phi^2}.$$

Hence δ_i can be interpreted as an EBE for $\theta_i, i = 1, \ldots, k$. In this case, with a common σ^2 JSE and EBE coincide.

4.2.3 Applications to Sampling of Similar Groups

Suppose there are k mutually exclusive population groups of sizes N_i supposed to be closely related from which samples of sizes n_i are taken, yielding sample means

$$\overline{y}_i = \frac{1}{n_i}\sum_{j=1}^{n_i} Y_{ij}, \; i = 1, \ldots, k,$$

Y_{ij} denoting the value of jth unit of ith group. Let

$$Y_{ij} \sim N(\theta_i, \sigma^2), \theta_i \sim N(\mu, \phi^2),$$

(with θ_i's independent of $\varepsilon_{ij} = Y_{ij} - \theta_i$ for every $j = 1, \ldots, n_i$). Define $\psi = \sigma^2/\phi^2$ and

$$B_i = \frac{\psi}{\psi + n_i}, \; W_i = 1 - (1 - f_i)B_i, \, f_i = \frac{n_i}{N_i}, \quad \textit{for } i = 1, \ldots, k.$$

Then, the BE of $\sum_1^{N_i} Y_{ij} = T_i$ is

$$\begin{aligned} t_{Bi} &= n_i\overline{y}_i + (N_i - n_i)\left[B_i\mu + (1 - B_i)\overline{y}_i\right] \\ &= N_i\left[W_i\overline{y}_i + (1 - W_i)\mu\right]. \end{aligned}$$

Assuming $n_i \geq 2$ and writing $n = \sum_1^k n_1$,

$$\overline{y}_{..} = \frac{1}{n}\sum_1^k n_i\overline{y}_i$$

$$BMS = \frac{1}{k-1}\sum_1^k n_i(\overline{y}_i - \overline{y})^2$$

$$WMS = \frac{1}{n-k}\sum_1^k\sum_{j=1}^{n_i}(y_{ij} - \overline{y}_i)^2$$

$$g = g(n_1, \ldots, n_k) = n - \sum_1^k n_i^2/n$$

one may estimate, following GHOSH and MEEDEN (1986),

$$1/\Psi \text{ by} \left[\frac{\hat{1}}{\Psi}\right] = \max\left\{0, \left[\frac{(k-1)BMS}{(k-3)WMS} - 1\right]\frac{k-1}{g}\right\} \text{ assuming } k \geq 4$$

$$B_i \text{ by } \hat{B}_i = \frac{1}{1 + n_i\left[\frac{\hat{1}}{\Psi}\right]}$$

$$\mu \text{ by } \hat{\mu} = \sum_1^k (1 - \hat{B}_i)\overline{y}_i \Big/ \sum_1^k (1 - \hat{B}_i) \text{ if } \hat{\Psi}^{-1} \neq 0$$

$$= \frac{1}{k}\sum_1^k \overline{y}_i \text{ if } \hat{\Psi}^{-1} = 0.$$

Then the EBE for T_i, the total of the ith group, is

$$t_{EBi} = N_i[\hat{W}_i\bar{y}_i + (1 - \hat{W}_i)\hat{\mu}]$$

writing $\hat{W}_i = 1 - (1 - f_i)\hat{B}_i$, $i = 1, \ldots, k$.

Again, suppose that t_i are estimators of parameters θ_i based on independent samples or on the same sample but θ_i's supposed closely similar. Then further improvements on t_i's may be desired and achieved if additional information is available through auxiliary well-correlated variables in the following way. First, let us postulate that $t_i \sim N(\theta_i, \sigma^2)$, $i = 1, \ldots, k$. Let $x_1, \ldots, x_p$ be $p(\geq 1)$ auxiliary variables with known values $X_{ji}(j = 1, \ldots, p; i = 1, \ldots, k)$ such that it is further postulated that $\theta_i \sim N(\underline{x}_i\underline{\beta}, \phi^2)$, θ_i independent of $t_i - \theta_i$, $i = 1, \ldots, k$, $\underline{x}_i = (X_{1i}, \ldots, X_{pi})'$, $\underline{\beta} = (\beta_1, \ldots, \beta_p)'$, a p vector of unknown parameters, with $p \leq k - 3$. Assuming that the matrix $\underline{X}'\underline{X}$ of order $p \times p$, with $\underline{X}' = (\underline{x}_1, \ldots, \underline{x}_N)$ has a full rank, the regression estimator for θ_i is $t_i^* = \underline{x}_i'[(\underline{X}'\underline{X})^{-1}\underline{X}'\underline{t}]$, writing $\underline{t} = (t_1, \ldots, t_k)'$. Then the BAYES estimator of θ_i is

$$\theta_{Bi}^* = t_i^* + \left[1 - \frac{\sigma^2}{\sigma^2 + \phi^2}\right](t_i - t_i^*)$$

$$= \left[\frac{\sigma^2}{\sigma^2 + \phi^2}\right]t_i^* + \left[\frac{\phi^2}{\sigma^2 + \phi^2}\right]t_i.$$

Writing $S^* = \sum_1^k (t_i - t_i^*)^2$, we have $E\,[\frac{k-p-2}{S*}] = \frac{\phi^2}{\sigma^2+\phi^2}$ yielding the JSE of θ_i as

$$\delta_i^* = t_i^* + \left[1 - \frac{k-p-2}{S*}\right](t_i - t_i^*)$$

$$= (k-p-2)\,\frac{\sigma^2}{S*}\,t_i^* + \left\{1 - (k-p-2)\,\frac{\sigma^2}{S*}\right\}t_i$$

which is, of course, an EBE. In particular, if $p = 1, X_i = 1$, $i = 1, \ldots, k$, then

$$\frac{1}{k}\sum_1^k t_i = \bar{t},\ \text{say},\ S^* = \sum (t_i - \bar{t})^2 \ \text{and}$$

$$\delta_i^* = \left[\frac{k-3}{S*}\,\sigma^2\right]\bar{t} + \left[1 - \frac{k-3}{S*}\,\sigma^2\right]t_i.$$

Further generalizations allowing σ^2 to vary with i as σ_i^2 render JSEs unavailable, but EBEs are yet available in the literature provided σ_i^2 are known. This latter condition is not very restrictive because from samples that are usually large σ_i^2 may be accurately estimated.

The BAYES estimators, as we have seen, are completely design-free, and in assessing their performances design-based properties are never invoked. The JAMES–STEIN estimators, whenever applicable, and their adaptations as empirical BAYES estimators, may start with design-based estimators, model-based estimators, or design-cum-model-based estimators, but these estimators get their final forms exclusively from considerations of postulated models. Also, only their model-based properties like model bias, model MSE, and related characteristics are studied in the literature. Details omitted here may be found in works by GHOSH and MEEDEN (1986) and GHOSH and LAHIRI (1987, 1988). Their design-based properties are not yet known to have been seriously examined. In the context of sample surveys, the question of robustness of BAYES estimators, JAMES–STEIN estimators, and empirical BAYES estimators is not yet known to have been seriously taken up or examined in the literature.

4.2.4 Applications to Multistage Sampling

Let us suppose, following LITTLE (1983), that a finite population U of N units with mean $\bar{Y}$ is divided into C mutually exclusive groups U_g with sizes N_g and group means $\bar{Y}_g$. Then, with $P_g = N_g/N$,

$$\sum_1^C N_g = N, \ \bar{Y} = \sum P_g \bar{Y}_g.$$

Let a sample s of size n be taken and denote by s_g the sample of n_g units selected from group U_g and $\bar{y}_g$ the corresponding mean. Then

$$\sum_1^C n_g = n; \ \bar{y} = \frac{1}{n}\sum_1^C n_g \bar{y}_g.$$

Let Y_{gi} denote the y variable value for the ith unit of the gth group and assume that all Y_{gi} are independently distributed with

$$Y_{gi} \sim N(\mu_g, \sigma^2 V_g)$$

where $V_1, V_2, \ldots, V_C > 0$ are known, $\sigma > 0$ and $\mu_1, \mu_2, \ldots, \mu_C$ are unknown. In practice n_g's are quite small for many of the groups and even $n_g = 0$ for several groups. One solution is to reduce the number of groups by coalescing several similar ones and thus ensure enough n_g per group with the number of groups reduced. Another alternative is to employ multistage sampling designs or clustered designs where several n_g's are taken to be zero deliberately. We may turn to such designs and see how an extension of the above approach may be achieved, yielding fruitful results.

Following SCOTT and SMITH (1969), we assume

$$\mu_g \sim N(\mu, \delta^2)$$

where μ_g and $Y_{gi} - \mu_g; g = 1, 2, \ldots, C$ are independent and μ is given a noninformative prior. Then one may derive the BLUP for Y as

$$\begin{aligned} t &= \sum_g \left[(n_g \bar{y}_g) + (N_g - n_g)\left\{\lambda_g \bar{y}_g + (1-\lambda_g)\bar{y}\right\}\right] \\ &= \sum_g \left[n_g(1-\lambda_g)(\bar{y}_g - \bar{y}) + N_g\left\{\lambda_g \bar{y}_g + (1-\lambda_g)\bar{y}\right\}\right] \end{aligned}$$

writing

$$\lambda_g = \frac{\delta^2}{\delta^2 + \sigma^2 \dfrac{V_g}{n_g}}$$

for $n_g > 0$ and $\lambda_g = 0$ for $n_g = 0$,

$$\tilde{y} = \left(\sum_g \lambda_g \bar{y}_g\right) \Big/ \left(\sum_g \lambda_g\right).$$

Note that $\tilde{\mu}_{gi} = \lambda_g \bar{y}_g + (1 - \lambda_g)\tilde{y}$ is a predicted value for unit i in group g. Thus, in this case only some of the groups are sampled and from each selected group only some of the units are selected. The units observed have values known and for them no prediction is needed. For those units that are not observed but belong to groups that are represented in the sample, there is one type of prediction utilizing the sampled group means, but there is a third type of unit with values not observed and not within groups represented in the sample, and hence they are predicted differently in terms of overall weighted sample group means.

This t is really a BAYES estimator and is not usable unless δ^2 and σ^2 are known. Since δ, σ are always unknown they have to be estimated from the sample; if they are estimated by $\hat{\delta}^2$, $\hat{\sigma}^2$ respectively t becomes an EBE. Writing $\hat{\lambda}_g(\tilde{y}_e)$ for $\lambda_g(\tilde{y})$ with δ^2, σ^2, therein replaced by $\hat{\delta}^2$, $\hat{\sigma}^2$, one gets the EBE as

$$\hat{t} = \sum_g [n_g(1 - \hat{\lambda}_g)(\bar{y}_g - \tilde{y}_e) + N_g\{\hat{\lambda}_g \bar{y}_g + (1 - \hat{\lambda}_g)\tilde{y}_e\}].$$

If $\frac{n_g}{N_g} \cong 0$, then

$$\hat{t} \cong \sum_g N_g\{\hat{\lambda}_g \bar{y}_g + (1 - \hat{\lambda}_g)\tilde{y}_e\}$$

which is a combination of shrinkage estimators. If $n_g = 0$ for a group, then $\lambda_g = 0$; hence $\hat{\lambda}_g = 0$, too.

Now, assume

$$Y_{gi} \sim N(\beta_{og} + \beta_1 X_{gi}, \sigma^2 V_g)$$
$$\beta_{og} \sim N(\beta_o, \delta^2)$$

where X_{gi} is the value of an auxiliary variable x for unit i of group U_g and the notation and independence assumptions are analogous to the above considerations. Then an unobserved value is predicted by

$$\hat{\mu}_{gi} = \lambda_g\{\bar{y}_g + \hat{\beta}_1(x_{gi} - \bar{x}_g)\} + (1 - \lambda_g)\{\tilde{y} + \hat{\beta}_1(x_{gi} - \tilde{x})\}$$

where

$$\lambda_g = \frac{\delta^2}{(\delta^2 + \frac{\sigma^2}{n_g} V_g)},$$

$$\tilde{y} = \frac{\sum \lambda_g \bar{y}_g}{\sum \lambda_g}, \quad \tilde{x} = \frac{\sum \lambda_g \bar{x}_g}{\sum \lambda_g}$$

and

$$\hat{\beta}_1 = \left[\sum_g \sum_{s_g} Y_{gi}(X_{gi} - \bar{x}_g)/V_g\right] \bigg/ \left[\sum_g \sum_{s_g} (X_{gi} - \bar{x}_g)^2/V_g\right].$$

Then the BLUP is

$$\begin{aligned} t &= \sum_g [n_g\bar{y}_g + (N_g - n_g)[\lambda_g\{\bar{y}_g + \hat{\beta}_1(\bar{x}_{rg} - \bar{x}_g)\} \\ &\quad + (1 - \lambda_g)\{\bar{y}_g + \hat{\beta}_1(\bar{x}_{rg} - \tilde{x})\}]] \\ &= \sum_g [n_g(1 - \lambda_g)\bar{y}_g + N_g\{\lambda_g\bar{y}_g + (1 - \lambda_g)\tilde{y}\} \\ &\quad + (N_g - n_g)\hat{\beta}_1\{\lambda_g(\bar{x}_{rg} - \bar{x}_g) + (1 - \lambda_g)(\bar{x}_{rg} - \tilde{x})\}] \end{aligned}$$

writing $\bar{x}_{rg}$ for the mean of units of group g that do not appear in the sample.

Chapter 5

Asymptotic Aspects in Survey Sampling

5.1 INCREASING POPULATIONS

It may be of interest to know the properties of a strategy as the population and sample sizes increase. To investigate these properties we follow ISAKI and FULLER (1982) and consider a sequence of increasing populations

$$U_1 \subset U_2 \subset U_3 \subset \ldots$$

of sizes $N_1 < N_2 < \ldots$ and a sequence of increasing sample sizes $n_1 < n_2 < \ldots$. The units of U_T are labeled

$$1, 2, \ldots, N_T$$

with values

$$Y_1, Y_2, \ldots, Y_{N_T}$$

of a variable y of interest and, possibly, with K vectors

$$\underline{x}_1, \underline{x}_2, \ldots, \underline{x}_{N_T}$$

defined by K auxiliary variables $x_1, \ldots, x_K$.

The discussion of the sequence of populations is greatly simplified by appropriate additional assumptions. To formulate such an assumption we define

$$U(1) = \{1, 2, \ldots, N_1\}$$

$$U(2) = \{N_1 + 1, N_1 + 2, \ldots, N_2\}$$

$$U(3) = \{N_2 + 1, N_2 + 2, \ldots, N_3\}$$

$$\vdots$$

Assumption A: *$U(1), U(2), \ldots$ are of the same size, that is,*

$$N_T = TN_1$$

and

$$n_T = Tn_1$$

for $T = 1, 2, \ldots$. In addition, for $i = 1, 2, \ldots, N_1$

$$Y_i = Y_{i+N_1} = Y_{i+2N_1} = \cdots$$

$$\underline{x}_i = \underline{x}_{i+N_1} = \underline{x}_{i+2N_1} = \cdots$$

According to this assumption $U(2), U(3), \ldots$ are copies of $U(1)$; U_T is the union of $U(1)$ with its first $T-1$ copies.

Note that Assumption **A** implies that

$$\overline{Y}_T = \frac{1}{TN_1} \sum_{1}^{TN_1} Y_i$$

$$\sigma_{yyT} = \frac{1}{TN_1} \sum_{1}^{TN_1} (Y_i - \overline{Y}_T)^2$$

are free of T and, similarly, for moments of the K vectors. So we may drop the index T and write

$$\overline{Y}, \sigma_{yy}$$

without ambiguity as long as Assumption **A** is true.

5.2 CONSISTENCY, ASYMPTOTIC UNBIASEDNESS

For $T = 1, 2, \ldots$ let (p_T, t_T) be a strategy for estimating $\overline{Y}_T$ by selecting a sample s_T of size n_T from U_T.

p_T and t_T may depend on auxiliary variables; however, p_T does not depend on the variable of interest y and t_T does not involve Y_i's with i outside s_T.

Let

$$\underline{Y} = (Y_1, Y_2, \cdots)$$

be a sequence of y values subject to Assumption **A**, but otherwise arbitrary. Given $\underline{Y}$,

$$t_T\,(s_T, \underline{Y}) - \overline{Y};\; T = 1, 2, \ldots \tag{5.1}$$

is a sequence of random variables with distributions defined by

$$p_T;\; T = 1, 2, \ldots$$

t_T is **asymptotically design unbiased** or more fully **asymptotically design unbiased** (ADU) if

$$\lim_{T \to \infty} E_{p_T}(t_T - \overline{Y}) = 0.$$

Exact unbiasedness of t_T of course ensures its asymptotic unbiasedness.

By describing the sequence Eq. (5.1) of random variables as converging in probability to 0 we mean

$$\lim_{T \to \infty} P_T\left\{\left|t_T - \overline{Y}\right| > \varepsilon\right\} = 0$$

for all $\varepsilon > 0$; here P_T is the probability defined by p_T.

In this case t_T is called **consistent** for $\overline{Y}$ (with respect to p_T) or more fully **asymptotically design consistent** (ADC).

This type of consistency is to be distinguished from COCHRAN's (1977) well-known finite consistency for a finite population parameter, meaning that the estimator and the estimand coincide if the sample is coextensive with the population.

EXAMPLE 5.1 *Accept condition A and let p_T denote SRSWOR of size*

$$n = T n_1$$

from a population of size

$$N = T N_1 .$$

For a sample $s = s_T$ define

$$t_T = t_T(s, \underline{Y}) = \frac{1}{n} \sum_s Y_i .$$

Then,

$$E_{p_T} t_T = \overline{Y}$$

$$V_{p_T}(t_T) = \frac{\sigma_{yy}}{n} \frac{N - n}{N - 1} .$$

Hence,

$$\lim_{T \to \infty} V_{p_T}(t_T) = \lim_{T \to \infty} \frac{\sigma_{yy}}{T n_1} \frac{T N_1 - T n_1}{T N_1 - 1} = 0$$

and it follows that t_T is a consistent estimator of $\overline{Y}$.

5.3 BREWER'S ASYMPTOTIC APPROACH

Looking for properties of a strategy as population and sample sizes increase presumes some relation between $p_1, p_2, \ldots$ on one hand and between $t_1, t_2, \ldots$ on the other hand.

In this and the next section relations on the design and estimator sequence, respectively, are introduced.

Consistency of an estimator t_T is easy to decide on if Assumption **A** is true and p_T satisfies a special condition considered by BREWER (1979):

Assumption B: *Using Assumption* **A** *and starting with an arbitrary design p_1 of fixed size n_1 for $\mathcal{U}(1)$, then p_T is as follows:*

Apply p_1 not only to $\mathcal{U}(1)$ but also, independently, to $\mathcal{U}(2)$, $\ldots$, $\mathcal{U}(T)$ and amalgamate the corresponding samples

$$s(1), s(2), \ldots, s(T)$$

to form

$$s_T = s(1) \cup s(2) \cup \cdots \cup s(T).$$

A design satisfying Assumption **B** to give the selection probability for s_T is appreciably limited in scope and application.

Some authors have considered such restrictive designs, notably HANSEN, MADOW and TEPPING (1983). However, interesting results have been derived under less restrictive assumptions as well as by alternative approaches.

We mention ISAKI and FULLER (1982) proving the consistency of the HT estimator under rather general conditions on p_T. In fact, they even drop Assumption **A**, a condition that seems quite rational to us.

BREWER's approach should be adequate where it is advisable to partition a large population $\mathcal{U}_T$ into subsets of similar size and structure and to use these subsets as strata in the selection procedure. This is acceptable only if there is no loss in efficiency. But it is doubtful that this may always be the case.

We plan to enlarge BREWER's class of designs and obtain a class containing the designs in common use and with the same technical amenities as BREWER's class.

Assumption $\mathbf{B_0}$: *Using Assumption* **A** *and letting*

$$\pi_1, \pi_2, \ldots, \pi_{N_1}$$

be the inclusion probabilities of first order for p_1, we have

$$\pi_i = \pi_{i+N_1} = \ldots, \pi_{i+(T-1)N_1};\ i = 1, \ldots, N_1. \tag{5.2}$$

The inclusion probabilities of second order π_{ij} satisfy the condition

$$\pi_{ij} - \pi_i \pi_j \leq 0 \tag{5.3}$$

for all $i, j = 1, 2, \ldots, TN_1$ with

$$|i - j| = N_1, 2N_1, \ldots. \tag{5.4}$$

Assumption $\mathbf{B_0}$ is obviously less restrictive than Assumption **B**. We want to motivate it more fully.

It is natural to give units with identical/similar K-vectors the same/nearly the same chance of being selected. If a design p_T is of this type, the first-order inclusion probabilities $\pi_1, \pi_2, \ldots$ of the population units are made to satisfy the condition

$$\underline{x}_i = \underline{x}_j \Rightarrow \pi_i = \pi_j \tag{5.5}$$

implying Eq. (5.2) as a consequence of Assumption **A**.

In addition, it is desirable not to select too many units with the same or similar K vectors implying

$$\underline{x}_i = \underline{x}_j \Rightarrow \pi_{ij} - \pi_i \pi_j < 0. \tag{5.6}$$

and, therefore, Eq. (5.3).

5.4 MOMENT-TYPE ESTIMATORS

To establish meaningful results of asymptotic unbiasedness and consistency, the estimators $t_1, t_2, \ldots$ of a sequence to be considered must be somehow related to each other. Subsequently, a relation is assumed that is based on the concept of a moment estimator we define as follows: Let $A_i, B_i, C_i, \ldots$ be values associated with $i \in U$. Then, for $s \subset U$ with $n(s) = n$

$$\frac{1}{n}\sum_s A_i, \quad \frac{1}{n}\sum_s A_i B_i, \quad \frac{1}{n}\sum_s A_i B_i C_i \tag{5.7}$$

are sample moments. Examples are

$$\frac{1}{n}\sum_s \frac{Y_i}{\pi_i}, \quad \frac{1}{n}\sum_s X_{i1} Y_i, \quad \frac{1}{n}\sum_s \frac{X_{i1} X_{i2}}{\pi_i}$$

where Y_i, X_{i1}, X_{i2} are values of variables y, x_1, x_2, respectively, and π_i inclusion probabilities defined by a design for $i \in U$.

$$\frac{1}{N}\sum_1^N A_i, \quad \frac{1}{N}\sum_1^N A_i B_i, \quad \frac{1}{N}\sum_1^N A_i B_i C_i$$

are population moments corresponding to the sampling moments Eq. (5.7).

A **moment estimator** t is an estimator that may be written as a function of sample moments $m^{(1)}, m^{(2)}, \ldots, m^{(\nu)}$:

$$t = f\,(m^{(1)}, m^{(2)}, \ldots, m^{(\nu)}). \tag{5.8}$$

Obvious examples of moment estimators are the sample mean, the HT-estimator, the HH-estimator, and the ratio estimator.

Now, let t_1 be a moment estimator, that is,

$$t_1 = f\left(m_1^{(1)}, \ldots, m_1^{(\nu)}\right)$$

where $m_1^{(1)}, \ldots, m_1^{(\nu)}$ are sample moments for s_1.

Then, t_T may be defined in a natural way:

$$t_T = f\left(m_T^{(1)}, m_T^{(2)}, \ldots, m_T^{(\nu)}\right) \tag{5.9}$$

where $m_T^{(j)}$ is the sample moment for s_T corresponding to $m_1^{(j)}$, $j = 1, 2, \ldots, \nu$. As an example, we mention the ratio estimator

$$t_1 = \frac{\sum_{s_1} Y_i}{\sum_{s_1} X_i} \overline{X}$$

for which

$$t_T = \frac{\sum_{s_T} Y_i}{\sum_{s_T} X_i} \overline{X}.$$

From this example it is clear that t_1 may depend on population moments also (here $\overline{X}$). These need not be noted explicitly in Eq. (5.9) because, according to Assumption **A**, population moments are free of T.

Of considerable importance are QR predictors, consistency and asymptotic unbiasedness of which are discussed in chapter 6.

5.5 ASYMPTOTIC NORMALITY AND CONFIDENCE INTERVALS

Let p denote SRSWR of size n and t the sample mean, that is, with $s = (i_1, \ldots, i_n)$

$$t(s, \underline{Y}) = \frac{1}{n}(Y_{i_1} + Y_{i_2} + \cdots + Y_{i_n}) = \overline{y}, \textit{say}.$$

$Y_{i_1}, \ldots, Y_{i_n}$ are independent and identically distributed (iid) with expectation $\overline{Y}$ and variance σ_{yy}. Hence, according to the central limit theorem

$$\frac{\overline{y}-\overline{Y}}{\sqrt{\frac{\sigma_{yy}}{n}}}$$

is asymptotically standard-normal.

$$s_{yy}=\frac{1}{n-1}\sum_{i\in s}(Y_i-\overline{y})^2$$

is consistent for σ_{yy}, hence by SLUTSKY's Theorem (cf. VALLIANT, DORFMAN and ROYALL, 2000, p. 414)

$$\frac{\overline{y}-\overline{Y}}{\sqrt{\frac{s_{yy}}{n}}}$$

is also standard-normal and confidence intervals may be derived. For the confidence level 95% we derive, for example, the interval

$$\left[\overline{y}-1,96\sqrt{\frac{s_{yy}}{n}};\ \overline{y}+1,96\sqrt{\frac{s_{yy}}{n}}\right].$$

Note that there is no need to consider a sequence of populations in connection with SRSWR. This is different for SRSWOR.

Let p_T denote SRSWOR of size n_T and $t_T=\overline{y}_T$ the sample mean.

Then,

$$E_{p_T}t_T=\overline{Y}_T$$

$$V_{p_T}(t_T)=\frac{\sigma_{yyT}}{n_T}\,\frac{N_T-n_T}{N_T-1}$$

HÁJEK (1960) and RÉNYI (1966) have proved under weak conditions (by far less restrictive than Assumption **A**)

$$\frac{\overline{y}_T-\overline{Y}_T}{\sqrt{\frac{\sigma_{yyT}}{n_T}\,\frac{N_T-n_T}{N_T-1}}}\qquad T=1,2,\cdots$$

is asymptotically standard-normal. Here σ_{yyT} may be replaced by a consistent estimator

$$s_{yyT}=\frac{1}{n_T-1}\sum_{i\in s_T}(Y_i-\overline{y}_T)^2$$

It should not be misleading to write N_T, n_T, $\overline{Y}_T$, $\overline{y}_T$, s_{yyT} without subscript T. A 95% confidence interval is then given as

$$\left[\overline{y} - 1,96\sqrt{\frac{s_{yy}}{n}\left(1-\frac{n}{N}\right)};\ \overline{y} + 1,96\sqrt{\frac{s_{yy}}{n}\left(1-\frac{n}{N}\right)}\right].$$

To have one more example of practical importance, consider the ratio strategy (p_T, t_T). Here, p_T is SRSWOR of size n_T and

$$t_T\left(s_T, \underline{Y}_T\right) = \frac{\overline{y}_T}{\overline{x}_T}\overline{X}_T.$$

We have

$$t_T\left(s_T, \underline{Y}_T\right) - \overline{Y}_T = \frac{\overline{X}_T}{\overline{x}_T}\left(\overline{y}_T - \frac{\overline{Y}_T}{\overline{X}_T}\overline{x}_T\right)$$

where

$$\overline{X}_T / \overline{x}_T$$

is consistent with limit 1. Further,

$$\left(\overline{y}_T - \frac{\overline{Y}_T}{\overline{X}_T}\overline{x}_T\right) \Big/ \sqrt{V_{p_T}\left(\overline{y}_T - \frac{\overline{Y}_T}{\overline{X}_T}\overline{x}_T\right)}$$

$$= \sqrt{n}\left(\overline{y}_T - \frac{\overline{Y}}{\overline{X}}\overline{x}_T\right) \Big/ \sqrt{\frac{N-n}{N-1}\left(\sigma_{yy} - 2\frac{\overline{Y}}{\overline{X}}\sigma_{yx} + \left(\frac{\overline{Y}}{\overline{X}}\right)^2\sigma_{xx}\right)}$$

is asymptotically standard-normal under the weak conditions stated by HÁJEK (1960) and RÉNYI (1966). Hence, according to SLUTSKY's Theorem

$$\sqrt{n}\left(t_T\left(s_T, \underline{Y}_T\right) - \overline{Y}_T\right) \Big/ \sqrt{\frac{N-n}{N-1}\left(\sigma_{yy} - 2\frac{\overline{Y}}{\overline{X}}\sigma_{yx} + \left(\frac{\overline{Y}}{\overline{X}}\right)^2\sigma_{xx}\right)}$$

is asymptotically standard-normal.

Now, the expression

$$\sigma_{yy} - 2\frac{\overline{Y}}{\overline{X}}\sigma_{yx} + \left(\frac{\overline{Y}}{\overline{X}}\right)^2\sigma_{xx}$$

may be estimated consistently by its sample analogy such that confidence intervals are derived in a straightforward way.

For strategies with designs of varying selection probabilities it is easy to derive confidence intervals under Assumptions **A** and **B**. However, the relevance of these intervals may be questionable. For a central limit theorem proved under much weaker assumptions for the HT estimator, we refer to FULLER and ISAKI (1981).

Chapter 6

Applications of Asymptotics

6.1 A MODEL-ASSISTED APPROACH

6.1.1 QR Predictors

In section 3.1.3 we saw that the generalized difference estimator (GDE)

$$t_A = \sum_s \left[\frac{Y_i - A_i}{\pi_i}\right] + \sum_1^N A_i$$

is a design-unbiased estimator of Y with $\underline{A} = (A_1, \ldots, A_i, \ldots, A_N)'$ as a vector of known quantities and that it has optimal superpopulation model-based properties in case $A_i = \mu_i = E_m(Y_i), i = 1, \ldots, N$. But the μ_i's are usually unknown in practice.

If one gets estimates $\hat{\mu}_i$ for μ_i then a possible estimator for Y is

$$t_{\hat{\mu}} = \sum_s \left[\frac{Y_i - \hat{\mu}_i}{\pi_i}\right] + \sum_1^N \hat{\mu}_i.$$

Consider the model

$$\underline{Y} = \underline{X}\,\underline{\beta} + \underline{\varepsilon}$$

with

$$E_m(\underline{\varepsilon}) = \underline{0}$$
$$V_m(\underline{\varepsilon}) = V, \quad V \quad \text{diagonal.}$$

Write for $i = 1, 2, \ldots, N$

$$\underline{x}_i = (X_{i1}, \ldots, X_{iK})'$$
$$\mu_i = \underline{x}'_i \underline{\beta}.$$

Then a natural choice of $\hat{\mu}_i$ would be

$$\hat{\mu}_i = \underline{x}'_i \, \hat{\underline{\beta}}$$

with the BLU estimator

$$\hat{\underline{\beta}} = \left(\underline{X}'_s V_{ss}^{-1} \underline{X}_s\right)^{-1} \left(\underline{X}'_s V_{ss}^{-1} \underline{Y}_s\right)$$

for $\underline{\beta}$. If V is not known, a suitably chosen $n \times n$ diagonal matrix Q_s with positive diagonal entries Q_i might be used to define

$$\hat{\underline{\beta}}_Q = (\underline{X}'_s Q_s \underline{X}_s)^{-1} (\underline{X}'_s Q_s \underline{Y}_s)$$
$$= \left(\sum_s Q_i \underline{x}_i \underline{x}'_i\right)^{-1} \left(\sum_s Q_i \underline{x}_i Y_i\right)$$
$$\hat{\mu}_i = \underline{x}'_i \hat{\underline{\beta}}_Q.$$

Note that, in spite of the unbiasedness of t_A, $t_{\hat{\mu}}$ will be p biased in general. Alternatively, in view of the model, we might be willing to use the predictor

$$\sum_{i \in s} Y_i + \sum_r \hat{\mu}_i = \sum_s (Y_i - \hat{\mu}_i) + \sum_1^N \hat{\mu}_i$$

with $\hat{\mu}_i = \underline{x}'_i \, \hat{\underline{\beta}}$, or, more generally, $\hat{\mu}_i = \underline{x}'_i \, \hat{\underline{\beta}}_Q$, which is m unbiased but p biased in general. In both cases we are concerned with functions of Y_i, $i \in s$, having the following structure

$$t_{QR} = \sum_s R_i (Y_i - \hat{\mu}_i) + \sum_1^N \hat{\mu}_i$$
$$= \sum_s R_i e_i + \sum_1^N \hat{\mu}_i$$

where

$$\hat{\mu}_i = \underline{x}_i' \, \hat{\underline{\beta}}_Q \, , \; e_i = Y_i - \hat{\mu}_i$$

with a diagonal matrix Q, $Q_i > 0$, and real numbers R_1, $R_2, \ldots, R_N$. These moment-type functions are called **QR predictors** and may finally be written as

$$\begin{aligned} t_{QR} = t_{QR}(s, \underline{Y}) &= \sum_s R_i Y_i + \left[\sum_1^N \underline{x}_i' - \sum_s R_i \, \underline{x}_i'\right] \hat{\beta}_Q \\ &= \sum_s R_i Y_i + \left[\sum_1^N \underline{x}_i' - \sum_s R_i \, \underline{x}_i'\right] \left(\sum_s Q_i \, \underline{x}_i \underline{x}_i'\right)^{-1} \left(\sum_s Q_i \underline{x}_i Y_i\right). \end{aligned}$$

EXAMPLE 6.1 *The choice $R_i = 1$ for all i yields the* ***linear predictor*** *(LPRE)*

$$t_{Q1} = \sum_s Y_i + \sum_r \hat{\mu}_i .$$

If $Q_i = 1/V_{ii}$, in addition, we obtain the BLUP, namely,

$$\begin{aligned} t_{BLUP} &= \sum_s Y_i + \sum_r \underline{x}_i' \hat{\underline{\beta}}_{BLU} \\ &= \sum_s Y_i + \sum_r \underline{x}_i' \left(\sum_s \underline{x}_i \, \underline{x}_i' / V_{ii}\right)^{-1} \left(\sum_s \underline{x}_i \, Y_i / V_{ii}\right). \end{aligned}$$

If $R_i = 0$, then

$$t_{Q0} = \sum_1^N \hat{\mu}_i ,$$

is called the ***simple projection predictor*** *(SPRO). If $R_i = 1/\pi_i$, then*

$$\begin{aligned} t_{Q1/\pi} &= \sum_s \frac{1}{\pi_i} (Y_i - \hat{\mu}_i) + \sum_1^N \hat{\mu}_i \\ &= \sum_s \frac{Y_i}{\pi_i} + \left(\sum_1^N \underline{x}_i' - \sum_s \frac{1}{\pi_i} \underline{x}_i'\right) \hat{\underline{\beta}}_Q \end{aligned}$$

with

$$\hat{\underline{\beta}}_Q = (\underline{X}_s' Q_s \underline{X}_s)^{-1} \, \underline{X}_s' Q_s \underline{Y}_s$$

is the GREG predictor.

A suitable choice for Q_i is not easy to make, but usual choices are

$$Q_i = \frac{1}{V_{ii}} \quad or \quad \frac{1}{\pi_i} \quad or \quad \frac{1}{\pi_i V_{ii}}.$$

REMARK 6.1 *For later reference we give QR predictors in matrix notation.*

Define

$$R = diag\,(R_1, \ldots, R_N)$$
$$\Pi = diag\,(\pi_1, \ldots, \pi_N)$$

and let R_s, π_s be the submatrices corresponding for s. Then

$$t_{QR} = \underline{1}_n' R_s \underline{Y}_s + (\underline{1}_N' \underline{X} - \underline{1}_n' R_s \underline{X}_s)\underline{\hat{\beta}}_Q$$

and especially

$$t_{Q1/\pi} = \underline{1}_n' \Pi_s^{-1} \underline{Y}_s + (\underline{1}_N' \underline{X} - \underline{1}_n' \Pi_s^{-1} \underline{X}_s)\underline{\hat{\beta}}_Q$$

6.1.2 Asymptotic Design Consistency and Unbiasedness

Introducing the indicator variable I defined by

$$I_{si} = \begin{cases} 1 & if \quad i \in s \\ 0 & if \quad i \notin s \end{cases}$$

we may write t_{QR}/N in the form

$$t = t\,(s, \underline{y}) = \frac{1}{N}\left(\sum_1^N \underline{x}_i' - \sum_1^N I_{si}\, R_i\, \underline{x}_i'\right) \cdot \left(\sum_1^N I_{si}\, Q_i\, \underline{x}_i \underline{x}_i'\right)^{-1} \cdot \left(\sum_1^N I_{si}\, Q_i\, \underline{x}_i Y_i\right) + \frac{1}{N} \cdot \sum_1^N I_{si}\, R_i\, Y_i.$$

We want to prove the consistency of this estimator and use Assumption **A**. Obviously,

$$t_T = t_T(s_T, \underline{Y}) = \frac{1}{N_T}\left(\sum_1^{N_T} \underline{x}_i' - \sum_1^{N_T} I_{s_T i} R_i \underline{x}_i'\right) \cdot \left(\sum_1^{N_T} I_{s_T i}\, Q_i \underline{x}_i \underline{x}_i'\right)^{-1} \cdot \left(\sum_1^{N_T} I_{s_T i}\, Q_i\, \underline{x}_i\, Y_i\right) + \frac{1}{N_T} \sum_1^{N_T} I_{s_T i}\, R_i Y_i$$

where for $i = 1, 2, \ldots, N_1$

$$Q_i = Q_{i+N_1} = Q_{i+2N_1} = \cdots$$
$$R_i = R_{i+N_1} = R_{i+2N_1} = \ldots$$

and, for the sample s_T,

$$I_{s_T i} = \begin{cases} 1 & if \quad i \in s_T \\ 0 & if \quad i \notin s_T. \end{cases}$$

Defining

$$f_{iT} = \frac{1}{T}(I_{s_T i} + I_{s_T\, i+N_i} + \ldots + I_{s_T i+(T-1)N_1})$$

we have

$$t_T = \frac{1}{N_1}\left(\sum_1^{N_1} \underline{x}_i' - \sum f_{iT} R_i \underline{x}_i'\right)\left(\sum_1^{N_1} f_{iT} Q_i \underline{x}_i \underline{x}_i'\right)^{-1}$$
$$\cdot\left(\sum_1^{N_1} f_{iT} Q_i \underline{x}_i Y_i\right) + \frac{1}{N_1}\sum_1^{N_1} f_{iT} R_i Y_i.$$

Now, let p_T be of type $\mathbf{B_0}$. Then

$$I_{s_T i}, I_{s_T\, i+N_1}, \ldots$$

are identically distributed with a common expectation π_i and a common variance $\pi_i(1 - \pi_i)$. Hence,

$$V_{p_T}(f_{i_T}) = V_{p_T}\left(\frac{1}{T}[I_{s_T i} + \ldots]\right) \leq \frac{1}{T^2}\, T\, \pi_i(1 - \pi_i)$$
$$= \frac{\pi_i(1 - \pi_i)}{T}$$

because of the assumption of nonpositivity of

$$C_{p_T}(I_{s_T i}, I_{s_T i+N_1,}) = \pi_{ii+N_1} - \pi_i \pi_{i+N_1}$$

for a $\mathbf{B_0}$-type design p_T. From CHEBYSHEV's inequality follows that f_{i_T} converges in probability to π_i. Also according to the consistency theorem, t_T is consistent (ADC) for

$$\frac{1}{N_1}\left(\sum_1^{N_1} \underline{x}_i' - \sum_1^{N_1} \pi_i R_i \underline{x}_i'\right)\left(\sum \pi_i Q_i \underline{x}_i \underline{x}_i'\right)^{-1} \sum \pi_i Q_i \underline{x}_i Y_i$$
$$+ \frac{1}{N_1}\sum \pi_i R_i Y_i.$$

The last expression is equal to $\overline{Y}$ if, for $j = 1, 2, \ldots, N_1$,

$$\frac{1}{N_1}\left(\sum \underline{x}_i' - \sum \pi_i R_i \underline{x}_i'\right)\left(\sum \pi_i Q_i \underline{x}_i \underline{x}_i'\right)^{-1} \pi_j Q_j \underline{x}_j + \frac{1}{N_1}\pi_j R_j = \frac{1}{N_1}$$

which may be written

$$\begin{aligned} 1 &= \left(\sum \underline{x}_i' - \sum \pi_i R_i \underline{x}_i'\right)\left(\sum \pi_i Q_i \underline{x}_i \underline{x}_i'\right)^{-1} \pi_j Q_j \underline{x}_j + \pi_j R_j \\ &= \underline{a}' \pi_j Q_j \underline{x}_j + \pi_j R_j, \text{ say,} \end{aligned}$$

with $\underline{a} = (a_1, a_2, \ldots, a_K)'$. This condition is equivalent to

$$\underline{a}' \underline{x}_j = \frac{1 - \pi_j R_j}{\pi_j Q_j} = u_j, \text{ say,}$$

for $j = 1, 2, \ldots, N_1$. Defining

$$\underline{X} = \begin{pmatrix} \underline{x}_1' \\ \vdots \\ \underline{x}_{N_1}' \end{pmatrix}$$

the last equation gives

$$\underline{X}\,\underline{a} = \underline{u}$$

that is, $\underline{u}$ is an element of the column space $M(\underline{X})$ of $\underline{X}$:

$$\underline{u} \in M(\underline{X}).$$

For the special case $K = 1$, x denoting a single auxiliary variable with values $X_1, X_2, \ldots > 0$, we derive that t_T is consistent (ADC) if and only if

$$u_j = \frac{1 - \pi_j R_j}{\pi_j Q_j}, \propto X_j.$$

RESULT 6.1 *Consider a sequence of populations satisfying condition A with K-vectors*

$$\begin{pmatrix} X_i \\ Q_i \\ R_i \end{pmatrix}; \; i = 1, 2, \ldots.$$

Let p_T be of type B_0 with inclusion probabilities $\pi_1, \pi_2, \ldots$ such that

$$\frac{1 - \pi_i R_i}{\pi_i Q_i} \propto X_i.$$

Then, the QR predictor

$$\frac{1}{N}\left(\sum_1^N X_i - \sum_s R_i X_i\right)\frac{\sum_s Q_i X_i Y_i}{\sum_s Q_i X_i^2} + \frac{1}{N}\sum_s R_i Y_i$$

(with x as a single auxiliary variable) is consistent (ADC) for $\overline{Y}$.

EXAMPLE 6.2 *We follow* LITTLE *(1983) and consider an arbitrary design p with inclusion probabilities $\pi_1, \pi_2, \ldots, \pi_N$. Writing $\pi_{(1)}$ for the smallest inclusion probability, $\pi_{(2)}$ for the next larger one, etc., we define*

$$U_{(g)} = \{i \in U : \pi_i = \pi_{(g)}\}.$$

Assume that $Y_1, Y_2, \ldots, Y_N$ are independently distributed but for $i \in U_{(g)}$, alternatively,

$$\begin{aligned} Y_i &\sim N(\alpha\,;\,\sigma^2 V_{(g)}) \\ &\sim N(\alpha + \beta X_i\,;\,\sigma^2 V_{(g)}) \\ &\sim N(\alpha_{(g)}\,;\,\sigma^2 V_{(g)}) \\ &\sim N(\alpha_{(g)} + \beta X_i\,;\,\sigma^2 V_{(g)}) \\ &\sim N(\alpha_{(g)} + \beta_{(g)} X_i\,;\,\sigma^2 V_{(g)}) \end{aligned}$$

where $V_{(g)}$ and X_i are known and $\sigma^2, \alpha, \alpha_{(g)}, \beta, \beta_{(g)}$ are unknown parameters.

According to RESULT 4.3 the BLU predictors are of the QR type. They are ADC in the first two cases if all

$$V_{(g)}\frac{1 - \pi(g)}{\pi(g)};\; g = 1, 2, \ldots$$

are equal. Assume this is not true. The BLU predictor is nevertheless consistent in the second alternative if

$$X_i = X_{(g)} \quad \textit{for all } i \in \mathcal{U}_{(g)}$$

and a_1, a_2 *exist with*

$$V(g)\,\frac{1-\pi_{(g)}}{\pi_{(g)}} = a_1 + a_2 X_{(g)}.$$

In the other three cases the BLU predictors are at any rate consistent according to the general criterion above. So, the presence of a non-zero intercept term $\alpha_{(g)}$ *in these regression models really ensures the ADC property of the BLUPs; hence* LITTLE *(1983) recommends basing BLUPs on such models. But the intercept term must be estimated for each group, and this requires large enough samples from all groups that are not always available.*

6.1.3 Some General Results on QR Predictors

In the sequel we present some results given by WRIGHT (1983) and SÄRNDAL and WRIGHT (1984).

It is easily seen that the ADC condition is always true for

$$R_i = \frac{1}{\pi_i} \quad \text{for} \quad i = 1, 2, \ldots, N.$$

Therefore,

RESULT 6.2 *All GREG predictors are consistent and ADU.*

Let t_{QR} be an arbitrary QR predictor that is consistent; that is,

$$\frac{1-\pi_i R_i}{\pi_i Q_i} = \underline{a}'\underline{x}_i \quad \text{for} \quad i = 1, 2, \ldots, N.$$

Consider the associated GREG predictor $t_{Q1/\pi}$ for which

$$\begin{aligned} t_{Q1/\pi} - t_{QR} &= \sum_s \frac{1}{\pi_i}(Y_i - \underline{x}_i'\hat{\beta}_Q) - \Sigma_s R_i(Y_i - \underline{x}_i'\hat{\beta}_Q) \\ &= \sum_s \frac{1-\pi_i R_i}{\pi_i Q_i} Q_i(Y_i - \underline{x}_i'\hat{\beta}_Q) \\ &= \sum_s \underline{a}'\underline{x}_i Q_i(Y_i - \underline{x}_i'\hat{\beta}_Q) \\ &= \underline{a}'\left(\sum Q_i\underline{x}_i Y_i - \sum Q_i\underline{x}_i\underline{x}_i'\hat{\beta}_Q\right). \end{aligned}$$

According to the definition of $\hat{\underline{\beta}}_Q$ the last difference equals 0; hence

RESULT 6.3 *Let t_{QR} be consistent. Then,*

$$t_{QR} = t_{Q1/\pi}$$

The following is easily seen:

RESULT 6.4 *Let $\underline{\theta} \in \mathbb{R}^k$ be such that $\underline{x}_i'\underline{\theta} > 0$ and define*

$$Q_i \propto \frac{1}{\pi_i \underline{x}_i'\underline{\theta}}$$

$$\tilde{Q}_i \propto \left[\frac{1}{\pi_i} - 1\right] \Big/ \underline{x}_i'\underline{\theta}$$

($i = 1, 2, \ldots, N$). Then the SPRO predictor t_{Q0} and the LPRE $t_{\tilde{Q}1}$ are consistent and hence ADU. For the special case $K = 1$, taking

$$Q_i^* \propto \frac{1}{X_i}\left[\frac{1}{\pi_i} - 1\right]$$

one gets the LPRE proposed by BREWER *(1979).*

REMARK 6.2 *Let us write*

$$\underline{B} = \left[\sum_1^N Q_i \underline{x}_i \underline{x}_i'\right]^{-1} \left[\sum_1^N Q_i \underline{x}_i Y_i\right] = (\underline{X}'Q\underline{X})^{-1}(\underline{X}'Q\underline{Y})$$

which is an estimate of $\underline{\beta}$ based on all the values Y_i; $i = 1, 2, \ldots, N$, an analogue of $\hat{\underline{\beta}}_Q$ both coinciding for $s = U$. This $\underline{B}$ is called a **census-fitted** *estimator for $\underline{\beta}$ and*

$$\hat{\mu}_{ci} = \underline{x}_i'\underline{B}$$

a census-fitted estimator of $\mu_i = E_m(Y_i)$. The residual

$$E_i' = Y_i - \hat{\mu}_{ci}$$

for a census fit obviously cannot be ascertained from a sample at hand. But for a consistent t_{QR}, an asymptotic formula for the design variance $V_p(t_{QR})$ or design mean square error $M_p(t_{QR})$ is available, as given by SÄRNDAL *(1982)*

$$V = \sum\sum_{i<j}(\pi_i\pi_j - \pi_{ij})\left[\frac{E_i}{\pi_i} - \frac{E_j}{\pi_j}\right]^2$$

where

$$E_i = Y_i - \underline{x}_i'\underline{B}_\pi$$

writing

$$\underline{B}_\pi = \left(\sum_1^N \pi_i Q_i \underline{x}_i \underline{x}_i'\right)^{-1} \sum_1^N \pi_i Q_i \underline{x}_i Y_i.$$

REMARK 6.3 *For $\tilde{Q}$ defined in RESULT 6.4 consider*

$$\begin{aligned} t_{\tilde{Q}1} &= \underline{1}_n'\underline{Y}_s + (\underline{1}_N'\underline{X} - \underline{1}_n'\underline{X}_s)\hat{\beta}_{\tilde{Q}} \\ t_{\tilde{Q}1/\pi} &= \underline{1}_n \Pi_s^{-1}\underline{Y}_s + (\underline{1}_N'\underline{X} - \underline{1}_n'\Pi_s^{-1}\underline{X}_s)\hat{\beta}_{\tilde{Q}}. \end{aligned}$$

where Π_s is the diagonal matrix with diagonal elements

$$\pi_i,\ i \in s.$$

$t_{\tilde{Q}1}$ is attractive in a model-based approach, $t_{\tilde{Q}1/\pi}$ in a design-based approach.

Now, BREWER *(1999a) shows*

$$t_{\tilde{Q}1} = t_{\tilde{Q}1/\pi} = t,\ say$$

and calls t a ***cosmetic*** *estimator.*

6.1.4 Bestness under a Model

To choose among different Q_i's satisfying the ADC and equivalently ADU requirement in case $R = 1$, BREWER (1979) recommended as a criterion

$$L = \lim_{T\to\infty} E_m E_p \left\{ \left[t_{Q1T}(s_T, \underline{Y}_T) - Y_T \right]^2 / T \right\}$$

where $Y_i = \underline{x}_i'\underline{\beta} + \varepsilon_i$ is assumed with

$$\begin{aligned} E_m(\varepsilon_i) &= 0 \\ C_m(\varepsilon_i, \varepsilon_j) &= \sigma_i^2, \quad \text{if } j = i \\ &= 0, \quad \text{if } j \neq i \end{aligned} \tag{6.1}$$

$(i, j = 1, 2, \ldots, TN)$. He has shown that

$$L \geq \sum \sigma_i^2 \left[\frac{1}{\pi_i} - 1 \right]$$

holds with equality for the LPRE defined by Q_i^* (see RESULT 6.4).

Now, every QR predictor with the consistency and ADU property is a GREG predictor, $t_{Q1/\pi}$, and

$$t_{Q1/\pi} - Y = \left[\sum_1^N \underline{x}_i' - \sum I_{si}\frac{1}{\pi_i}\underline{x}_i'\right]\left[\sum_1^N I_{si}Q_i\underline{x}_i\underline{x}_i'\right]^{-1}\left[\sum_1^N I_{si}Q_i\underline{x}_iY_i\right]$$
$$+ \sum_1^N I_{si}\frac{1}{\pi_i}Y_i - \sum_1^N I_{si}Y_i - \sum_1^N (1 - I_{si})Y_i$$
$$= \sum_1^N I_{sj}\left\{\left[\sum_1^N \underline{x}_i' - \sum I_{si}\frac{1}{\pi_i}\underline{x}_i'\right]\left[\sum_1^N I_{si}Q_i\underline{x}_i\underline{x}_i'\right]^{-1} Q_j\underline{x}_j\right.$$
$$\left. + \left[\frac{1}{\pi_j} - 1\right]\right\}Y_j - \sum_1^N (1 - I_{sj})Y_j.$$

With s replaced by s_T and N by NT we obtain

$$t_{Q1/\pi T} - Y_T.$$

It is easily checked that $E_m(t_{Q1/\pi T} - Y_T) = 0$ and under Eq. (6.1)

$$E_m\left[t_{Q1/\pi T} - Y_T\right]^2 = V_m\left[t_{Q1/\pi T} - Y_T\right]$$
$$= \sum_1^{NT} I_{s_Tj}\left\{\left[\sum_1^{NT} \underline{x}_i' - \sum I_{s_Ti}\frac{1}{\pi_i}\underline{x}_i'\right]\left[\sum_1^{NT} I_{s_Ti}Q_i\underline{x}_i\underline{x}_i'\right]^{-1} Q_j\underline{x}_j\right.$$
$$\left. + \left[\frac{1}{\pi_i} - 1\right]\right\}^2 \sigma_j^2 + \sum_1^{NT}(1 - I_{s_Tj})\sigma_j^2.$$

Hence

$$E_m\left(\left[t_{Q1/\pi T} - Y_T\right]^2 / T\right)$$
$$= \sum_1^N f_{jT}\left\{\left[\sum \underline{x}_i' - \sum f_{iT}\frac{1}{\pi_i}\underline{x}_i'\right]\left[\sum_1^N f_{iT}Q_i\underline{x}_i\underline{x}_i'\right]^{-1} Q_j\underline{x}_j\right.$$
$$\left. + \left[\frac{1}{\pi_i} - 1\right]\right\}^2 \sigma_j^2 + \sum (1 - f_{jT})\sigma_j^2$$

and

$$\lim_{T\to\infty} E_p E_m \left([t_{Q1/\pi T} - Y_T]^2 / T \right)$$

$$= \sum_i^N \pi_j \left[\frac{1}{\pi_i} - 1 \right]^2 \sigma_j^2 + \sum_i^N (1 - \pi_j) \sigma_j^2$$

$$= \sum \sigma_j^2 \left[\frac{1}{\pi_j} - 1 \right]$$

that is, every QR predictor with the consistency property has the common limiting value

$$\sum \sigma_j^2 \left[\frac{1}{\pi_j} - 1 \right]$$

which is equal to the lower bound of BREWER's (1979) L.

Restricting to p_n designs, the minimum value of BREWER's lower bound is

$$\frac{[\sum \sigma_j]^2}{n} - \sum \sigma_j^2.$$

If, in particular, $\sigma_j = \sigma f_j, j = 1, \ldots, N$ with $\sigma(>0)$ unknown but $f_j(>0)$ known, so that $\Sigma = \sigma^2 V$ with $V = diag(f_1^2, \ldots, f_N^2)$, the strategy (p_{nf}, e_Q) is regarded as best when

$$e_Q = \underline{1}_s' \underline{\Pi}_s^{-1} \underline{Y}_s + (\underline{1}' \underline{X} - \underline{1}_s' \underline{\Pi}_s^{-1} \underline{X}_s) \hat{\beta}(Q_s)$$

is based on the p_n design p_{nf} for which

$$\pi_i = \frac{n f_i}{\sum_1^N f_i}, \quad i = 1, \ldots, N.$$

By best we mean a strategy involving an ADU predictor for which the above minimal value is attained.

TAM (1988a) has shown that

(a) $\underline{1}_s' \underline{X}_s = \underline{1}' \underline{X}$
(b) $Q_s^{-1}(\underline{1}_s - k V_{ss}^{-1/2} \underline{1}_s) \in M(\underline{X}_s)$

are sufficient conditions for a strategy (p_n, e_L) with $e_L = \underline{1}_s' \underline{Y}_s$ to be best in estimating Y. Here $k = \frac{1}{n} \sum f_j$ and

$$V = diag\left(f_1^2, \ldots, f_N^2\right) = \begin{pmatrix} V_{ss} & \underline{0} \\ \underline{0} & V_{rr} \end{pmatrix}$$

It may be noted that (a) here is a condition of model unbiasedness. This is relevant in prescribing conditions for robustness. If a working model differs from a true model one may go wrong in misspecifying the design parameters π_i and/or misspecifying V. As long as both the conditions (a) and (b) are satisfied by a strategy the latter is robust even if one goes wrong in postulating the right model in other respects. TAM (1988a, 1988b) and BREWER, HANIF and TAM (1988) give further results useful in fixing conditions on design parameters, on the features of models in achieving the ADU property and/or in bestowing optimality properties on several alternative design-cum-model-based predictors and related strategies. One may consult further the references cited in the above two, especially the works due to SÄRNDAL and his colleagues.

6.1.5 Concluding Remarks

For a fuller treatment and alternative approaches by asymptotic analyses in survey sampling along with their interpretations, one may refer to BREWER (1979), SÄRNDAL (1980), FULLER and ISAKI (1981), ISAKI and FULLER (1982), ROBINSON and SÄRNDAL (1983), HANSEN, MADOW and TEPPING (1983), and CHAUDHURI and VOS (1988). We omit the details to avoid a too technical discussion.

Robustness has been on the focus relating to LPREs. GREG predictors by virtue of their forms acquire robustness from design considerations in the sense of asymptotic design unbiasedness, as we noticed in the previous section. At this stage let us turn again to them to examine their robustness.

An LPRE is of the form $t_L = \Sigma_s Y_i + \Sigma_r \hat{\mu}_i$ where $E_m(Y_i) = \mu_i$. If μ_i is a polynominal in an auxiliary variable x, for samples balanced up to a certain order every t_{BLU} is bias robust, that is, $E_m(t_{BLU} - Y) = 0$, and asymptotically so for large samples selected by SRSWOR, preferably with appropriate stratifications. But t_{BLU} is not usually MSE robust, by which we mean the following: Let us write $t_{m'}$ for the predictor, which is BLU under a model m'; its bias, MSE, and variance under a true model, m, are, respectively, $B_m(t_{m'})$, $M_m(t_{m'})$, and $V_m(t_{m'} - Y)$. Then, $M_m(t_{m'}) = V_m(t_{m'} - Y) + B_m^2(t_{m'})$ and

$M_m(t_m) = V_m(t_m - Y)$ because $B_m(t_m) = 0$. Even if $|B_m(t_{m'})|$ is negligible, $V_m(t_{m'} - Y)$ may be too far away from $V_m(t_m - Y)$ and so may be $M_m(t_{m'})$ from $M_m(t_m)$. So $t_{m'}$, even if bias robust, may be quite fragile in respect to MSE.

Very little with practical utility is known about MSE robustness of LPREs. More importantly, nobody knows what the true model is; even with a polynomial assumption it is hard to know its degree, and in large-scale surveys diagnostic analysis to fix a correct model is a far cry. So, it is being recognized that even for model-based LPREs robustness should be examined with respect to design, that is, one should examine the magnitude of

$$M_p(t_L) = E_p(t_L - Y)^2 = V_p(t_L) + B_p^2(t_L).$$

Since the sample size is usually large, we may presume $V_p(t_L)$ to be suitably under control and we should concentrate on $|B_p(t_L)|$. In section 4.1.2 we saw how a restriction $B_p(t) = 0$ may lead to loss of efficiency, especially if a model is accurately postulated. An accepted criterion for robustness studies is therefore to demand that t_L be ADC. Similar are the desirable requirements for any other estimator or predictor.

6.2 ASYMPTOTIC MINIMAXITY

In practice it is difficult to find a strategy (p^*, t^*) which is minimax in the strict sense, that is, with the property

$$\sup_{\underline{Y}\in\Omega} M_{p^*}(t^*) = \inf_{(p,t)\in\Delta} \sup_{\underline{Y}\in\Omega} M_p(t) = r^*, \text{ say}$$

where Ω is the set of all relevant parameters $\underline{Y}$ and Δ the set of all strategies available in a situation. So, CHENG and LI (1983) have reported how one may derive strategies (p', t') that are approximately minimax in the sense that

$$\sup_{\underline{Y}\in\Omega} M_{p'}(t')$$

comes close to r^*.

A more satisfactory approach is to aim at strategies that are asymptotically minimax. In describing this approach we follow STENGER (1988, 1989, 1990) to show, for example, that the ratio estimator, when based on SRSWOR, is asymptotically minimax. The RHC strategy, however, which is approximately minimax in the sense defined by CHENG and LI (1983), is not minimax in our asymptotic setup.

6.2.1 Asymptotic Approximation of the Minimax Value

For a population U and a size measure x with $X_1, X_2, \ldots, X_N > 0$ we define (c.f. section 3.4.2)

$$\Omega_x = \{\underline{Y} \in \mathbb{R}^N : 0 \leq Y_i \leq X_i \quad \text{for all } i = 1, 2, \ldots, N\}$$

$$\Delta_n = \left\{(p, t) : p \text{ a design of fixed size } n, t = \sum_{i \in s} b_{si} Y_i\right\}$$

Define, as in section 5.1, $X_{N+1}, X_{N+2}, \ldots, X_{NT}$ with

$$X_i = X_{i+N} = X_{i+2N} = \ldots$$

for $i = 1, 2, \ldots, N$, which may be interpreted as reproducing $T - 1$ times the population U with the known x values leading to an extended population $(1, 2, \ldots, NT)$ and $\underline{X}_T = (X_1, \ldots, X_{NT})$.

Define $\underline{Y}_T = (Y_1, Y_2, \ldots, Y_{NT})$ where Y_i is the value of the variate under study for the unit i. We assume the parameter space

$$\Omega_{x_T} = \left\{\underline{Y}_T \in \mathbb{R}^{NT} : 0 \leq Y_i \leq X_i \text{ for } i = 1, 2, \ldots, NT\right\}$$

It is worth noting that $\underline{Y}_T \in \Omega_{x_T}$ is assumed, but not $Y_i = Y_{i+N} = \ldots$.

RESULT 6.5 *Let Δ_{nT} be the class of all strategies (p_T, t_T) where p_T is a design of size Tn used to select a sample s_T from U_T and*

$$t_T = t_T(s_T, \underline{Y}_T) = \sum_{i \in s_T} b_{s_T i} Y_i$$

a homogeneously linear estimator. Then, assuming

$$n\frac{X_i}{X} \leq 1 \quad \textit{for } i = 1, 2, \ldots, N \tag{6.2}$$

we have

$$\lim_{T\to\infty} nTr_T = \frac{1}{4}\left[\overline{X}^2\left(1-\frac{n}{N}\right)-\frac{n}{N}\sigma_{xx}\right]$$

where

$$r_T = \inf_{\Delta_{nT}} \sup_{\Omega_{x_T}} M_{p_T}(t_T)$$

$$\sigma_{xx} = \frac{1}{N}\sum(X_i-\overline{X})^2.$$

Hence,

$$\frac{1}{4n}\left[\overline{X}^2\left(1-\frac{n}{N}\right)-\frac{n}{N}\sigma_{xx}\right] = r_x,\ say$$

approximates Tr_T.

PROOF: *Define for* $i = 1, 2, \ldots, N$

$$U_i = (i, i+N, i+2N, \ldots, i+(T-1)N)$$

and consider a design p_T *of size* nT *selecting a sample* s_T *that is composed of samples* $s_1, s_2, \ldots, s_N$ *of sizes* $Tf_1, Tf_2, \ldots$ Tf_N *from* $U_1, U_2, \ldots, U_N$*, respectively.* $\underline{f} = (f_1, f_2, \ldots, f_N)'$ *may be a random vector; we assume that, conditional on* $\underline{f}$, s_i *is selected by SRSWOR of size* Tf_i.

The MSE of the estimator

$$\sum \tau_i(\underline{f})\overline{y}_i$$

where $\overline{y_i}$ *is the mean of the y values of all* Tf_i *units of* U_i *in the sample is then*

$$M_0 = E_f\left\{\sum \tau_i^2(\underline{f})\frac{\sigma_{iyy}}{f_i}\frac{1-f_i}{T-1} + \left[\sum \tau_i(\underline{f})\overline{Y}_i - \frac{1}{N}\sum \overline{Y}_i\right]^2\right\}$$

where the expectation operator E_f *refers to* $\underline{f}$ *and* $\overline{Y}_i(\sigma_{iyy})$ *is the mean (variance) of the y values of all units in* U_i.

Now, under condition (6.2) the design may be chosen such that

$$nT\cdot\frac{X_i}{X} - 1 < Tf_i \le nT\cdot\frac{X_i}{X} + 1 \quad for\ i = 1, 2, \ldots, N$$

with Tf_i *an integer and* $\Sigma f_i = n$*, provided* T *is large enough. Setting* $\tau_i(\underline{f}) = 1/N$ *and taking into account* $\sigma_{iyy} \le X_i^2/4$

we derive

$$r_T \leq \sum_1^N \frac{1}{N^2}\frac{X_i^2}{4}\left[\frac{1}{nT\frac{X_i}{X}-1}\frac{T}{T-1}-\frac{1}{T}\right]$$

$$\overline{\lim_{T\to\infty}}\, T r_T \leq r_x.$$

Assume $(p,t)\in\Delta_{nT}$ *exists with*

$$T\sup_{\Omega x_T} M_p(t) < r_x.$$

Define for $j=1,2,\ldots,N$ *a vector* $\underline{Y}^{(j)}$ *with*

$$Y_j = Y_{j+N} = Y_{j+2N} = \ldots = X_j$$

and $Y_i = 0$ *for* $i \neq j,\ j+N,\ j+2N,\ldots$. *Then* $\underline{Y}^{(j)}\in\Omega_{x_T}$ *and*

$$E\left[\tau_j(\underline{f})X_j - \frac{X_j}{N}\right]^2 < \frac{r_x}{T}$$

which implies

$$\frac{X_j}{N} - \sqrt{\frac{r_x}{T}} < E\tau_j(\underline{f})X_j < \frac{X_j}{N} + \sqrt{\frac{r_x}{T}}$$

$$E\tau_j^2(\underline{f})X_j^2 < \left[\frac{X_j}{N} + \sqrt{\frac{r_x}{T}}\right]^2.$$

Therefore, by Cauchy's inequality

$$E\frac{\tau_j^2(\underline{f})X_j^2}{f_j} \geq \frac{[E\tau_j(\underline{f})X_j]^2}{Ef_j} > \frac{1}{Ef_j}\left[\frac{X_j}{N} - \sqrt{\frac{r_x}{T}}\right]^2$$

and because of $\sup\sigma_{iyy} \geq X_i^2(T-1)/(4T)$

$$\sup_{\Omega_{x_T}} M_0 \geq E\left\{\sum \tau_i^2(\underline{f})\frac{X^2(T-1)}{4T}\left[\frac{1}{f_i}\frac{1}{T-1} - \frac{1}{T-1}\right]\right\}$$

$$\geq \frac{1}{4T}\left\{\sum\frac{1}{Ef_i}\left[\frac{X_i}{N} - \sqrt{\frac{r_x}{T}}\right]^2 - \sum\left[\frac{X_i}{N} + \sqrt{\frac{r_x}{T}}\right]^2\right\}.$$

From $n = \Sigma Ef_i$ *we derive, therefore,*

$$T\inf\sup M_0 \geq \frac{1}{4n}\left[\sum\left[\frac{X_i}{N} - \sqrt{\frac{r_x}{T}}\right]\right]^2 - \frac{1}{4}\sum\left[\frac{X_i}{N} + \sqrt{\frac{r_x}{T}}\right]^2.$$

Obviously, the right-hand side converges to r_x *and the desired result follows.*

In a similar way, asymptotic approximations may be derived for the minimax value with respect to other parameter spaces introduced in section 3.4.1. By equating x and z in Ω_{xz} we obtain

$$\Omega_{xx} = \left\{ \underline{Y} \in \mathbb{R}^N : \frac{1}{X} \sum \frac{1}{X_i} \left[Y_i - \frac{Y}{X} X_i \right]^2 \leq c^2 \right\}$$

and by defining $X_i = Z_i^2$

$$\Omega_{z^2z} = \left\{ \underline{Y} \in \mathbb{R}^N : \frac{1}{\sum Z_i^2} \sum \left[Y_i - \frac{Y}{Z} Z_i \right]^2 \leq c^2 \right\}.$$

The asymptotic approximations of the minimax values (with respect to Δ_n) are

$$r_{xx} = \frac{c^2}{n} \overline{X} \cdot \zeta \quad \text{and}$$

$$r_{z^2z} = \frac{c^2}{n} \left[1 - \frac{n}{N} \right] \frac{1}{N} \sum Z_i^2$$

respectively, as has been shown by STENGER (1989); here ζ is the unique solution of

$$\sum X_i \bigg/ \left[\zeta + \frac{n}{N} X_i \right] = N$$

and satisfies

$$\zeta \leq \overline{X} \left[1 - \frac{n}{N} \right]$$

with equality if and only if $X_1 = X_2 = \ldots = X_N$.

6.2.2 Asymptotically Minimax Strategies

To introduce the notion of asymptotic minimaxity of a strategy we consider the following modification of Ω_{z^2z}:

$$\Omega^{(L)} = \left\{ \underline{Y} \in \mathbb{R}^N : 0 < Y_i < L \quad \text{for } i = 1, 2, \ldots, N \quad \text{and} \right.$$
$$\left. \frac{1}{\sum Z_i^2} \sum \left(Y_i - \frac{Y}{Z} Z_i \right)^2 \leq c^2 \right\}$$

where $L > 0$ is given. $\Omega_T^{(L)}$ is correspondingly defined by $\underline{Z}_T$ instead of $\underline{Z}$ and Δ_{nT} has the same meaning as earlier. Suppose a sample of size nT is selected by SRSWOR and denote by $\overline{y}_T, \overline{z}_T$ the sample means of the y and z values, respectively. For the MSE M_T of the ratio estimator

$$\overline{Z}\frac{\overline{y}_T}{\overline{z}_T}$$

we then have (cf. STENGER, 1990)

$$T \sup_{\Omega_T^{(L)}} M_T \leq \frac{c^2}{n}\left[1 - \frac{n}{N}\right]\frac{1}{N}\sum_1^N Z_i^2 + \frac{A}{\sqrt{T}}$$

with A free of T. Hence

$$\lim_{T\to\infty} T \sup_{\Omega_T^{(L)}} M_T \leq \frac{c^2}{n}\left[1 - \frac{n}{N}\right]\frac{1}{N}\sum_1^N Z_i^2 = r_{z^2z}$$

such that the ratio strategy achieves the asymptotic approximation of the minimax value with respect to $\Omega^{(L)}$ and Δ_n in an asymptotic sense and may be called an asymptotically minimax strategy.

To give a more general definition of asymptotic minimaxity let Ω be any parameter space defined by a vector $\underline{X}$ (or vectors $\underline{X}$ and $\underline{Z}$). Ω_T is the subset of $\mathbb{R}^{NT}$ given by $\underline{X}_T$ (or $\underline{X}_T$ and $\underline{Z}_T$). Let a design p_T of fixed size nT and an estimator t_T be defined by $\underline{X}_T$ (and $\underline{Z}_T$) without T appearing explicitly. Then (p_1, t_1) may be called **asymptotically minimax** if for the MSE M_T of (p_T, t_T)

$$\lim_{T\to\infty} T \sup_{\Omega_T} M_T$$

equals the asymptotic approximation of the minimax value with respect to Ω and Δ_n.

It is easily seen that the MSE M_T of the RHC strategy of size n_T satisfies

$$T \sup_{\Omega_{xx}} M_T = \frac{c^2}{n}\left[1 - \frac{n}{N}\right]\frac{NT}{NT - 1}\overline{X}^2$$

Hence,

$$\lim_{T\to\infty} T \sup M_T = \frac{c^2}{n}\left[1 - \frac{n}{N}\right]\overline{X}^2 > r_{xx}$$

and the RHC strategy is not asymptotically minimax with respect to Ω_{xx} and Δ_n.

6.2.3 More General Asymptotic Approaches

In an asymptotic theory the actual population U is usually treated as an element of a sequence of populations $U_1, U_2, \ldots$ with increasing sizes $N_1, N_2, \ldots$ and the vector $\underline{X}$ of values of an auxiliary variable x as an element of a sequence of vectors $\underline{X}_1, \underline{X}_2, \ldots$ associated with $U_1, U_2, \ldots$. In section 6.2.1, U and $\underline{X}$ are the first elements of sequences defined in a very special way such that doubts may arise on the relevance of the results.

Therefore, more general approaches will be described.

Define for $\xi \in \mathbb{R}$

$$G(\xi) = \frac{1}{N}\,[\text{number of } X_i \text{ in } \underline{X} \text{ with } X_i \leq \xi].$$

Replacing N and $\underline{X}$ in the definitions of Ω_x and G by N_T and $\underline{X}_T$ we obtain

$$\Omega_{xT}, G_T(\xi).$$

Consider sample sizes $n_1, n_2, \ldots$ such that

$$\lim_{T\to\infty} \frac{n_T}{N_T} = f$$

exists and define

$$r_T = \inf_{\Delta_{n_T}} \sup_{\Omega_{xT}} M_{p_T}(t_T).$$

Now, imposing suitable conditions on G_T; $T = 1, 2, \ldots$ the limit of $n_T \cdot r_T$ for $T \to \infty$ should exist. In fact, let

$$\lim_{T\to\infty} G_T(\xi) = \Gamma(\xi)$$

be a distribution function. Then, as has been shown by STENGER (1989), weak additional assumptions are sufficient for the existence of

$$\lim_{T\to\infty} n_T r_T = \rho(\Gamma, f), \text{ say.} \tag{6.3}$$

Hence,

$$\frac{1}{n_T}\rho\left(G_T, \frac{n_T}{N_T}\right)$$

is an approximation of r_T and

$$\frac{1}{n}\rho\left(G, \frac{n}{N}\right)$$

is an approximation of the minimax value of interest

$$r^* = \inf_{\Delta_n} \sup_{\Omega} M_p(t).$$

If Eq. (6.3) is taken for granted, $\rho(G, n/N)/n$ may be determined by the simple procedure described in section 6.2.1.

Chapter 7

Design- and Model-Based Variance Estimation

In estimating Y by a design-based estimator, a choice among competing strategies (p, t_p) is made on considerations of the magnitudes of $|B_p(t_p)|$, $V_p(t_p)$, and $M_p(t_p)$, each required to be small. Once a choice is made and a sample is drawn and surveyed, it is customary to report an estimated value v_p of $V_p(t_p)$ along with the value of t_p.

A variance estimator indicates the level of accuracy attained by the estimator actually employed but, more importantly, it provides a measure of the variability of the estimator over conceptual repeated sampling. Planning of future surveys is aided by indicating, among other things, a sample size needed to achieve a desired level of precision by adopting a similar strategy. Moreover, it helps in making confidence statements. If v_p is an estimator for $V_p(t_p)$, then the following standardized error (SZE)

$$(t_p - Y)/\sqrt{v_p}$$

is supposed to have STUDENT's t distribution with a number of degrees of freedom (df) determined by the sample size n.

This supposition is valid under many usual situations when the distribution of the SZE is considered over all possible samples s with $p(s) > 0$. For large n and N its distribution is often found close to that of the standardized normal deviate τ. Writing

$$Pr(\tau > \tau_\alpha) = \alpha, \quad 0 < \alpha < 1,$$

the interval $(t_p - \tau_{\alpha/2}\sqrt{v_p}, t_p + \tau_{\alpha/2}\sqrt{v_p})$, or briefly $(t_p \pm \tau_{\alpha/2}\sqrt{v_p})$, is supposed to be a $100(1-\alpha)\%$ **confidence interval** for Y. The interpretation here is that for the fixed $\underline{Y} = (Y_1, \ldots, Y_i, \ldots, Y_N)'$ the probability to obtain a sample s with an interval $(t_p \pm \tau_{\alpha/2}\sqrt{v_p})$ covering Y is $100(1-\alpha)\%$.

We have also considered a linear predictive approach based on least squares that involves treatment of model-based predictors t_m and their biases $B_m(t_m) = E_m(t_m - Y)$, mean square errors (MSE) $M_m(t_m) = E_m(t_m - Y)^2$, and variances $V_m = V_m(t_m - Y) = E_m[(t_m - Y) - E_m(t_m - Y)]^2$. It is also important to consider estimators v_m of V_m for the purposes of assessing the level of accuracy attained for a predictor t_m actually employed for Y, gaining insight into how a future survey should be planned for predictions and in making confidence statements.

In this case it is desirable to have

$$B_m(v_m) = E_m(v_m - V_m) \quad \text{and}$$
$$M_m(v_m) = E_m(v_m - V_m)$$

under control. Here the SZE is taken as

$$(t_m - Y)/\sqrt{v_m}$$

which is supposed to have student's t distribution and approximately the $N(0,1)$ distribution for large n, N. But here a $100(1-\alpha)\%$ confidence interval $(t_m \pm \tau_{\alpha/2}\sqrt{v_m})$ or $(t_m \pm t_{\alpha/2}\sqrt{v_m})$ is constructed with the interpretation that if $\underline{Y}$ is generated as hypothesized through a postulated model, then for $100(1-\alpha)\%$ of $\underline{Y}s$ so generated, the intervals will cover the unknown Y with the sample actually drawn held fixed.

In this context the main problem is robustness. Both the actual sample drawn and the estimation procedures are required to be so chosen that t_m may continue to predict Y well,

v_m may estimate $V_m(t_m - Y)$ well, and the SZE above may continue to yield confidence intervals with coverage probabilities close to the nominal value $1-\alpha$ even if the model on which t_m, v_m are based may be wrong, that is, some other model may underlie the process that generates $\underline{Y}$. Keeping this in mind, it is often necessary to examine several alternative but plausible formulae for v_m for a given t_m with respect to their biases, MSEs, that is, $E_m(v_m - V_m)^2$, and coverage probabilities of the confidence intervals they lead to. In this context, also, asymptotic analyses are necessary, and discussion of rigorous treatment of asymptotic studies here is again beyond our scope and aim. But we shall illustrate a few developments in a somewhat simplistic manner.

Innumerable strategies for estimating Y or $\bar{Y}$ are available. RAO and RAO (1971), WOLTER (1985), CHAUDHURI and VOS (1988), J. N. K. RAO (1986, 1988), P. S. R. S. RAO (1988), and ROYALL (1988) give accounts of many such along with variance estimators. But we shall cover only a few, our own interest drawing especially on the works mainly of ROYALL and EBERHARDT (1975), ROYALL and CUMBERLAND (1978a, 1978b, 1981a, 1981b, 1985), CUMBERLAND and ROYALL (1988), WU (1982), WU and DENG (1983), DENG and WU (1987), SÄRNDAL (1982, 1984), and, only in passing, SÄRNDAL and HIDIROGLOU (1989), SÄRNDAL, SWENSSON and WRETMAN (1992), and KOTT (1990), among others.

7.1 RATIO ESTIMATOR

The ratio estimators for Y, $\bar{Y}$, $R = \frac{Y}{X} = \frac{\bar{Y}}{\bar{X}}$, respectively, are

$$t_R = X\frac{\bar{y}}{\bar{x}},\ \bar{t}_R = \bar{X}\frac{\bar{y}}{\bar{x}} \quad \text{and} \quad r = \frac{\bar{y}}{\bar{x}}.$$

When based on the LMS scheme (cf. section 2.4.5) t_R is p unbiased for Y, but it is more popularly based on SRSWOR. Then it is biased, but its design bias is considered negligible for large n because the coefficient of variation (CV) of $N\bar{x}$ is small for large n and

$$|B_p(t_R)/\sigma_p(t_R) \le CV(N\,\bar{x})$$

(cf. RAO, 1986).

7.1.1 Ratio- and Regression-Adjusted Estimators

Although an exact formula for $V_p(\bar{t}_R)$ based on SRSWOR, along with one for its unbiased estimator, is given in section 2.4.1, it is traditional to turn to their respective approximations

$$\bar{M}' = \frac{1-f}{n}\frac{1}{N-1}\sum_1^N (Y_i - RX_i)^2$$

$$v_0 = \frac{1-f}{n}\frac{1}{n-1}\sum_s (Y_i - r\,X_i)^2.$$

J. N. K. RAO (1968, 1969) found empirically for $n \le 12$ that $\Delta = \bar{M}' - V_p(\bar{t}_R) < 0$ for many actual populations, but later, WU and DENG (1983) found both positive and negative values of Δ for $n = 32$, but none appreciably high in magnitude with more extensive empirical investigations. So it is considered adequate in practice to estimate $\bar{M}'$ rather than $V_p(\bar{t}_R)$ if n is not too small.

Since $\bar{M}'/\bar{X}^2$ is an approximation for $V_p(r)$ an estimator for it, in case $\bar{X}$ is unknown, is usually taken as

$$v_0/\bar{x}^2.$$

In case $\bar{X}$ is known, an alternative customary estimator for $\bar{M}'$ is therefore

$$v_2 = \left(\frac{\bar{X}}{\bar{x}}\right)^2 v_0.$$

WU (1982) suggests instead a ratio adjustment to v_0 to propose another alternative estimator for $\bar{M}'$ as

$$v_1 = \left(\frac{\bar{X}}{\bar{x}}\right) v_0$$

and goes a step further to propose a class of estimators

$$v_g = \left(\frac{\bar{X}}{\bar{x}}\right)^g v_0$$

and recommends choosing a suitable g in the following way:

Let $E_i = Y_i - RX_i$ with $\sum E_i = 0$ be the residual in fitting a straight line through the origin and the point $(\bar{X}, \bar{Y})$ in the scatter diagram of $(X_i, Y_i), i = 1, \ldots, N$ and let $e_i = Y_i - r\,X_i$

be taken as estimated residuals. Let

$$Z_i = E_i^2 - 2\,E_i \sum_1^N X_j\,E_j/X\,,\ \bar{Z} = \frac{1}{N}\sum_1^N Z_i\,.$$

Then, WU (1982) recommends (a) the optimal choice of g as

$$g_{opt} = \text{the regression coeffizient of } Z_i/\bar{Z} \text{ on}$$
$$X_i/\bar{X},\ \text{based on } (X_i,\ Y_i),\ i = 1, \ldots,\ N$$

and (b), because it is unavailable, replacing g_{opt} by

$$\hat{g} = \text{the sample analogue of } g_{opt} \text{ based on } (X_i, Y_i, e_i), i \in s.$$

To arrive at these recommendations WU (1982) carried out an asymptotic analysis to evaluate $V_p(v_g)$ using TAYLOR series expansion. They found it expedient to omit terms too small for large n and N and showed the term retained in the expansion of $V_p(v_g)$, called the **leading term**, to be minimum if g is taken as g_{opt}.

Another choice of g suggested by WU (1982) is $\tilde{g}$, which is the sample analogue of the regression coefficient of $E_i^2/\frac{1}{N}\sum_1^N E_i^2$ on $X_i/\bar{X}$. This is intended only to find a simpler substitute for $\hat{g}$.

Just as v_1 is a ratio adjustment on v_o, FULLER (1981) proposed a regression adjustment to propose another alternative estimator for $\bar{M}'$ as

$$v_{reg} = v_o + \frac{1-f}{n}\,\hat{b}(\bar{X} - \bar{x}).$$

Here $\hat{b}$ is the regression coefficient of e_i^2 on X_i evaluated from (X_i, Y_i); $i \in s$.

Although $v_{g_{opt}}$ is asymptotically optimal, it is not known how it may fare compared to v_o, v_1, v_2 in specific situations with given N, n and it is more important to examine the performance of $v_{\hat{g}}$, $v_{\tilde{g}}$, and v_{reg} vis-à-vis v_o, v_1, v_2 using empirical data at hand. Also, if one restricts g for simplicity to 0, 1, 2, one should be curious about how in practice to choose among these three competitors.

Even with the design-based approach it is known that one will be well off to use $\bar{t}_R$ based on SRSWOR to estimate $\bar{Y}$ if from the sample observations $(X_i, Y_i), i \in s$ one is justified to believe that a straight line passing closely through the origin gives an adequate fit to the scatter of all (x, y) values in the population to which the values $(X_i, Y_i), i = 1, \ldots, N$ belong.

In fact, the use of $\bar{t}_R$ to estimate $\bar{Y}$ is well known to be appropriate if a model M_{1_γ} (cf. section 4.1.2) may be correctly postulated for the $(X_i, Y_i), i = 1, \ldots, N$ under investigation, for which

$$E_m(Y_i) = \beta X_i, \; V_m(Y_i) = \sigma^2 X_i^\gamma, \; C_m(Y_i, Y_j) = 0, \; i \neq j$$

and more specifically, if $\gamma = 1$.

By dint of his asymptotic analysis without model postulations, WU (1982) concludes that among v_0, v_1, v_2 as estimators of $\bar{M}'$

$$\begin{aligned} &v_0 \text{ is the best if } g_{opt} \leq 0.5 \\ &v_1 \text{ is the best if } 0.5 \leq g_{opt} \leq 1.5 \\ &v_2 \text{ is the best if } g_{opt} \geq 1.5 . \end{aligned}$$

But postulating the model $M_{1\gamma}$ he concludes that among v_g

$$\begin{aligned} &v_0 \text{ is optimal if } \gamma = 0 \\ &v_1 \text{ is optimal if } \gamma = 1 \\ &v_1, v_2 \text{ are better than } v_0 \text{ if } \gamma \geq 1 \end{aligned}$$

as estimator of $\overline{M}'$. He further observed that for large n the squared p bias of v_g is inconsequential relative to $\bar{M}'$ and so one need not bother about the p bias in employing a v_g. But for sample size actually at hand, correcting for the bias may be useful, and a large-sample approximation formula for $E_p(v_g - \bar{M}')$ has been given by WU (1982), who suggests using an estimator for it to correct for the p bias of v_g.

Incidentally, if the model M_{21} is postulated instead (cf. section 4.1.2), demanding independence of estimating equations (cf. section 3.3) to the multiparameter cases, GODAMBE and THOMPSON (1988a, 1988b) lay down estimating equations for β and γ^2 in this case as

$$\sum_1^N (Y_i - \beta X_i) = 0 \quad \text{and} \quad \sum_1^N \{(Y_i - \beta X_i)^2 - \sigma^2 X_i\} = 0.$$

From these the solutions are

$$\beta_0 = \frac{Y}{X} \quad \text{and} \quad \sigma_o^2 = \frac{1}{X}\sum_1^N \left(Y_i - \frac{Y}{X}X_i\right)^2$$

and their estimators based on SRSWOR are

$$\hat{\beta} = \sum_s Y_i / \sum_s X_i = r \quad \text{and} \quad \hat{\sigma}^2 = \sum_s (Y_i - r\, X_i)^2 / \sum_s X_i.$$

So they propose

$$\frac{X}{\sum_s X_i}\sum_s (Y_i - r\, X_i)^2 \text{ as an estimator for } \sum_1^N \left(Y_i - \frac{Y}{X}X_i\right)^2$$

and hence

$$\frac{N}{n(N-1)}\,\frac{1-f}{N}\,\frac{X}{\bar{x}}\sum_s (Y_i - r\, X_i)^2$$

as an estimator for $\bar{M}'$. This variance estimator is obviously quite close to v_1.

7.1.2 Model-Derived and Jackknife Estimators

For a decisive choice among the estimators of $\bar{M}'$ keeping in mind their p biases, design MSEs (often called measures of stability of variance estimators), and efficacy in yielding design-based confidence intervals one recognized approach is to examine empirical evidences of their relative performances. Before briefly narrating some such exercises reported in the literature, let us mention some more competitive variance estimators that have emerged through the model-based predictive approach in the context of applicability of ratio predictor.

If the model $\mathcal{M}_{11}$ (cf. section 4.1.2) is true, $\bar{t}_R$ is the BLUP for $\bar{Y}$ with

$$B_m(\bar{t}_R) = E_m(\bar{t}_R - \bar{Y}) = 0$$

$$V_m = V_m(\bar{t}_R - \bar{Y}) = \frac{1-f}{n}\,\frac{\bar{X}\,\bar{x}_r}{\bar{x}}\,\sigma^2 = g(s)\sigma^2, \text{ say.}$$

Since

$$\hat{\sigma}^2 = \frac{1}{n-1}\sum_s \left[\frac{e_i^2}{X_i}\right]$$

has

$$E_m(\hat{\sigma}^2) = \sigma^2$$

under $\mathcal{M}_{11}$,

$$v_L = g(s)\hat{\sigma}^2$$

is an m-unbiased estimator for V_m, no matter how a sample s of size n is drawn.

A sample of size n containing the largest X_i's, a so-called extreme sample, yields the minimal value of V_m and hence is the optimal.

Suppose $\mathcal{M}_{11}$ is incorrect but $\mathcal{M}'_{11}$ holds, that is,

$$E_m(Y_i) = \alpha + \beta X_i, \alpha \neq 0$$
$$V_m(Y_i) = \sigma^2 X_i.$$

Then $\bar{t}_R$ is still m unbiased if based on a balanced sample for which $\bar{x} = \bar{X} = \bar{x}_r$ and v_L is m unbiased for V_m. Since from a study of the sample α may not be conclusively ignored, a balanced rather than an extreme sample is preferred in practice in using $\bar{t}_R$ and v_L.

But if $\mathcal{M}_{12}$ is true, that is, $E_m(Y_i) = \beta X_i$ and

(a) $V_m(Y_i) = \sigma^2 X_i^2$,

then

$$V_m(\bar{t}_R - \overline{Y}) = \sigma^2 \left[\left(\frac{1-f}{n} \right)^2 \left(\frac{\overline{x}_r}{\overline{x}} \right)^2 \sum_s X_i^2 + \frac{1}{N^2} \sum_r X_i^2 \right]$$

while

$$E_m(v_L) = \frac{1-f}{n} \frac{\overline{X}\overline{x}_r}{\overline{x}^2} \frac{\sigma^2}{n-1} \left(n\overline{x}^2 - \frac{1}{n} \sum_s X_i^2 \right)$$

and the relative bias

$$\frac{E_m(v_L - V_m)}{V_m} \quad \text{is approximately} \quad -\frac{\sum_s (X_i - \overline{x})^2}{\sum_s X_i^2}.$$

If we have $\mathcal{M}_{10}$, i.e., $E_m(Y_i) = \beta X_i$ and

(b) $V_m(Y_i) = \sigma^2$,

then the relative bias of v_L is approximately

$$\frac{\bar{x}}{n}\sum\left[\frac{1}{X_i}-1\right].$$

These biases cannot be neglected in practice whether a sample is balanced, extreme, or random. The actual coverage probability for a model-based confidence interval ($\bar{t}_R \pm \tau_{\alpha/2}\sqrt{v_L}$) will be less than or greater than the nominal value if $B_m(v_L)$ is negative or positive, respectively. So, variance estimation using v_L is not a robust procedure.

If $\mathcal{M}_{11}$ is true and v_0 is used as a variance estimator for $\bar{t}_R$, then

$$\frac{B_m(v_0)-V_m(\bar{t}_R-\bar{Y})}{V_m(\bar{t}_R-\bar{Y})}=\frac{\bar{x}^2}{\bar{X}\,\bar{x}_r}\left(1-\frac{C_s^2}{n}\right)-1$$

writing

$$C_s^2=\frac{1}{n}\sum_s(X_i-\bar{x})^2/\bar{x}^2=(CV \text{ of } X_i,\ i\in s)^2.$$

Observing this, ROYALL and EBERHARDT (1975) propose the alternative variance estimator

$$v_H=v_0\frac{\bar{X}\,x_r}{\bar{x}^2}\Big/\left(1-\frac{C_s^2}{n}\right)$$

and they find its m bias negligible in samples balanced or not even if the condition

$$V_m(Y_i)\propto X_i$$

is violated.

Keeping the prerequisite of robustness in mind, ROYALL and CUMBERLAND (1978a) proposed another variance estimator, namely,

$$v_D=\frac{1-f}{n}\left[\frac{\bar{X}\,\bar{x}_r}{\bar{x}}\right]\sum_s e_i^2/(n\bar{x}-X_i).$$

Another competitor receiving attention, although not from the predictive approach, is the **jackknife** estimator (cf. section 9.2)

$$v_J=\frac{1-f}{n}\,\overline{X}^2\sum_{j\in s}D^2(j).$$

Here

$$D(j) = r(j) - \frac{1}{n}\sum_{i\in s} r(i)$$

$$r(j) = \frac{n\bar{y} - Y_j}{n\bar{x} - X_j}$$

and j is a unit in s.

ROYALL and CUMBERLAND (1978a) presented results based on asymptotic analyses relating to the comparative performances of v_L, v_H, v_D, and v_J with respect to their model-based biases, MSEs, and the convergence in law of the associated SZEs in examining the efficacy of the corresponding confidence intervals. In this context the questions of robustness and efficacy of balanced sampling and the role of large SRSWORs in achieving balance have also been taken up by them. Their main findings are that

(a) v_L is unsuitable because of its lack of robustness even if the sample is balanced.
(b) It is difficult to choose from v_H, v_D, and v_J, each of which seems serviceable.

CUMBERLAND and ROYALL (1988), however, have cast doubt on the efficacy of large SRSWORs in achieving rapid convergence to normality of SZEs even if balance is preserved for an increasing proportion of sample with increasing sizes.

7.1.3 Global Empirical Studies

Fortunately, considerable empirical studies have been reported by ROYALL and CUMBERLAND (1978b, 1981a, 1981b, 1985) and also by WU and DENG (1983), in light of which the following brief comments seem useful concerning comparative performances of $v_0, v_1, v_2, v_{\hat{g}}, v_{\tilde{g}}, v_{reg}, v_H, v_D, v_J$, and v_{gopt} leaving out v_L, which is generally disapproved as a viable competitor.

Keeping in mind three key features namely, (1) linear trend, (2) zero intercept, and (3) increasing squared residuals with x in the scatter diagram of (x, y), ROYALL et al. studied appropriate actual populations including one with $N = 393$ hospitals with x as the number of beds and y as the number of patients discharged in a particular month. They took $n = 32$ for (1) extreme samples, (2) balanced samples with $|\bar{x} - \bar{X}|$ suitably bounded above, (3) SRSWOR samples, (4) **best fit** samples with a minimal discrepancy among sample- and population-based cumulative distribution functions. WU and DENG (1983), however, considered only SRSWORs with $n = 32$ from the same populations and also from a few others, purposely violating one or the other of the above three characteristics.

Two types of studies have been made. Simulating 1000 SRSWORs of $n = 32$ from each population the values of $\bar{t}_R$ and the above 10 variance estimators v, in general, are calculated. The MSE of $\bar{t}_R$ is taken as

$$M = \frac{1}{1000} \sum{}'(\bar{t}_R - \bar{Y})^2.$$

and the bias of v is taken as

$$B = \frac{1}{1000} \sum{}' v - M$$

and the root MSE of v is taken as

$$RM = \left[\frac{1}{1000} \sum{}'(v - M)^2\right]^{1/2}.$$

Each sum Σ' is over the 1000 simulated samples. Also, for each of the 1000 simulated samples the SZEs $\tau = (\bar{t}_R - \bar{Y})/\sqrt{v}$ and the intervals $\bar{t}_R \pm \tau_{\alpha/2}\sqrt{v}$ are calculated to examine the closeness of t to τ in terms of mean, standard deviation, skewness, and kurtosis. The df of t is taken as $n - 1 = 31$.

With respect to RM,

(a) v_{gopt} is found the best, with $v_{\hat{g}}$, $v_{\tilde{g}}$, v_{reg} closely behind.
(b) Among v_0, v_1, v_2 the one closest to v_{gopt} is found the best.
(c) v_H is found to be close to v_2 and fairly good, but v_D is found to be poor, and v_J is found to be the worst.

The biases of $v_0, v_1, v_2, v_{\hat{g}}$, $v_{\tilde{g}}$, and v_{reg} are negative, but v_J is positively biased, and the biases of v_H, v_D are erratic; among v_0, v_1, and v_2, those with small RM are more biased.

The intervals $\bar{t}_R \pm \tau_{\alpha/2}\sqrt{v}$ are wider for v_J but narrower for $v_0, v_1, v_2, v_{\hat{g}}$, $v_{\tilde{g}}$, and v_{reg}, and those for v_H, v_D are in between. The actual coverage probabilities are mostly less than the nominal $(1-\alpha)$, and pronouncedly so for v_0. In this respect v_J is the best, with v_D closely behind; v_H does not lag far behind. Among v_0, v_1, and v_2 the best is v_2 and v_0 is the worst. But v_1, v_2, $v_{\hat{g}}$, $v_{\tilde{g}}$, and v_{reg} are close to each other, and each is behind v_H.

7.1.4 Conditional Empirical Studies

From these global studies, where the averages are taken over all of the 1000 simulated samples, it is apparent that different variance estimators may suit different purposes. For example, one with a small MSE may yield a poor coverage probability, while one with a coverage probability close to the nominal value may not be stable, bearing an unacceptably high MSE. To get over this anomaly, these investigators adopt a conditional approach, which seems to be promising.

In a variance estimator alternative to v_0 the term $\bar{x}$ occurs as a prominent factor and its closeness to or deviation from $\bar{X}$ seems to be a crucial factor in determining its performance characteristics. This $\bar{x}$ is an *ancillary statistic,* that is, the distribution of $\bar{x}$ is free of $\underline{Y}$, and it seems proper to examine how each v performs for a given value of $\bar{x}$ or over several disjoint intervals of values of $\bar{x}$. In other words, for conditional biases, conditional MSEs, and conditional confidence intervals, given $\bar{x}$ may be treated as suitable criteria for judging the relative performances of these variance estimators.

With this end in view, in their empirical studies ROYALL and CUMBERLAND (1978b, 1981a, 1981b, 1985) and WU and DENG (1983) divided the 1000 simulated samples each of size $n = 32$ into 20 groups of 50 each in increasing order of $\bar{x}$ values for the samples. Thus, the first 50 smallest $\bar{x}$ values are placed in the first group, the next 50 larger $\bar{x}$ values are taken in the second group, and so on. Then they calculate

(a) the average of $\bar{x}$, $A_{\bar{x}} = \frac{1}{50}\Sigma'\,\bar{x}$ for respective groups
(b) the conditional MSE of $\bar{t}_R$ within respective groups as $M_{\bar{x}} = \frac{1}{50}\Sigma''\,(\bar{t}_R - \bar{Y})^2$
(c) averages $v_{\bar{x}} = \frac{1}{50}\Sigma' v$ of each of the v's within respective groups where Σ' denotes summation over 50 samples within respective groups.

Graphs are then plotted for $\sqrt{v_{\bar{x}}}/\sqrt{M_{\bar{x}}}$ against $A_{\bar{x}}$ to see how closely the trajectories for respective v's track the one for the

MSEs, that is, for $M_{\bar{x}}$ across the groups. For an overall comparison WU and DENG (1983) propose the distance measure

$$d_v = \left[\frac{1}{20}\Sigma'''(\sqrt{v_{\bar{x}}} - \sqrt{M_{\bar{x}}})^2\right]^{1/2}$$

the sum Σ''' being over the 20 groups. A variance estimator with a small d_v value is regarded to be close to the conditional MSE.

In terms of this criterion for performance, the variance estimators rank as follows in decreasing order. Those within parentheses are tied in rank and v_{gopt} is excluded:

$$(v_H, v_D), (v_J, v_2, v_{\tilde{g}}), (v_{\hat{g}}, v_{reg}), v_1, v_0.$$

With this conditional approach, it is remarkable that they find that the variance estimators that are good point estimators for conditional (given $\bar{x}$) MSE of t_R also yield good interval estimates in terms of achieving conditional coverage probabilities close to the nominal values respectively for each group of $\bar{x}$ values.

An important message from these empirical evidences with both global and conditional approaches is that, in spite of recommendations in many textbooks, v_0 does not fare well with respect to its bias, MSE, and coverage probabilities associated with the confidence interval based on it.

Behaviors of some of the variance estimators when based on simulated balanced, best fit, or extreme samples rather than random samples are also reported in the literature.

Many modifications of the ratio estimator based on SRSWOR and variance estimators for the latter also occur in the literature. An interested reader may consult RAO (1986), CHAUDHURI and VOS (1988), and the references cited therein.

7.1.5 Further Measures of Error in Ratio Estimation

CHAUDHURI and MITRA (1996) introduced additional estimators for the measures of error of the ratio estimator

$$\bar{t}_R = \overline{X}\frac{\overline{y}}{\overline{x}}$$

based on SRSWOR utilizing models and asymptotics.

They considered the standard model (a) $\mathcal{M}$ for which

$$Y_i = \beta X_i + \varepsilon_i$$

ε_i's independent with

$$E_m(\varepsilon_i) = 0$$
$$V_m(\varepsilon_i) = \sigma^2 X_i$$

$i \in \mathcal{U}$, its modifications (b) $\mathcal{M}'$ with

$$V_m(\varepsilon_i) = \sigma_i^2$$

and a second modification (c) $\mathcal{M}_\theta$ for which

$$Y_i = \theta + \beta X_i + \varepsilon_i$$

without changes for ε_i's in $\mathcal{M}$.

For the TAYLOR approximation-based variance of $\bar{t}_R$, namely

$$V_T = \frac{1-f}{n(N-1)} \sum (Y_i - RX_i)^2$$

they calculated

$$M_T = E_m(V_T) \text{ under } \mathcal{M}.$$

They also calculated

$$M' = \lim E_p E_m(\bar{t}_R - \overline{Y})^2 \text{ under } \mathcal{M} \text{ and}$$
$$M'' = \lim E_p E_m(\bar{t}_R - \overline{Y})^2 \text{ under } \mathcal{M}'.$$

In order to work out estimators υ and

$$\upsilon(\alpha) = \sum_{i \in s} \alpha_i \left(\frac{Y_i}{X_i} - \frac{1}{n} \sum_{i \in s} \frac{Y_i}{X_i} \right)^2 = \sum_{i \in s} \alpha_i (r_i - \bar{r})^2, \text{ say,}$$

with suitable coefficients α_i ($i \in s$), they equated

(a) $E_m(\upsilon)$ to M_T
(b) $\lim E_p E_m(\upsilon)$ to M_T and M' with a suitable initial function υ of $(Y_i, X_i, i \in s), \bar{x}$
(c) $E_m \upsilon(\alpha)$ to M''
(d) $\lim E_p E_m \upsilon(\alpha)$ to M''.

The approaches in mean square error (MSE) estimation by BREWER (1999a) and SUNDBERG (1994) are also worthy of attention in this context. Writing

$$S_x^2 = \frac{1}{N-1}\sum(X_i - \overline{X})^2$$

$$C_0^2 = S_x^2/\overline{X}^2$$

$$s_x^2 = \frac{1}{n-1}\sum_{i\in s}(X_i - \overline{X})^2$$

$$c_x^2 = s_x^2/\overline{x}^2$$

some of the MSE estimators for $\bar{t}_R$ introduced by CHAUDHURI and MITRA (1996) are

$$\upsilon_{01} = \frac{1 - C_0^2/N}{1 - C_0^2/n}\upsilon_0, \ \upsilon_{21} = \left(\frac{\overline{X}}{\overline{x}}\right)^2 \upsilon_{01}$$

$$\upsilon_{02} = \frac{\overline{X}}{\overline{x}}\frac{1 - C_0^2/N}{1 - c_x^2/n}\upsilon_0,$$

$$\upsilon_{03} = \frac{\upsilon_0}{1 - C_0^2/n}, \ \upsilon_{23} = \left(\frac{\overline{X}}{\overline{x}}\right)^2 \upsilon_{03}$$

$$\upsilon_{04} = \frac{\overline{x}_r}{\overline{x}}\upsilon_H$$

$$m_1 = \frac{1-f}{n(n-2)}\sum_{i\in s}(r_i - \overline{r})^2\left(X_i^2 - \frac{\sum X_i^2}{N(n-1)}\right)$$

$$m_2 = \frac{1-f}{n(n-2)}\sum_{i\in s}(r_i - \overline{r})^2\left(X_i^2 - \frac{\sum_{i\in s} X_i^2}{n(n-1)}\right)$$

$$m_3 = \frac{n(n-2)}{N(n-1)}\frac{\sum X_i^2}{\sum_{i\in s} X_i^2 - \frac{n}{N(n-1)}\sum X_i^2}m_1$$

$$m_4 = f\,\frac{\sum X_i^2}{\sum_{i\in s} X_i^2}m_2.$$

Drawing samples from artificial populations conforming to the models $\mathcal{M}$, $\mathcal{M}'$, $\mathcal{M}_\theta$ with various choices of N, n, β, σ^2, σ_i^2, θ, CHAUDHURI and MITRA (1996) studied numerical data, giving the relative performances of the confidence intervals (CI) for $\overline{Y}$ both conditionally, as in section 7.1.4, and unconditionally, as in section 7.1.3, based on $\bar{t}_R$ and these MSE estimators, along with the others like υ_0, υ_1, υ_2, υ_L, υ_H, υ_J, and υ_D. Many of the newly proposed ones, especially m_1 and m_2, were illustrated to yield better CIs.

7.2 REGRESSION ESTIMATOR

7.2.1 Design-Based Variance Estimation

When (X_i, Y_i) values are available for SRSWOR of size n an alternative to the ratio estimator for $\bar{Y}$ is the regression estimator

$$t_r = \bar{y} + b(\bar{X} - \bar{x}).$$

Here b is the sample regression coefficient of y on x. Its variance $V_p(t_r)$ and mean square error $M_p(t_r)$ are both approximated by

$$V = \frac{1-f}{n} \frac{1}{N-1} \sum_1^N D_i^2$$

where

$$D_i = (Y_i - \bar{Y}) - B(X_i - \bar{X})$$

$$B = \sum_1^N (Y_i - \bar{Y})\ (X_i - \bar{X}) \Big/ \sum_1^N (X_i - \bar{X})^2.$$

The errors in these approximations are neglected for large n and N although for n, N, and $\underline{X}$ at hand it is difficult to guess the magnitudes of these errors. However, there exists evidence that t_r may be more efficient than the ratio estimator $\bar{t}_R$ in many situations in terms of mean square error (cf. DENG and WU, 1987).

Writing

$$d_i = (Y_i - \bar{y}) - b(X_i - \bar{x}),$$

$$v_{lr} = \frac{1-f}{n(n-2)} \sum_s d_i^2$$

is traditionally taken as an estimator for V. DENG and WU (1987) consider a class of generalized estimators

$$v_g = \left[\frac{\bar{X}}{\bar{x}}\right]^g v_{lr}$$

They work out an asymptotic formula for $V_p(v_g)$ using TAYLOR series expansions and neglecting terms therein supposed to be small for large n relative to the term they retain, called the **leading term**. They find the leading term to be minimal if one chooses g equal to

$$g_{opt} = \text{regression coefficient of } D_i^2 \Big/ \left[\frac{1}{N}\sum_1^N D_i^2\right]$$

on $X_i/\bar{X}$, $i = 1, 2, \ldots, N$.

Since g_{opt} is unavailable they recommend the variance estimator $v_{\hat{g}}$ with $\hat{g}$ as the sample analogue of g_{opt} calculated using (Y_i, X_i, d_i), $i \in s$.

7.2.2 Model-Based Variance Estimation

Besides these ad hoc variance estimators, hardly any others are known to have been proposed as estimators for V with a design-based approach. However, some rivals have emerged from the least squares linear predictive approach.

Suppose $\underline{Y}, \underline{X}$ are conformable to the model $\mathcal{M}'_{10}$ (cf. section 4.1.2) for which the following is tenable:

$$E_m(Y_i) = \alpha + \beta X_i,\ \alpha \neq 0,\ V_m(Y_i) = \sigma^2,$$
$$C_m(Y_i,\ Y_j) = 0,\ i \neq j.$$

Then the BLUP for $\bar{Y}$ is t_r and

$$B_m(t_r) = E_m(t_r - \bar{Y}) = 0$$

$$V_m(t_r - \bar{Y}) = \frac{1-f}{n}\left[1 + \frac{(\bar{X} - \bar{x})^2}{(1-f)\,g(s)}\right]\sigma^2 = \phi(s)\,\sigma^2, \text{ say,}$$

writing

$$g(s) = \frac{1}{n}\sum_s (X_i - \bar{x})^2.$$

Then, for

$$\hat{\sigma}^2 = \frac{1}{(n-2)}\sum_s d_i^2$$

we have

$$E_m(\hat{\sigma}^2) = \sigma^2.$$

Consequently,

$$v_L = \phi(s)\,\hat{\sigma}^2 = \frac{1-f}{n(n-2)}\left[1+\frac{(\bar{X}-\bar{x})^2}{(1-f)\,g(s)}\right]\sum_s d_i^2$$

is an m-unbiased estimator for $V_m(t_r - \bar{Y})$ under $\mathcal{M}'_{10}$. The term

$$h(s) = \frac{(\bar{X}-\bar{x})^2}{(1-f)\,g(s)}$$

in v_L vanishes if the sample is balanced, that is, $\bar{x} = \bar{X}$, and for a balanced sample $V_m(t_r - \bar{Y})$ is the minimal under $\mathcal{M}'_{10}$.

In general,

$$v_L = (1+h(s))\,v_{lr} \geq v_{lr}$$

with equality only for a balanced sample. If a balanced sample is drawn, then the classical design-based estimator v_{lr} based on it becomes m-unbiased for $V_m(t_r - \bar{Y})$.

As usual with the predictive approach, the main problem is robustness. If the model $\mathcal{M}'_{10}$ is not correctly applicable to the $\underline{X}$, $\underline{Y}$ at hand, for example, if

$$E_m(Y_i) \neq \alpha + \beta X_i,$$

then $B_m(t_r)$ may not vanish for a realized sample and if $V_m(Y_i) \neq \sigma^2$, then $V_m(t_r - \bar{Y})$ does not equal $\phi(s)\sigma^2$ and one does not know the quantity that v_L may m-unbiasedly estimate. Consequently, the SZE, which is here

$$(t_r - \bar{Y})/\sqrt{v_L}$$

may not have a distribution close to that of a standardized normal variate as it may be supposed to be for large n, N if $\mathcal{M}'_{10}$ is correct. So, in fact one may not know to what extent the true coverage probability for the confidence interval $(t_r \pm \tau_{\alpha/2}\sqrt{v_L})$ matches the nominal value $(1-\alpha)$.

For example, if the correct model is $\mathcal{M}'_{11}$ (cf. section 4.1.2) for which $V_m(Y_i) = \sigma^2 X_i$, then

$$V_m(t_r - \bar{Y}) = \frac{\sigma^2}{n}[(2-f)\bar{X} - \bar{x} + (\bar{X}-\bar{x})^2 C(s)]$$

where

$$C(s) = \left(\sum_{x} X_i^3 - 2\bar{x}\sum_{s} X_i^2 + n\bar{x}^3\right) \Big/ ng^2(s).$$

But in this case

$$E_m(v_L) = \frac{1-f}{n}\sigma^2(1+h(s))[\bar{x} + \{\bar{x} - C(s)g(s)\}/(n-2)]$$

and

$$B_m(v_L) = E_m(v_L) - V_m(t_r - \bar{Y})$$

may not be negligible in general.

This only illustrates how v_L may not legitimately be treated as a robust estimator for $V_m(t_r - \overline{Y})$.

If one uses v_{lr} to estimate $V_m(t_r - \overline{Y})$ in this case, then obviously

$$B_m(v_{lr}) \neq 0$$

as one may check on noting that

$$E_m(v_{lr}) = E_m(v_L)$$

with $h(s) = 0$ in the latter.

So, even for a balanced sample v_{lr} is not m-unbiased for $V_m(t_r - \bar{Y})$ if $\mathcal{M}_{10}$ is inapplicable, that is, it is not robust.

However, ROYALL and CUMBERLAND (1978a) have proposed the following alternative estimators for $V_m(t_r - \bar{Y})$:

$$v_H = \frac{1-f}{n^2}\,\Sigma\, d_i^2\left[1 + (X_i - \bar{x})(\bar{x}_r - \bar{x})/g(s)\right]^2 \Big/ \left(1 - \frac{1}{n}\sum_{s} W_i K_i\right) + (N-n)\hat{\sigma}^2$$

where

$$W_i = [g(s) + (X_i - \bar{x})(\bar{x}_r - \bar{x})]^2 \Big/ \left[\sum_{s}\{g(s) + (X_i - \bar{x})(\bar{x}_r - \bar{x})\}^2\right]$$

$$K_i = 1 + (X_i - \bar{x})^2/g(s)$$

and

$$v_D = \frac{(1-f)^2}{n(n-1)}\sum_{s} d_i^2\left[\frac{\left[1 + (X_i - \bar{x})(\bar{x}_r - \bar{x})/g(s)\right]^2 + \frac{1-f}{f}}{\left[1 - \{(X_i - \bar{x})^2/(n-1)g(s)\}\right]}\right]$$

$$v_J = (1-f)\left[\frac{n-1}{n}\right]\sum_{j\in s}(\hat{T}_j - \hat{T})^2.$$

In v_J, $\hat{T}_j$ is t_r calculated from s omitting (Y_j, X_j) and $\hat{T} = \frac{1}{n}\sum_{j\in s}\hat{T}_j$.

These authors have noted that

(a) $E_m(v_H) = E_m(v_D) = E_m(v_J) = V_m(t_r - \bar{Y})$ if $\mathcal{M}'_{10}$ is true

(b) $B_m(v)$ is negligible if $V_m(Y_i)$ is not a constant for each i but $\frac{N}{n}$ is large provided $E_m(t_r - \bar{Y}) = 0$ for a sample at hand

(c) $|B_m(v)|$ is not negligible even for large n in case $|E_m(t_r - \bar{Y})|$ is not close to zero, when v is one of v_H, v_D, or v_J above.

7.2.3 Empirical Studies

ROYALL and CUMBERLAND (1981b, 1985) therefore made empirical studies in an effort to make a right choice of an estimator for $V_m(t_r - \bar{Y})$ because a model cannot be correctly postulated in practice. DENG and WU (1987) also pursued with an empirical investigation to rightly choose from these several variance estimators. But they also examined the design biases and design MSEs of all the above-noted estimators v, each taken by them as an estimator for V, considering SRSWOR only. The theoretical study concerning them is design based, and because of the complicated nature of the estimators their analysis is asymptotic. From their theoretical results v_D seems to be the most promising variance estimator from the design-based considerations and v_L and v_{lr} are both poor.

In the empirical studies undertaken by ROYALL and CUMBERLAND (1981b, 1985) and DENG and WU (1987) 1000 simple random samples of size $n = 32$ each are simulated from several populations including one of size $N = 393$. For each of these 1000 SRSWORs values of t_r, $\bar{x}$, v_0, v_1, v_2, $v_{\hat{g}}$, v_L, v_H, v_D, and v_J are calculated. The estimator v_{lr} is found too poor to be admitted as a viable competitor and is discarded by the authors mentioned. For each sample again for each of these variance estimators v, as above, the SZEs and confidence intervals are also calculated

$$\tau = (t_r - \bar{Y})/\sqrt{v} \quad \text{and} \quad t_r \pm \tau_{\alpha/2}\sqrt{v}$$

with $\tau_{\alpha/2}$ as the $100\alpha/2$ % point in the upper tail of the STUDENT's t distribution with $df = n - 2 = 30$ in this case.

First, from the study of the entire sample the unconditional behavior is reviewed using the overall averages to denote respectively by

$$\bar{M} = \frac{1}{1000}\Sigma'(t_r - \bar{Y})^2, \text{ the MSE}$$

$$B = \frac{1}{1000}\Sigma' v - \bar{M}, \text{ the bias,}$$

Σ' denoting the sum over the 1000 simulated samples. Again, taking $\bar{x}$ as the ancillary statistic conditional (given $\bar{x}$) behavior is examined on dividing the 1000 simulated samples into 10 groups, each consisting of 100 samples with the closest values of $\bar{x}$ within each, the groups being separated according to changes in the values of $\bar{x}$. For each group

$$\frac{1}{100}\Sigma'\bar{x}, \ \frac{1}{100}\Sigma' v,$$

are separately calculated, Σ' denoting the sum over the 100 samples in respective groups and the estimated coverage probabilities associated with the confidence intervals. Thus, the unconditional and the conditional behavior of variance estimators related to t_r are investigated, following the same two approaches as with variance estimation related to the ratio estimator $\bar{t}_R$ discussed in section 7.1. The estimators are compared with respect to MSE, bias, and associated conditional and unconditional coverage probabilities.

Empirical findings essentially show the following:

With respect to MSE:

(a) $v_{\hat{g}}$ is the best and v_J is the worst
(b) among v_0, v_1, and v_2 the one closest to $v_{\hat{g}}$ is the best
(c) between v_H, and v_D, the former is better but v_H is worse than $v_0, v_1, v_2, v_g, v_{\hat{g}}$ and v_L.

With respect to bias, v_J is positively biased, v_D has the least absolute bias, and v_L has less bias than v_0, v_1, v_2, and $v_{\hat{g}}$.

In terms of unconditional coverage probabilities:

(a) each coverage probability is less than the nominal value, v_0 giving the lowest but v_J the closest to it

(b) v_0, v_1, and v_2 rank in improving order
(c) v_H is worse than v_D.

In terms of conditional coverage probabilities:

(a) v_J is the most excellent and its associated coverage probabilities remain stable over variations of $\bar{x}$; those with v_H and v_D are also pretty stable but those with v_0, v_L, and $v_{\hat{g}}$ increase with $\bar{x}$
(b) among v_0, v_1, and v_2, the one with the most stable coverage probability across $\bar{x}$ is v_2
(c) v_D is better than v_H.

For nearly balanced samples all estimators perform similarly. One important message is that the traditional estimator v_{lr} is outperformed by each new competitor and the least squares estimator v_L is also inferior to the other alternatives from overall considerations.

7.3 HT ESTIMATOR

In section 2.4.4 we presented the formula for the variance of the HTE $\bar{t} = \sum_{i \in s} \frac{Y_i}{\pi_i}$ based on a fixed sample size design available due to YATES and GRUNDY (1953) and SEN (1953), along with an unbiased estimator v_{YG} thereof. For designs without restriction on sample size the corresponding formulae given by HORVITZ and THOMPSON (1952) themselves were also noted as

$$V_p(\bar{t}) = \sum_i \frac{Y_i^2}{\pi_i} + \sum_{i \neq j} Y_i Y_j \frac{\pi_{ij}}{\pi_i \pi_j} - Y^2$$

$$v_p(\bar{t}) = \sum_s Y_i^2 \frac{1-\pi_i}{\pi_i^2} + \sum\sum_{i \neq j \in s} Y_i Y_j \frac{\pi_{ij} - \pi_i \pi_j}{\pi_i \pi_j \pi_{ij}}.$$

It is well known that $v_p(\bar{t})$ has the defect of bearing negative values for samples with high selection probabilities. The estimator v_{YG} may also turn out negative for designs not subject to the constraints

$$\pi_i \pi_j \geq \pi_{ij} \quad \text{for all} \quad i \neq j$$

as may be seen in BIYANI's (1980) work. To get rid of this problem of negative variance estimators, JESSEN (1969) proposed the following variance estimator

$$v_J = \bar{W} \sum\sum_{i<j \in s} \left[\frac{Y_i}{\pi_i} - \frac{Y_j}{\pi_j} \right]^2$$

where

$$\bar{W} = \frac{n - \sum \pi_i^2}{N(N-1)},$$

with n as the fixed sample size.

This is uniformly non-negative and is free of π_{ij} and very simple in form.

KUMAR, GUPTA and AGARWAL (1985), following JESSEN (1969), suggest the following uniformly non-negative variance estimator for $V_p(\bar{t})$, namely,

$$v_0(\bar{t}) = K \sum\sum_{i<j \in s} \left(\frac{Y_i}{\pi_i} - \frac{Y_j}{\pi_j} \right)^2.$$

Their choice of K is

$$K = \frac{1}{(n-1)} \frac{\sum_1^N p_i^{\gamma-1}(1 - np_i)}{\sum p_i^{\gamma-1}}$$

from considerations of a fixed sample size n and the model $\mathcal{M}_{1\gamma}$ for which

$$Y_i = \beta p_i + \varepsilon_i$$

with $0 < p_i < 1$, $n\, p_i = \pi_i$, $\sum_1^N p_i = 1$, and

$$E_m(\varepsilon_i) = 0,\ V_m = (\varepsilon_i) = \sigma^2 p_i^\gamma,\ C_m(\varepsilon_i,\ \varepsilon_j) = 0 \text{ for } i \neq j$$

with $\gamma \geq 0$, $\sigma < 0$. Under this model

$$E_m\, V_p(\bar{t}) = \frac{\sigma^2}{n} \sum_1^N p_i^\gamma (1 - np_i)$$

to which $E_m v_0(\bar{t})$ agrees with the above choice of K. Thus, $v_0(\bar{t})$ is an m-unbiased estimator of $V_p(\bar{t})$. But since $\bar{t}$ is predominantly a p-based estimator, they also consider the magnitude of

$$\Delta = \left[\frac{E_p\, v_0(\bar{t})}{V_p(\bar{t})} - 1\right] \times 100$$

and also of

$$\delta = \frac{V_p(v_0(\bar{t}))}{\left[E_p(v_0(\bar{t})\right]^2}\,.$$

They also undertake a comparative study for the performances of v_J and v_{YG} in terms of criteria similar to Δ and δ for the latter. Their empirical study demonstrates that $v_0(\bar{t})$ may be quite useful in practice. BREWER (1990) recommends it from additional considerations we omit to save space.

SÄRNDAL (1996) mentioned two crucial shortcomings in the unbiased estimators v_{HT} and v_{YG} for $V_p(t_{HT}) = V_p(t_H)$, namely that (1) computation of π_{ij} is very difficult for many standard schemes of sampling, and for systematic sampling with a single random start it is often zero, and (2) for large-scale surveys the variation in

$$\frac{\pi_i\pi_j - \pi_{ij}}{\pi_{ij}} \quad \text{and} \quad \frac{\pi_{ij} - \pi_i\pi_j}{\pi_i\pi_j\pi_{ij}}$$

involved in the numerous cross-product terms of v_{YG} and v_{HT}, respectively, is so glaring that these variance estimators achieve little stability.

Motivated by this, DEVILLE (1999) and BREWER (1999a, 2000) are inclined to offer the following approximations by way of getting rid of the cross-product terms in $V_p(t_H)$ and in estimators thereof.

Confirming the sampling schemes for which $\nu(s)$, the effective size of a sample s, that is, the number of the distinct units in it, is kept fixed at an integer n ($2 \le n < N$), BREWER (2000) gives the formula for $V_p(t_H)$ as

$$V_{Br}(t_H) = \sum \pi_i(1-\pi_i)\left(\frac{Y_i}{\pi_i} - \frac{Y}{n}\right)^2 + \sum\sum_{i\neq j}(\pi_{ij} - \pi_i\pi_j)\left(\frac{Y_i}{\pi_i} - \frac{Y}{n}\right)\left(\frac{Y_j}{\pi_j} - \frac{Y}{n}\right).$$

He then recommends approximating π_{ij} by

$$\pi_{ij}^{*} = \pi_i \pi_j \frac{c_i + c_j}{2}$$

choosing c_i as one of

(a) $c_i = \frac{n-1}{n-\pi_i}$

(b) $c_i = \frac{n-1}{n-2\pi_i+\frac{\sum \pi_i^2}{n}}$

(c) $c_i = \frac{n-1}{n-\frac{1}{n}\sum \pi_i^2}$

from certain well-accounted-for considerations that we omit. The resulting approximate variance formula for t_H is then

$$V_{Br}^{*}(t_H) = \sum \pi_i (1 - c_i \pi_i) \left(\frac{Y_i}{\pi_i} - \frac{Y}{n} \right)^2$$

and BREWER (2000) calls it the **natural variance** of t_H free of π_{ij}'s. He proposes the approximately unbiased formula for an estimator of $V_p(t_H)$ as

$$v_4 = \sum_{i \in s} \left(\frac{1}{c_i} - \pi_i \right) \left(\frac{Y_i}{\pi_i} - \frac{t_H}{n} \right)^2 = v_{BR}.$$

For $V_4(t_H)$, DEVILLE's (1999) recommended estimator is

$$v_5 = \frac{1}{1 - \sum_{i \in s} a_i^2} \sum_{i \in s} (1 - \pi_i) \left(\frac{Y_i}{\pi_i} - A_s \right)^2 = v_{DE}, \text{ say,}$$

on writing

$$a_i = \frac{1 - \pi_i}{\sum_{i \in s} (1 - \pi_i)}, \quad A_s = \sum_{i \in s} a_i \frac{Y_i}{\pi_i}$$

also to get rid of π_{ij}'s.

STEHMAN and OVERTON (1994) recommended approximating π_{ij} by

(a) $\pi_{ij}^{(1)} = \frac{(n-1)\pi_i \pi_j}{n-\frac{1}{2}(\pi_i+\pi_j)}$ and

(b) $\pi_{ij}^{(2)} = \frac{(n-1)\pi_i \pi_j}{n-\pi_i-\pi_j+\frac{1}{n}\sum_i^N \pi_i^2}$

for the fixed sample size (n) scheme of HARTLEY and RAO (1962), which is a systematic sampling scheme with unequal selection probabilities with a prior random arrangement of the units in the population.

They empirically demonstrated these choices to be useful in retaining high efficiency even on getting rid of the cross-product terms in variance estimators.

HÁJEK's (1964, 1981) Poisson sampling scheme, however, is very handy to accommodate SÄRNDAL's (1996) viewpoint. To draw a sample s from $\mathcal{U} = (1, 2, \ldots, N)$ by this scheme one has to choose N suitable numbers π_i $(0 < \pi_i < 1,\ i \in \mathcal{U})$, associate them with i in $\mathcal{U}$, implement N independent Bernoullian trials with π_i as the probability of success for the ith trial $(i = 1, 2, \ldots, N)$, and take into s those units for which successes were achieved. For this scheme, of course, $0 < \nu(s) \leq N$, π_i is the inclusion probability of i,

$$E_p(\nu(s)) = \sum \pi_i$$

and $\pi_{ij} = \pi_i \pi_j$ for every $i \neq j$ $(= 1, 2, \ldots, N)$.

Consequently,

$$V_p(t_H) = \sum Y_i^2 \frac{1 - \pi_i}{\pi_i} \quad \text{and}$$

$$v_p = \sum_{i \in s} Y_i^2 \frac{1 - \pi_i}{\pi_i^2}$$

is an unbiased estimator for $V_p(t_H)$.

The most unpleasant feature here is that there is little control on the magnitude of $\nu(s)$ and hence it is difficult to plan a survey within a budget and aimed at efficiency level.

This topic is widely studied in the literature, especially because of its uses in achieving coordination and control on the choice of units over a number of time points when, for the sake of comparability, it is desired to partially rotate some fractions of the units over certain time intervals.

BREWER, EARLY and JOYCE (1972), BREWER, EARLY and HANIF (1984), and OHLSSON (1995) are among the researchers who explored its possibilities, especially by introducing the concept of **permanent random numbers** (PRN) to be associated with the **take-some** units of a survey population, namely those units with selection probabilities p_i $(0 < p_i < 1, i \in \mathcal{U})$

contrasted with the **take-all** units for which selection probabilities are q_i ($=1$ for $i \in \mathcal{U}_c$) when $\mathcal{U}$ is the union of $\mathcal{U}_s$ and $\mathcal{U}_c$, which are disjoint, and also with the units that are to be added on subsequent occasions, omitting the units that may be found irrelevant later.

These researchers also modified the Poisson scheme, allowing repeated drawing until $\nu(s)$ turns out positive, and also studied collocated sampling, which uses the PRNs effectively to keep the selection confined to desirable ranges of the units of $\mathcal{U}_s$.

The inclusion probabilities of units i and pairs of units (i, j) of course deviates for the modified Poisson and collocated Poisson schemes from those of the Poisson scheme, and they do not retain the requirements of SÄRNDAL (1996).

BREWER, EARLY and JOYCE (1972) and BREWER, EARLY and HANIF (1984) considered the ratio version of t_H based on the Poisson scheme, that is,

$$\begin{aligned} t_{HR} &= \frac{\sum \pi_i}{\nu(s)} \sum_{i \in s} \frac{Y_i}{\pi_i} \quad \text{if } \nu(s) > 0 \\ &= 0 \qquad\qquad\qquad \text{otherwise.} \end{aligned}$$

Writing

$$P_0 = Prob(\nu(s) = 0) = \prod_1^N (1 - \pi_i)$$

BREWER et al. (1972) approximated $V_p(t_{HR})$ by

$$V_{BEJ} = \sum_1^N \pi_i(1 - \pi_i)\left(\frac{Y_i}{\pi_i} - \frac{Y}{n}\right)^2 + P_0 Y^2$$

writing $n = \Sigma \pi_i$, and gave two estimators for it as

$$\begin{aligned} v_{1B} &= \sum_{i \in s}(1 - \pi_i)\left(\frac{Y_i}{\pi_i} - \frac{t_{HR}}{n}\right)^2 + P_0\, t_{HR}^2 \\ v_{2B} &= \frac{\sum \pi_i}{\nu(s)}\left[\sum_{i \in s}(1 - \pi_i)\left(\frac{Y_i}{\pi_i} - \frac{t_{HR}}{n}\right)^2 + P_0\, t_{HR}^2\right]. \end{aligned}$$

Observing that $\nu(s) = \Sigma I_{si}$ and $\Sigma \pi_i = E_p(\Sigma I_{si})$ and hence

$$\frac{\sum \pi_i}{\nu(s)} \sum_{i \in s} \frac{Y_i}{\pi_i}$$

may be treated as a ratio estimator for ΣY_i, the first terms of v_{1B} and v_{2B} are analogous to v_0 and v_1 of subsection 7.1.1.

BREWER et al. (1984), on the other hand, approximated $V(t_{HR})$ for this Poisson sampling scheme by

$$V_{BEH} = (1 - P_0) \sum \pi_i (1 - \pi_i) \left(\frac{Y_i}{\pi_i} - \frac{Y}{n} \right)^2 + P_0 Y^2$$

and proposed for it the estimator

$$v_{BEH} = \frac{1 - P_0}{1 + P_0} \frac{n}{\nu(s)} \sum_{i \in s} (1 - \pi_i) \left(\frac{Y_i}{\pi_i} - \frac{t_{HR}}{n} \right)^2 + P_0 Y^2.$$

Incidentally, SÄRNDAL (1996) also considered t_{HR} based on the Poisson scheme, but, in examining its variance on MSE and in proposing estimators thereof, did not care to take account of the possibility of $\nu(s)$ being zero, and simply considered t_{HR} as

$$t_{HR} = \frac{\sum \pi_i}{\nu(s)} \sum_{i \in s} \frac{Y_i}{\pi_i}.$$

In the next section we shall treat this case.

7.4 GREG PREDICTOR

Let y be the variable of interest and $x_1, \ldots, x_k$ be k auxiliary variables correlated with y. Let Y_i and X_{ij} be the values of y and x_j on the ith unit of $U = (1, \ldots, i, \ldots, N), i = 1, \ldots, N, j = 1, \ldots, k$. Let $\underline{\beta} = (\beta_1, \ldots, \beta_k)'$ be a $k \times 1$ vector of unknown parameters, $\underline{x}_i = (X_{i1}, \ldots, X_{ik})'$, $\underline{Y} = (Y_1, \ldots, Y_N)'$, $\underline{X} = (\underline{x}_1, \ldots, \underline{x}_N)'$ and $\mu_i = \underline{x}_i' \underline{\beta}, i = 1, \ldots, N$.

Let there be a model for which we may write

$$Y_i = \mu_i + \varepsilon_i,$$

with $E_m(\varepsilon_i) = 0$, $V_m(\varepsilon_i) = \sigma_i^2$, ε_i's independent. Let Q be an $N \times N$ diagonal matrix with non-zero diagonal entries Q_i, $i = 1, \ldots,$ and s a sample of n units of U chosen according to a design p with positive inclusion probabilities $\pi_i, i = 1, \ldots, N$.

Let

$$\underline{B} = (\underline{X}\, Q\, \underline{X}')^{-1} (\underline{X}\, Q\, \underline{Y})$$
$$E_i' = Y_i - \underline{x}_i'\, \underline{B}$$
$$\underline{\hat{B}}_s = \left[\sum_{i \in s} \frac{Q_i}{\pi_i}\, \underline{x}_i\, \underline{x}_i'\right]^{-1} \left[\sum_{i \in s} \frac{Q_i}{\pi_i}\, \underline{x}_i\, Y_i\right]$$
$$\hat{\mu}_i = \underline{x}_i'\, \underline{\hat{B}}_s,\ e_i = Y_i - \hat{\mu}_i.$$

Then the GREG predictor for $Y = \sum_1^N Y_i$ is

$$t_G = \sum_1^N \hat{\mu}_i + \sum_{i \in s} \frac{e_i}{\pi_i}.$$

With

$$\pi_{ij} = \sum_{s \ni i,j} p(s),\ \ \Delta_{ij} = \pi_i \pi_j - \pi_{ij}$$
$$\underline{B}_\pi = \left(\sum_1^N Q_i \underline{x}_i \underline{x}_i' \pi_i\right)^{-1} \left(\sum_1^N Q_i \underline{x}_i Y_i \pi_i\right)$$

and

$$E_i = Y_i - \underline{x}_i'\, \underline{B}_\pi$$

an asymptotic formula for the variance of t_G is given by SÄRNDAL (1982) as

$$V_G = \sum_{i<j}\sum \Delta_{ij} \left[\frac{E_i}{\pi_i} - \frac{E_j}{\pi_j}\right]^2$$

and an approximately design-unbiased estimator for V_G as

$$v_G = \sum_{i<j \in s}\sum \frac{\Delta_{ij}}{\pi_{ij}} \left[\frac{e_i}{\pi_i} - \frac{e_j}{\pi_j}\right]^2$$

provided $\pi_{ij} > 0$ for all i, j.

SÄRNDAL (1984) and SÄRNDAL and HIDIROGLOU (1989) give details about its performances which we omit. The simple projection (SPRO) estimator for Y given by $t_{sp} = \sum_1^N \underline{x}_i' \underline{\hat{B}}_s$ can be expressed in the form

$$t_{sp} = \sum_s g_{si} \frac{Y_i}{\pi_i},$$

writing $Q_i = 1/C_i\ \pi_i$, for $C_i \neq 0$ and

$$g_{si} = \left[\sum_1^N \underline{x}_i'\right]\left[\sum_s \underline{x}_i\underline{x}_i'/C_i\pi_i\right]^{-1}(\underline{x}_i/C_i).$$

SÄRNDAL, SWENSSON and WRETMAN (1989) propose

$$v_{sp} = \sum\sum_{i<j} \frac{\Delta_{ij}}{\pi_{ij}}\left[\frac{g_{si}e_i}{\pi_i} - \frac{g_{sj}\,e_j}{\pi_j}\right]^2$$

as an approximately unbiased estimator for $V_p(t_{sp})$ and examine its properties valid for large samples.

KOTT (1990), on the other hand, proposes the estimator

$$t_K = \sum_s \frac{Y_i}{\pi_i} + \left(\sum_1^N \underline{x}_i - \sum_{i\in s} \underline{x}_i/\pi_i\right)' \underline{b}$$

where $\underline{b} = (b_1, \ldots, b_k)'$ is a suitable estimator of $\underline{\beta}$. Writing

$$T_1 = \sum\sum_{i<j\in s} \frac{\Delta_{ij}}{\pi_{ij}}\left[\frac{e_i}{\pi_i} - \frac{e_j}{\pi_j}\right]^2$$

$$T_2 = V_m(t_K - Y)$$

$$T_3 = E_m(T_1)$$

KOTT (1990) proposes

$$v_K = \frac{T_1 T_2}{T_3}$$

as an estimator for $V_p(t_K)$.

Letting $k = 2, \underline{x}_i' = (1, X_i), \underline{\beta}' = (\beta_1, \beta_2)$ and $\underline{b}$ the least squares estimator for $\underline{\beta}$ and postulating the appropriate model $\mathcal{M}_{10}'$ for the use of the regression estimator $t_r = N\,\bar{t}_r$ based on SRSWOR for Y, it is easy to check that t_{sp} and t_K both coincide with t_r. CHAUDHURI (1992) noted that in this particular case (a) v_G closely approximates v_D and (b) v_K coincides with v_L considered in section 7.2. Since from DENG and WU (1987) we know that v_D is better than v_L, at least in this particular case we may conclude that v_G is better than v_K, although in general it is not easy to compare them.

With a single auxiliary variable x for which the values X_i are positive and known for every i in $\mathcal{U}$ with a total X, it is of interest to pursue with a narration of some aspects of the GREG predictor t_G because of the attention it is receiving, especially since the publication of the celebrated text *Model Assisted Survey Sampling* by SÄRNDAL, SWENSSON and WRETMAN (SSW, 1992).

In this context it is common to write t_G as

$$t_G = \sum_{i \in s} \frac{Y_i}{\pi_i} + \left(X - \sum_{i \in s} \frac{X_i}{\pi_i} \right) b_Q = \sum_{i \in s} \frac{Y_i}{\pi_i} g_{si}$$

where

$$b_Q = \frac{\sum_{i \in s} Y_i X_i Q_i}{\sum_{i \in s} X_i^2 Q_i}$$

with $Q_i(>0)$ arbitrarily assignable constants free of $\underline{Y} = (Y_1, \ldots, Y_N)'$ but usually as

$$\frac{1}{X_i}, \frac{1}{X_i^2}, \frac{1 - \pi_i}{\pi_i X_i}, \frac{1}{\pi_i X_i}, \frac{1}{X_i^g}, \quad (0 < g < 2) \quad \text{etc.}$$

and

$$g_{si} = 1 + \left(X - \sum_{i \in s} \frac{X_i}{\pi_i} \right) \frac{X_i Q_i \pi_i}{\sum_{i \in s} X_i^2 Q_i}.$$

Letting

$$B_Q = \frac{\sum Y_i X_i Q_i \pi_i}{\sum X_i^2 Q_i \pi_i}$$

$$E_i = Y_i - X_i B_Q$$

$$e_i = Y_i - X_i b_Q$$

SÄRNDAL (1982), essentially employing first-order TAYLOR series expansion, gave the following two approximate formulae for the MSE of t_G about Y as

$$M_1(t_G) = \sum_i \frac{1 - \pi_i}{\pi_i} E_i^2 + \sum_{i \neq j} \sum \frac{\pi_{ij} - \pi_i \pi_j}{\pi_i \pi_j} E_i E_j$$

for general designs and

$$M_2(t_G) = \sum_{i<j} \sum (\pi_i \pi_j - \pi_{ij}) \left(\frac{E_i}{\pi_i} - \frac{E_j}{\pi_j} \right)^2$$

for a design of fixed size $\nu(s)$. To these CHAUDHURI and PAL (2002) add a third as

$$M_3(t_G) = M_2(t_G) + \sum \alpha_i \frac{E_i^2}{\pi_i}$$

for a general design where

$$\alpha_i = 1 + \frac{1}{\pi_i} \sum_{j \neq i} \pi_{ij} - \sum \pi_i.$$

For $M_1(t_G)$, recommended estimators are, writing $a_{1i} = 1, a_{2i} = g_{si}$,

$$m_{1k}(t_G) = \sum_{i \in s} a_{ki}^2 \frac{1-\pi_i}{\pi_i} \frac{e_i^2}{\pi_i} + \sum \sum_{i \neq j \in s} a_{ki} a_{kj} \frac{\pi_{ij} - \pi_i \pi_j}{\pi_i \pi_j \pi_{ij}} e_i e_j; \; k = 1, 2$$

and for $M_2(t_G)$ estimators are

$$m_{2k}(t_G) = \sum \sum_{i < j \in s} \frac{\pi_i \pi_j - \pi_{ij}}{\pi_{ij}} \left(\frac{a_{ki} e_i}{\pi_i} - \frac{a_{kj} e_j}{\pi_j} \right)^2; \; k = 1, 2$$

as given by SÄRNDAL (1982). For $M_3(t_G)$ the estimators as proposed by CHAUDHURI and PAL (2002) are

$$m_{3k}(t_G) = m_{2k} + \sum_{i \in s} \frac{\alpha_i}{\pi_i} (a_{ki} e_i)^2; \; k = 1, 2.$$

In order to avoid instability in $m_{jk}(t_G); j = 1, 2, 3; k = 1, 2$ due to (a) the preponderance of numerous cross-product terms involving exorbitantly volatile terms

$$\frac{\pi_{ij} - \pi_i \pi_j}{\pi_i \pi_j \pi_{ij}}, \quad \frac{\pi_i \pi_j - \pi_{ij}}{\pi_{ij}}$$

in them and (b) the terms π_{ij}, which are hard to spell out and compute accurately for many sampling schemes, SÄRNDAL (1996) recommends approximating MSE(t_G) by

$$M_S(t_G) = \sum \frac{1-\pi_i}{\pi_i} E_i^2$$

and estimating it by

$$m_{Sk}(t_G) = \sum_{i \in s} \frac{1-\pi_i}{\pi_i}(a_{ki}e_i)^2;\ k = 1, 2$$

possibly with a slight change in the coefficient of E_i^2 in $M_S(t_G)$ when ΣE_i equals zero at least approximately.

He illustrated the two specific sampling schemes, namely (1) stratified simple random sampling without replacement, STSRS in brief, and (2) stratified sampling with sampling from each stratum by the special case of the Poisson sampling scheme for which π_i is a constant for every unit within the respective strata. He showed $m_{Sk}(t_G)$ for these two schemes composed with variance estimators for certain unequal probability sampling schemes illustratively chosen by them as the RAO, HARTLEY and COCHRAN (RHC) scheme.

Incidentally, choosing (1) $Q_i = 1/\pi_i X_i$ and (2) $X_i = \pi_i$ the estimator t_G takes the form

$$t_G = \frac{\sum \pi_i}{\nu(s)} \sum_{i \in s} \frac{Y_i}{\pi_i}.$$

Let this be based on a Poisson scheme and ignore the possibility of $\nu(s)$ equalling 0. Then

$$m_{11}(t_G) = \sum_{i \in s} \frac{1-\pi_i}{\pi_i^2} \left(Y_i - \frac{\sum_{i \in s} \frac{Y_i}{\pi_i}}{\nu(s)} \pi_i \right)^2$$

$$m_{12}(t_G) = \left(\frac{\sum \pi_i}{\nu(s)} \right)^2 m_{11}(t_G)$$

consistently with the formulae for v_0 and v_2 of section 7.1.

CHAUDHURI and MAITI (1995) and CHAUDHURI, ROY and MAITI (1996) considered a generalized regression version of the RAO, HARTLEY, COCHRAN (RHC) estimator as

$$t_{GR} = \sum_{i=1}^{n} Y_i \frac{Q_i}{P_i} + \left(X - \sum_{i=1}^{n} X_i \frac{Q_i}{P_i} \right) b_R = \sum_{i=1}^{n} Y_i \frac{Q_i}{P_i} h_{si}$$

where $R_i (> 0)$ is a suitably assignable constant like

$$R_i = \frac{1}{X_i}, \frac{1}{X_i^2}, \frac{1}{X_i^g}, \frac{Q_i}{P_i X_i}, \frac{1-P_i/Q_i}{X_i P_i/Q_i} \text{ etc. } (0 < g < 2)$$

and

$$b_R = \frac{\sum_{i=1}^{n} Y_i X_i R_i}{\sum_{i=1}^{n} X_i^2 R_i}$$

$$h_{si} = 1 + \left(X - \sum_{i=1}^{n} X_i \frac{Q_i}{P_i} \right) \frac{X_i R_i \frac{P_i}{Q_i}}{\sum_{i=1}^{n} X_i^2 R_i}.$$

Clearly, here R_i corresponds to Q_i, P_i/Q_i to π_i, and b_R to b_Q in t_G.

Accordingly, writing

$$B_R = \frac{\sum Y_i X_i R_i \frac{P_i}{Q_i}}{\sum X_i^2 R_i \frac{P_i}{Q_i}} \quad \text{that parallels } B_Q$$

$$F_i = Y_i - X_i B_R$$

$$f_i = Y_i - X_i b_R$$

and using first-order TAYLOR series expansion we may write the approximate MSE of t_{GR} about Y as

$$M(t_{GR}) = c \sum \sum_{1 \le i < j \le n} P_i P_j \left(\frac{F_i}{P_i} - \frac{F_i}{P_j} \right)^2$$

where

$$c = \frac{\sum_1^n N_i^2 - N}{N(N-1)}$$

and two reasonable estimators for it as

$$m_k(t_{GR}) = D \sum \sum_{1 \le i < j \le n} Q_i Q_j \left(\frac{b_{ki} f_i}{P_i} - \frac{b_{kj} f_j}{P_j} \right)^2 ; \; k = 1, 2$$

all analogous to $M_1(t_G)$, $M_2(t_G)$, $m_{1k}(t_G)$, $m_{2k}(t_G)$; here

$$b_{1i} = 1; \; b_{2i} = h_{si}$$

$$D = \frac{\sum_1^n N_i^2 - N}{N^2 - \sum N_i^2}.$$

We emphasize the importance of this t_{GR}, especially because SÄRNDAL (1996) compared t_G based on STSRS and STBE with t_{RHC}, but it would have been fairer if, instead of t_{RHC}, t_{GR} was brought under a comparison to keep the contestants under a common footing.

Finally, remember that DEVILLE and SÄRNDAL (1992) derived t_G as a calibration estimator on modifying the sample weight $a_k = 1/\pi_k(> 0)$ in

$$HTE = \sum_{k \in s} a_k Y_k$$

into w_k so as to (a) keep the revised weight w_k close to a_k, (b) taking account of the calibration constraint (CE)

$$\sum_{k \in s} w_k X_k = \sum_{k=1}^{N} X_k$$

by minimizing the distance function

$$\sum_{k \in s} a_k (w_k - a_k)^2 / Q_k, \text{ with } Q_k > 0$$

subject to the above CE. By the same approach one may derive t_{GR} as a calibration estimator by modifying t_{RHC} as well.

7.5 SYSTEMATIC SAMPLING

Next we consider variance estimation in systematic sampling where we have a special problem of unbiased variance estimation because a necessary and sufficient condition for the existence of a p-unbiased estimator for a quadratic form with at least one product term $X_i X_j$ is that the corresponding pair of units (i, j) has a positive inclusion probability π_{ij}. But systematic sampling is a cluster sampling where the population is divided into a number of disjoint clusters, one of which is selected with a given probability. Thus a pair of units belonging to different clusters has a zero probability of appearing together in a sample. Hence the problem of p-unbiased estimation of variance. Let us turn to it.

Let us consider the simplest case of **linear systematic sampling** with equal probabilities where in choosing a sample of size n from the population of N units it is supposed that $\frac{N}{n}$ is an integer K. Then, the population is divided into K mutually exclusive clusters of n units each and one of them is selected at random, that is, with probability $\frac{1}{K}$. If the ith cluster is selected then one takes $\bar{y}_i$, the mean of the n units of the ith cluster, $i = 1, \ldots, K$ as the unbiased estimator for the population mean $\overline{Y}$. Then,

$$V(\bar{y}_i) = \frac{1}{K} \sum_{i=1}^{K} (\bar{y}_i - \bar{Y})^2 = \frac{S^2}{n} \{1 + (n-1)\rho\}$$

writing $S^2 = \frac{1}{nK} \sum_1^K \sum_{j=1}^n (Y_{ij} - \bar{Y})^2$, Y_{ij} = the value of y for the jth member of ith cluster and

$$\rho = \frac{1}{K n (n-1) S^2} \sum_1^K \sum\sum_{j \neq j'} (Y_{ij} - \bar{Y})(Y_{ij'} - \bar{Y}).$$

For the reasons mentioned above one cannot have a p-unbiased estimator for $V(\bar{y}_i)$ for the sampling scheme employed as above. However, there are several approaches to bypass this problem.

One procedure is to postulate a model characterizing the nature of the y_{ij} values when they are arranged in K clusters as narrated above and then work out an estimator based on the sample, for example, v such that $E_m(v)$ equals $E_m V(\bar{y}_i)$, which therefore becomes a DM approach (cf. SÄRNDAL, 1981).

Second, the N elements are arranged in order, a number r is found out so that $\frac{n}{r}$ is an integer m. Then, $Kr = L$, clusters are formed, and an SRSWOR of r clusters is chosen. Each of these L clusters has m units and so a required sample of size $n = mr$ is thus realized. This is distinct from the original systematic sampling. To distinguish between the two they are respectively called **single-start** and **multiple-start** systematic sampling schemes. For the latter, one may suppose to have drawn r different systematic samples each of size m and the sample mean of each provides an unbiased estimator for the population mean. Denoting them by $\bar{y}_1, \bar{y}_2, \ldots, \bar{y}_r$ one may use $\bar{\bar{y}} = \frac{1}{r} \sum_1^r \bar{y}_i$ as an unbiased estimator for $\bar{Y}$ and $\frac{1}{r(r-1)} \sum_1^r (\bar{y} - \bar{\bar{y}})^2$ as an unbiased estimator for $V_p(\bar{\bar{y}})$. Two variations of this procedure are (a) to choose by SRSWOR method 2 or more clusters out of the K original clusters or (b) to divide the chosen cluster into a number of subsamples, and in either case obtain several unbiased estimators for $\bar{Y}$ and from them get an unbiased estimator of the variance of the pooled mean of these unbiased estimators.

A third approach is to first choose a systematic sample from the population and supplement it with an additional SRSWOR or another systematic sample from the remainder of the population. A variation of this is given by SINGH and SINGH (1977), who first make a random start out of all the N units arranged in a certain order, select a few successive units, and then follow up by choosing later units at a constant interval in a circular order until a required effective sample size is realized. They call it **new systematic sampling**, derive certain conditions on its applicability, show that $\pi_{ij} > 0$ for every i, j for this scheme and hence derive a Yates–Grundy-type variance estimator.

COCHRAN's (1977) standard text gives several estimators following the first model-based approach. GAUTSCHI (1957), TORNQVIST (1963), and KOOP (1971) applied the second approach. HEILBRON (1978) also gives model-based optimal estimators of Var (systematic sample mean) as the conditional expectations of this variance given a systematic sample under various models postulated on the observations arranged in certain orders.

ZINGER (1980) and WU (1984) follow the third approach, taking a weighted combination of the unbiased estimators of $\bar{Y}$ based on the two samples and choosing the weights, keeping in mind the twin requirements of resulting efficiency and non-negativity of the variance estimators. For a review one may refer to BELLHOUSE (1988) and IACHAN (1982).

Finally, we present below a number of estimators for $V(\bar{y}_i)$ based on the single-start simple linear systematic sample as given by WOLTER (1984).

We consider first the following notations: For the ith ($i = 1, \ldots, K$) systematic sample supposed to have been chosen containing n units, let Y_{ij} be the sample values, $j = 1, \ldots, n$. Then,

$$\bar{y}_i = \frac{1}{n}\sum_{j=1}^{n} Y_{ij}.$$

Let further

$$
\begin{aligned}
a_{ij} &= Y_{ij} - Y_{i,j-1},\ j = 2, \ldots, n \\
b_{ij} &= Y_{ij} - 2Y_{i,\,j-1} + Y_{i,j-2} \\
c_{ij} &= \frac{1}{2}Y_{ij} - Y_{i,j-1} + Y_{i,j-2} - Y_{i,j-3} + \frac{1}{2}Y_{i,j-4} \\
d_{ij} &= \frac{1}{2}Y_{ij} - Y_{i,j-1} + \ldots + \frac{1}{2}Y_{i,j-8}
\end{aligned}
$$

and

$$
s^2 = \frac{1}{(n-1)}\sum_{1}^{n}(y_{ij} - \bar{y}_i)^2.
$$

Then WOLTER (1984) proposed the following estimators for $V(\bar{y}_i)$.

$$
\begin{aligned}
v_1 &= (1-f)\frac{s^2}{n} \\
v_2 &= \frac{1-f}{2n(N-1)}\sum_{j=2}^{n} a_{ij}^2 \\
v_3 &= \frac{1-f}{n}\frac{1}{n}\sum_{1}^{n/2} a_{i,\,2j}^2 \\
v_4 &= \frac{1-f}{n}\frac{1}{6(n-2)}\sum_{j=3}^{n} b_{ij}^2 \\
v_5 &= \frac{1-f}{n}\frac{1}{3\times 5(n-4)}\sum_{j=5}^{n} c_{ij}^2 \\
v_6 &= \frac{1-f}{n}\frac{1}{7\times 5(n-8)}\sum_{j=9}^{n} d_{ij}^2.
\end{aligned}
$$

For a multiple-start systematic sample with r starts, let $\bar{y}_\alpha$ denote the sample mean based on the αth replicate and

$$
\bar{y} = \frac{1}{r}\sum_{\alpha=1}^{r}\bar{y}_\alpha.
$$

Then for $V(\bar{y})$ the estimator is taken as

$$v_7 = \frac{1-f}{r(r-1)} \sum_{\alpha=1}^{r} (\bar{y}_\alpha - \bar{y})^2.$$

This is also applicable if the ith systematic sample is split up into r random subsamples (cf. KOOP, 1971). Writing

$$\hat{\rho}_K = \frac{1}{(n-1)s^2} \sum_{j=2}^{n} (Y_{ij} - \bar{y}_i)(Y_{i,j-1} - \bar{y}_i)$$

another estimator for $V(\bar{y}_i)$ is

$$v_8 = \frac{1}{(n-1)s^2} \sum_{j=2}^{n} (Y_{ij} - \bar{y}_i)(Y_{i,j-1} - \bar{y}_i).$$

WOLTER (1984) examined relative performances of these estimators considering $B_m(v) = E_m[E_p(v) - V(\bar{y})]$ and $B_m(v)/E_m V(\bar{y}_i)$ for v as $v_i, i = 1, \ldots, 8$ for several models usually postulated in the context of systematic sampling. He also examined how good these are in providing confidence intervals for $\bar{Y}$. His recommendations favor v_2, and v_3, and, to some extent, v_8.

The general varying probability systematic sampling is known as circular systematic sampling (CSS) with probabilities proportional to sizes (PPS). From MURTHY (1967) we may describe it as follows. Suppose positive integers $X_i (i = 1, \ldots N)$ with a total X are available as size measures and a sample of n units is required to be drawn from $\mathcal{U} = (1, \ldots, N)$. Then a member K is fixed as the integer nearest to X/n.

A random positive integer R is chosen between 1 and X. Then, let

$$a_r = (R + rK) \bmod (X), r = 0, \ldots, n-1$$

and

$$C_0 = 0\,,\; C_i = \sum_{j=1}^{i} X_j, i = 1, \ldots, N.$$

Then, a CSSPPS sample s is formed of the units i for which

$$C_{i-1} < a_r \leq C_i \quad \text{for} \quad r = 0, 1, \ldots, n-1$$

and the unit N if $a_r = 0$.

If $\nu(s)$ happens to equal n, the intended sample size (in practice it often falls short by 1, 2, or even more for arbitrary values of $P_i = X_i/X$), then for this scheme

$$\pi_i \text{ equals } nP_i$$

provided $nP_i < 1 \forall i \in \mathcal{U}$, a condition that also often fails.

If $nP_i > 1$, then calculation of π_i becomes a formidable task, especially if X is large and n is not too small. For many pairs (i, j), $i \neq j$, π_{ij} for CSSPPS scheme turns out to be zero and is also difficult to compute even if found positive.

Following DAS (1982) and RAY and DAS (1997) one may modify the scheme CSSPPS and (a) choose K above as a positive integer at random from 1 to $X - 1$ instead of (b) keeping it fixed as earlier. It is easy to check that for this scheme, CSSPPS (n),

$$\pi_{ij} > 0 \quad \forall i \neq j.$$

However, $\nu(s)$ need not then equal n nor may π_i equal nP_i. Nevertheless, the HT estimator may be calculated for this scheme. Importantly, CHAUDHURI's (2000a) unbiased estimator for its variance is available as

$$v_c = \sum\sum_{i<j} \frac{\pi_i\pi_j - \pi_{ij}}{\pi_{ij}} \left(\frac{Y_i}{\pi_i} - \frac{Y_j}{\pi_j}\right)^2 + \sum_{i \in s} \frac{Y_i^2}{\pi_i^2} \alpha_i$$

where

$$\alpha_i = 1 + \frac{1}{\pi_i} \sum_{j \neq i} \pi_{ij} - \sum \pi_i, \; i \in \mathcal{U}.$$

This is a vindication of the utility of v_c in practice.

If one heeds the recommodation of SÄRNDAL (1996) to get rid of any situation when one encounters (a) difficulty in calculating π_i's and (b) instability in

$$\frac{\pi_i\pi_j - \pi_{ij}}{\pi_{ij}} \quad \text{or} \quad \frac{\pi_{ij} - \pi_i\pi_j}{\pi_i\pi_j\pi_{ij}}$$

involved in numerous cross-product terms in $\hat{V}$(HTE), by employing the generalized regression estimator with its variance approximated by

$$V_{APP} = \sum \frac{1 - \pi_i}{\pi_i} E_i^2$$

and taking its estimator as

$$v_R = \sum_{i \in s} \frac{1 - \pi_i}{\pi_i} (a_{ki} e_i)^2,$$

then there is no problem with either the CSSPPS or CSSPPS(n) schemes except that computation of π_i is also not easy if $\pi_i \neq nP_i (< 1)$ or if X is large.

Chapter 8

Multistage, Multiphase, and Repetitive Sampling

8.1 VARIANCE ESTIMATORS DUE TO RAJ AND RAO IN MULTISTAGE SAMPLING: MORE RECENT DEVELOPMENTS

Suppose each unit of the population $U = (1, \ldots, i \ldots, N)$ consists of a number of subunits and hence may be regarded as a **cluster**, the ith unit forming cluster of M_i subunits with a total Y_i for the variable y of interest; $i = 1, \ldots, N$. For example, we may consider districts as clusters and villages in them as subunits or cluster elements. Then quantity of interest is $Y = \Sigma_1^N Y_i$ or

$$\overline{Y} = \frac{\sum_1^N Y_i}{\sum_1^N M_i} = \frac{\sum_1^N M_i \overline{Y}_i}{\sum_1^N M_i},$$

where Y_{ij} is the value of the jth element of the ith cluster and

$$\overline{Y}_i = \frac{Y_i}{M_i} = \sum_{j=1}^{M_i} \frac{Y_{ij}}{M_i}$$

is the ith cluster mean of y. Now, often it is not feasible to survey all the M_i elements of the ith cluster to ascertain Y_i.

Instead, a policy that may be implemented is to first take a sample s of n clusters out of U according to a suitable design p and then from each selected cluster, i, take a further sample, of m_i elements out of the M_i elements in it following another suitable scheme of selection of these elements; the selection procedures in all selected clusters have to be independent from each other. Then one may derive suitable unbiased estimators, say, T_i of Y_i for $i \in s$ and derive a final estimator for Y or $\overline{Y}$. This is **two-stage sampling**, the clusters forming the **primary** or **first-stage** units (psu or fsu) and the elements within the fsus being called the **second stage** units (ssu). Further stages may be added allowing the elements to consist of subelements, the **third-stage** units to be subsampled and so on, leading, in general, to multistage sampling. We will now discuss estimation of totals, or means and estimation of variances of estimators of totals, or means in multistage sampling.

8.1.1 Unbiased Estimation of Y

Let E_1, V_1 denote expectation variance operators for the sampling design in the first stage and E_L, V_L those in the later stages. Let R_i be independent variables satisfying

(a) $E_L(R_i) = Y_i$,

(b) $V_L(R_i) = V_i$ or

(c) $V_L(R_i) = V_{si}$

and let there exist (b)′ random variables v_i such that $E_L(v_i) = V_i$ or (c)′ random variables v_{si} such that $E_L(v_{si}) = V_{si}$.

Let $E = E_1 E_L = E_L E_1$ be the overall expectation and $V = E_1 V_L + V_1 E_L = E_L V_1 + V_L E_1$ the overall variance operators. CHAUDHURI, ADHIKARI and DIHIDAR (2000a, 2000b) have illustrated how these commutativity assumptions may be valid in the context of survey sampling.

Let

$$t_b = \sum b_{si} I_{si} Y_i,$$

$$M_1(t_b) = E_1(t_b - Y)^2 = \sum\sum d_{ij} y_i y_j,$$

$$d_{ij} = E_1(b_{si} I_{si} - 1)(b_{sj} I_{sj} - 1),$$

d_{sij} be constants free of Y such that

$$E_1(d_{sij}I_{sij}) = d_{ij}\forall_{i,j} \text{ in } U.$$

Let w_i's be certain non-zero constants. Then, one gets

$$M_1(t_b) = -\sum_{i<j}\sum d_{ij}w_iw_j\left(\frac{Y_i}{w_i} - \frac{Y_j}{w_j}\right)^2 + \sum \beta_i\frac{Y_i^2}{w_i} \text{ when } \beta_i = \sum_{j=1}^{N} d_{ij}w_j.$$

Let

$$m_1(t_b) = -\sum_{i<j}\sum d_{sij}I_{sij}w_iw_j\left(\frac{Y_i}{w_i} - \frac{Y_j}{w_j}\right)^2 + \sum \beta_i\frac{I_{si}}{\pi_i}\frac{Y_i^2}{w_i}.$$

Then, we have already seen that

$$E_1m_1(t_b) = M_1(t_b),$$

Let

$$e_b = t_b|_{\underline{Y}=\underline{R}} = \Sigma b_{si}I_{si}R_i,$$

writing

$$\underline{Y} = (Y_1, \ldots, Y_i, \ldots, Y_N)$$

and

$$\underline{R} = (R_1, \ldots, R_i, \ldots, R_N).$$

Then, it follows that (1) $E_L(e_b) = t_b$, (2) $E_1(e_b) = \Sigma R_i = R$ in case we assume that $E_1(t_b) = Y$, which means

$$E_1(b_{si}I_{si}) = 1\forall i \text{ in } U \tag{8.1}$$

So,

$$E(e_b) = E_1(t_b) = Y = E_L(R)$$

if Eq. (8.1) is assumed.

$$M_1(t_b)|_{\underline{Y}=\underline{R}} = E_1(e_b - R)^2.$$

Now, writing

$$M(e_b) = E_1E_L(e_b - Y)^2 = E_LE_1(e_b - Y)^2,$$

the overall mean square error of e_b about Y and $m_1(e_b) = m_1(t_b)|_{\underline{Y}=\underline{R}}$ we intend to find $m(e_b)$ such that

$$Em(e_b) = E_1 E_L m(e_b) = E_L E_1 m(e_b)$$

may equal $M(e_b)$.

First let us note that

$$\begin{aligned} E_1 m_1(e_b) &= E_1\left[-\sum\sum_{i<j} d_{sij} I_{sij} w_i w_j \left(\frac{R_i}{w_i} - \frac{R_j}{w_j}\right)^2 + \Sigma\beta_i \frac{I_{si}}{\pi_i}\frac{R_i^2}{w_i}\right] \\ &= -\sum\sum_{i<j} d_{ij} w_i w_j \left(\frac{R_i}{w_i} - \frac{R_j}{w_j}\right)^2 + \Sigma\beta_i \frac{R_i^2}{w_i} \\ &= E_1(e_b - R)^2 = M_1(e_b) \end{aligned}$$

Now,

$$\begin{aligned} M(e_b) &= E_L E_1 (e_b - Y)^2 \\ &= E_L E_1 \left[(e_b - R) + (R - Y)\right]^2 \\ &= E_L E_1 (e_b - R)^2 + E_L (R - Y)^2 \\ &= E_L M_1(e_b) + \Sigma V_i \end{aligned}$$

if (b) holds.

So,

$$m(e_b) = m_1(e_b) + \Sigma b_{si} I_{si} v_i$$

satisfies $Em(e_b) = M(e_b)$ if in addition to (b), Eq. (8.1) also holds.

Thus, treating R_i's as estimators of Y_i obtained through later stages of sampling and v_i's as their unbiased variance estimators, it follows that under the specified conditions we may state the following result.

RESULT 8.1 *$m(e_b)$ is an unbiased estimator for $M(e_b)$.*

REMARK 8.1 *This is a generalization of* RAJ's *(1968) result, which demands that $M_1(t_b)$ be expressed as a quadratic form in Y with $m_1(t_b)$ also expressed as a quadratic form in Y_i's for $i \in s$.*

But we know from the previous chapters that often variances of estimators for Y in a single stage of sampling and their unbiased estimators, for example, those for RHC (1962), MURTHY (1957) or RAJ's (1956) estimators, are not so expressed. Our Result (8.1) avoids the tedious steps of first re-expressing the variances of these estimators as quadratic forms in seeking their estimators. Second, we may observe that

$$E_L m_1(e_b) = \left[-\sum\sum_{i<j} d_{sij} w_i w_j \left(\frac{Y_i}{w_i} - \frac{Y_j}{w_j} \right)^2 + \Sigma \beta_i \frac{I_{si}}{\pi_i} \frac{Y_i^2}{w_i} \right]$$
$$- \sum\sum_{i<j} d_{sij} I_{sij} w_i w_j \left(\frac{W_{si}}{w_i^2} + \frac{W_{sj}}{w_j^2} \right) + \Sigma \beta_i \frac{I_{si}}{\pi_i} \frac{W_{si}}{w_i},$$

writing W_{si} commonly for V_i or V_{si}, assuming either (b) or (b)' to hold:

$$M_1(t_b) = -\sum\sum_{i<j} d_{sij} I_{sij} w_i w_j \left(\frac{W_{si}}{w_i^2} + \frac{W_{sj}}{w_j^2} \right) + \Sigma \beta_i \frac{I_{si}}{\pi_i} \frac{W_{si}}{w_i}$$

But

$$\begin{aligned} M(e_b) &= E_1 E_L (e_b - Y)^2 \\ &= E_1 E_L \left[(e_b - t_b) + (t_b - Y) \right]^2 \\ &= E_1 V_L (\Sigma b_{si} I_{si} R_i) + M_1(t_b) \\ &= E_1 \Sigma b_{si}^2 I_{si} W_{si} + M_1(t_b) \end{aligned}$$

So, we have

RESULT 8.2

$$m_2(e_b) = m_1(e_b) + \sum\sum_{i<j} d_{sij} I_{sij} w_i w_j \left(\frac{w_{si}}{w_i^2} + \frac{w_{sj}}{w_j^2} \right)$$
$$+ \Sigma \left(b_{si}^2 - \frac{\beta_i}{\pi_i^2} \right) I_{si} w_{si}$$

writing w_{si} commonly for v_{si} and v_i is an unbiased estimator for $M(e_b)$ when either (b) and (c) together or (b)′ and (c)′ together hold.

Here the condition (8.1) is not required.

REMARK 8.2 *Result 8.2 is somewhat similar to* RAO's *(1975a) result, which is also constrained by the quadratic form expressions for the variances of estimators t for Y.*

It is appropriate to briefly state below RAJ's (1968) and RAO's (1975a) results in this context to appreciate the roles for these changes. Relevant references are CHAUDHURI (2000) and CHAUDHURI, ADHIKARI and DIHIDAR (2000a, 2000b).

For $t_b = \Sigma b_{si} I_{si} Y_i$ subject to $E_1(b_{si} I_{si}) = 1 \; \forall i$ in U so that $E_1(t_b) = Y$ and its variance is

$$V_1(t_b) = \Sigma C_i Y_i^2 + \sum\sum_{i \neq j} C_{ij} Y_i Y_j$$

where

$$C_i = E_1(b_{si}^2 I_{si}) - 1$$

and

$$C_{ij} = E_1(b_{si} b_{sj} I_{sij}) - 1$$

if there exist C_{si}, C_{sij} free of $\underline{Y}$ such that $E_1(C_{si} I_{si}) = C_i$ and $E_1(C_{sij} I_{sij}) = C_{ij}$, it follows that $e_b = \Sigma b_{si} I_{si} R_i$ satisfies, assuming (a), (b), and (c) above,

$$E(e_b) = Y, V(e_b) = V_1(t_b) + E_1(\Sigma b_{si}^2 I_{si} V_i) = V,$$

and noting

$$v_1(t_b) = \Sigma C_{si} I_{si} Y_i^2 + \sum\sum_{i \neq j} C_{sij} I_{sij} Y_i Y_j$$

satisfies $E_1 v_1(t_b) = V_1(t_b)$, it follows on writing

$$v_1(e_b) = v_1(t_b)|_{Y=R} = \Sigma C_{si} I_{si} R_i^2 + \sum\sum_{i \neq j} C_{sij} I_{sij} R_i R_j$$

that one has for

$$\begin{aligned} v(e_b) &= v_1(e_b) + \Sigma b_{si} I_{si} v_i, \\ E v(e_b) &= V(e_b) = V \end{aligned} \tag{8.2}$$

This is due to RAJ (1968). If, instead of (b) and (c) we have (b)′ and (c)′, then RAO (1975a) has the following modifications to the above.

$$V(e_b) = V_1(t_b) + E_1(\Sigma b_{si}^2 I_{si} V_{si}) = V',$$

and

$$v'(e_b) = v_1(e_b) + \Sigma (b_{si}^2 - C_{si}) I_{si} v_{si}$$

satisfies $Ev'(e_b) = V'$. Thus, $v'(e_b)$ is another unbiased estimator for $V(e_b)$ as alternative to $v(e_b)$.

In particular, if $v(s)$ is a constant for every s with $p(s) > 0$, so that SEN (1953) and YATES and GRUNDY's (1953) unbiased estimator v_{syg} is available for the variance of the HTE in a single-stage sampling, RAJ (1968) has the following results. Under (a)–(b),

$$t_H = \sum_{i \in P} \frac{Y_i}{\pi_i}, \; e_H = \sum_{i \in S} \frac{R_i}{\pi_i}, \; E(e_H) = Y,$$

$$V(e_H) = \sum_{i<j}\sum (\pi_i \pi_j - \pi_{ij}) \left(\frac{Y_i}{\pi_i} - \frac{Y_i}{\pi_j} \right)^2 + \sum_i \frac{V_i}{\pi_i} = V',$$

For

$$v'(e_H) = \sum_{i<j}\sum_{\in s} (\frac{\pi_i \pi_j - \pi_{ij}}{\pi_{ij}} \left(\frac{R_i}{\pi_i} - \frac{R_j}{\pi_j} \right)^2 + \sum_{i \in s} \frac{v_i}{\pi_i}$$

one has

$$Ev'(e_H) = V(e_H) = V'.$$

In case, instead, (b)′ and (c)′ hold, then the above results change into less elegant results.

If (a), (b)′ and (c)′ hold, then

$$V(e_H) = \sum_{i<j}\sum \left(\frac{\pi_i \pi_j - \pi_{ij}}{\pi_{ij}} \right) \left(\frac{Y_i}{\pi_i} - \frac{Y_j}{\pi_j} \right)^2 + E_1 \left(\sum_{i \in s} \frac{V_{si}}{\pi_i^2} \right) = V'',$$

and

$$v''(e_H) = \sum_{i<j}\sum_{\in s} \left(\frac{\pi_i \pi_j - \pi_{ij}}{\pi_{ij}} \right) \left(\frac{R_i}{\pi_i} - \frac{R_j}{\pi_j} \right)^2 + \sum_{i \in s} \frac{v_{si}}{\pi_i^2}$$
$$+ \sum_{i<j}\sum_{\in s} \left(\frac{\pi_i \pi_j - \pi_{ij}}{\pi_{ij}} \right) \left(\frac{v_{si}}{\pi_i^2} + \frac{v_{sj}}{\pi_j^2} \right)$$

satisfies

$$Ev''(e_H) = V''.$$

If, in the single-stage sampling, one is satisfied to employ a biased estimator for Y like the generalized regression (GREG) estimator t_G or a version of it like t_{GR}, and is also satisfied to employ a not-unbiased estimator like $m_k(t_G)$ or $m_k(t_{GR})$ for the TAYLOR version of an approximate MSE for t_G or for t_{GR} as M_G or M_{GR}, then supposing that Y_i is not ascertainable but is required to be unbiasedly estimated by R_i, through sampling at later stages while X_i, an auxiliary positive value with total X, is available for every i in U, we may be satisfied with the results of the following types.

Let

$$e_G = t_G|_{\underline{Y}=\underline{R}} = \sum_{i \in s} \frac{R_i}{\pi_i} g_{si}.$$

Then,

$$\begin{aligned} M(e_G) &= E_1 E_L \left[e_G - Y\right]^2 \\ &= E_L E_1 \left[(e_G - R) + (R - Y)\right]^2 \\ &= E_L \left[M(t_G)|_{\underline{Y}=\underline{R}}\right] + \sum V_i \end{aligned}$$

assuming (a)–(c) to hold.

Then,

$$m_k(t_G)|_{\underline{Y}=\underline{R}} + \sum_{i \in s} b_{si} I_{si} v_i = v_k(e_G),\ k = 1, 2$$

provides a desirable estimator for $M(e_G)$ with a suitable choice of b_{si}, which may be subject to $E_1(b_{si} I_{si}) = 1\ \forall i$.

If instead of (b) and (c), only (b)$'$ and (c)$'$ are supposed to hold, elegant results are hard to come by.

An analogous treatment is recommended starting with t_{GR}. Suppose one needs to estimate instead of Y, the mean

$$\overline{Y} = \frac{\sum_1^N Y_i}{\sum_1^N M_i} = \frac{\sum_1^N \sum_{j=1}^{M_i} Y_{ij}}{\sum_1^N \sum_1^{M_i} 1_{ij}} = \frac{\sum_1^N \sum_1^{M_i} \sum_1^{T_{ij}} \sum_1^{L_{ijk}} \sum_1^{R_{ijkl}} Y_{ijklu}}{\sum_1^N \sum_1^{M_i} \sum_1^{T_{ij}} \sum_1^{L_{ijk}} \sum_1^{R_{ijkl}} 1_{ijklu}}$$

writing $1_{ijklu} = 1$ if uth 5th-stage unit of lth 4th-stage unit of kth 3rd-stage unit of jth 2nd-stage unit of ith first stage unit has a y value, for example, with a 5-stage sampling.

Here both $\sum_1^N Y_i$ and $\sum_1^N M_i$ are unknown and both are to be estimated, and $\overline{Y}$ is to be estimated by the ratio of an estimator $\hat{Y}_N$ for $Y = \sum_1^N Y_i$ to the estimator $\hat{M}$, for $M = \sum_1^M M_i$. Then, $\hat{R} = \frac{\hat{Y}}{\hat{M}}$ is clearly a ratio estimator for the ratio

$\overline{Y} = \frac{Y}{M}$. Then, supposing a suitable estimator $\hat{V}(\hat{Y})$ for the variance or MSE of $\hat{Y}$ is employed, then $\hat{V}(\hat{R})$ is to be taken as

$$\hat{V}(\hat{R}) = \frac{1}{(\hat{M})^2}\left[\hat{V}(\hat{Y})|_{y_{ijklu}=y_{ijklu}-\hat{R}I_{ijklu}} + \sum_{i\in s} b_{si}^2 w_{si}\right], \quad (8.3)$$

applying the usual procedure involved for ratio estimation.

This is because writing $\hat{y}_i$ as an unbiased estimator for $y_i = \sum_j^{M_i}\sum_k^{T_{ij}}\sum_l^{L_{ijk}}\sum_k^{R_{ijkl}} y_{ijklu}$ and w_{si} as an estimator for $Var\,(\hat{y}_i) = V_L(\hat{y}_i)$

$$\hat{Y} = \sum_{i\in s} b_{si}\hat{y}_i,\ \hat{M} = \sum_{i\in s} b_{si}M_i,\ \hat{\overline{Y}} = \frac{\hat{Y}}{\hat{M}},$$

$$E_1E_L(\hat{\overline{Y}} - \overline{Y})^2 \simeq E_1\left[\sum_{i\in s} b_{si}^2 V_L(\hat{y}_i)/(\hat{M})^2\right]$$

$$+ E_1\left[\frac{\sum_{i\in s} b_{si}y_i}{\sum_{i\in s} b_{si}M_i} - \frac{Y}{M}\right]^2$$

$$\simeq E_1E_L\left[\frac{\sum_{i\in s} b_{si}^2 w_{si}}{(\hat{M})^2}\right]$$

$$+ \frac{1}{M^2}V\left[\sum_{i\in s} b_{si}\left(y_i - \frac{Y}{M}M_i\right)\right]$$

An estimator for this may therefore be taken as Eq. (8.3) above.

It may be in order at this stage to elaborate on the concept of Rao-Blackwellization, relevant in the context of survey sampling.

Let from a survey population $U = (1, \ldots, i, \ldots, N)$ a sample sequence $s = (i_1, \ldots, i_j, \ldots, i_n)$ of n units of U be drawn that are not necessarily distinct and where the order in which the units are drawn is maintained as the 1st, 2nd, …, nth.

Let $s^* = \{j_1, \ldots, j_i, \ldots, j_k\}$ be the set of distinct elements $(1 \le k \le n)$ in s ignoring the order of their occurrence with no repetition of the elements in s^*. Let $\sum_{s\to s^*}$ denote the sum over the sequences s for each of which s^* is the set of distinct units with no repetitions therein. Let $p(s)$ be the probability of selecting s and $p(s^*) = \sum_{s\to s^*} p(s)$ that of s^*.

Let $t = t(s, \underline{Y})$ be any estimator for a parameter θ which is a function of $\underline{Y} = (y_1, \ldots, y_i, \ldots, y_N)$. Then, let

$$t^* = t^*(s, \underline{Y}) = \frac{\sum_{s \to s^*} t(s, \underline{Y}) p(s)}{\sum_{s \to s^*} p(s)}$$

$= t^*(s^*, \underline{Y})$ for every s to which s^* corresponds as the set of all the distinct units therein with no repetitions.

Then,

$$\begin{aligned} E_p(t) &= \sum_s p(s) t(s, \underline{Y}) \\ &= \sum_{s^*} \sum_{s \to s^*} p(s) t(s, \underline{Y}) \\ &= \sum_{s^*} \left[\frac{\sum_{s \to s^*} t(s, Y) p(s)}{\sum_{s \to s^*} p(s)} \right] p(s^*) \\ &= \sum_{s^*} t^*(s^*, \underline{Y}) p(s^*) \\ &= E_p(t^*) \end{aligned}$$

Also,

$$\begin{aligned} E_p(tt^*) &= \sum_s p(s) t(s, Y) t^*(s, \underline{Y}) \\ &= \sum_{s^*} t^*(s^*, \underline{Y}) \left[\frac{\sum_{s \to s^*} t(s, \underline{Y}) p(s)}{\sum_{s \to s^*} p(s)} \right] p(s^*) \\ &= \sum_{s^*} p(s^*) \left[t^*(s^*, \underline{Y}) \right]^2 = E_p(t^*)^2 \end{aligned}$$

So,

$$\begin{aligned} 0 \le E_p(t - t^*)^2 &= E_p(t^2) - E_p(t^*)^2 \\ &= V_p(t) - V_p(t^*) \end{aligned}$$

Thus,

$$\begin{aligned} V_p(t) &= V_p(t^*) + E_p(t - t^*)^2 \\ &\ge V_p(t^*) \end{aligned}$$

equality holding only in case $t(s, \underline{Y}) = t^*(s, \underline{Y})$ for every s with $p(s) > 0$.

So, the statistic t^* free of order and/or repetition of units in a sample is better than t as an estimator for θ, both having the same expectation but t^* having a less variance than t.

The operation of deriving t^* from t may be regarded as one of Rao-Blackwellization, which consists of deriving an estimator based on a sufficient statistic, rather the minimal sufficient statistic, from another statistic and showing that the former has the same expectation as the latter, but with a smaller variance.

In order to further elaborate on this let us write

$$d = ((i_1, y_{i1}), \ldots, (i_n, y_{in}))$$

to denote survey data on choosing a sample s with probability $p(s)$ and observing the values of y as $\underline{y} = (y_{i1}, \ldots, y_{in})$ for the respective sampled units $(i_1, \ldots, i_n) \doteq s$. Let $\Omega = \{\underline{Y} \mid -\infty < a_i \le y_i \le b_i < +\infty\}$ be the parametric space, of which $\underline{Y}$ is an element and $\Omega_d = \{\underline{Y} \mid -\infty < a_i \le y_i \le b_i + \infty$ for $i = 1, \ldots, N (\neq i_1, \ldots, i_n)$ but $y_{i1}, \ldots, y_{in}$ are as observed, be the subset of Ω that is consistent with d. It follows that $\Omega_d = \Omega_{d^*}$ where

$$d^* = \{(j_1, y_{j1}), \ldots (j_k, y_{jk})\}.$$

Then the probability of observing d is $P_{\underline{Y}}(d) = p(s) I_{\underline{Y}}(d)$, where $I_{\underline{Y}}(d) = 1$ if $\underline{Y} \in \Omega_d$, $= 0$ otherwise and that of observing d^* is

$$P_{\underline{Y}}(d^*) = p(s^*) I_{\underline{Y}}(d^*)$$

where

$$I_{\underline{Y}}(d^*) = 1 \quad \text{if} \quad \underline{Y} \in \Omega_d, = 0 \text{ else}.$$

Then, $I_{\underline{Y}}(d) \doteq I_{\underline{Y}}(d^*)$ and assuming $p(\cdot)$ as a noninformative design, it follows that the conditional probability of observing d, given d^* is

$$P_{\underline{Y}}(d \mid d^*) = \frac{P_{\underline{Y}}(d \cap d^*)}{P_{\underline{Y}}(d^*)} = \frac{P_{\underline{Y}}(d)}{P_{\underline{Y}}(d^*)} = \frac{p(s)}{p(s^*)}$$

As the ratio $\frac{p(s)}{p(s^*)}$ is free of $\underline{Y}$, it follows that d^* is a sufficient statistic.

To prove that d^* is the minimal sufficient statistic, let $t = t(d)$ be another sufficient statistic.

Let d_1, d_2 be two separate survey data points and d_1^*, d_2^* the corresponding sufficient statistics of the form d^* as derived from d. We state below that

$$t(d_1) = t(d_2) \text{ will imply } d_1^* = d_2^*$$

and hence imply that d^* is a minimal sufficient statistic.

Letting p be a noninformative design, we may notice that

$$\begin{aligned} P_{\underline{Y}}(d_1) &= P_{\underline{Y}}(d_1 \cap t(d_1)) \\ &= P_{\underline{Y}}(t(d_1))P_Y(d_1|t(d_1)) \\ &= P_{\underline{Y}}(t(d_1))C_1, \end{aligned}$$

where C_1 is a constant free of $\underline{Y}$ because t is a sufficient statistic. Similarly,

$$\begin{aligned} P_{\underline{Y}}(d_2) &= P_{\underline{Y}}(t(d_2))C_2, \text{ say,} \\ &= P_{\underline{Y}}(t(d_1))C_2 \end{aligned}$$

because $t(d_1) = t(d_2)$ by hypothesis.

So,

$$P_{\underline{Y}}(d_2) = P_{\underline{Y}}(d_1)\frac{C_2}{C_1}$$

or

$$p(s_2)I_{\underline{Y}}(d_2) = p(s_1)I_{\underline{Y}}(d_1)C,$$

where C is a constant free of $\underline{Y}$ or

$$p(s_2^*)I_{\underline{Y}}(d_2^*) \propto p(s_1^*)I_{\underline{Y}}(d_1^*)$$

and this implies $d_2^* = d_1^*$ as is required to be shown.

8.1.2 PPSWR Sampling of First-Stage Units

First, from DES RAJ (1968) we note the following. Suppose a PPSWR sample of fsus is chosen in n draws from U using normed size measures $P_i(0 < P_i < i, \Sigma P_i = 1)$. Writing $y_r(p_r)$ for the $Y_i(p_i)$ value for the unit chosen on the rth draw, ($r = 1, \ldots, n$) the HANSEN–HURWITZ estimator

$$t_{HH} = \frac{1}{n}\sum_{n=1}^{n}\frac{y_r}{p_r}$$

might be used to estimate Y because $E_p(t_{HH}) = Y$ if Y_i could be ascertained. But since Y_i's are not ascertainable, suppose that each time an fsu i appears in one of the n independent draws by PPSWR method, an independent subsample of elements is selected in subsequent stages in such a manner that estimators $\hat{y}_r$ for y_r are available such that $E_L(\hat{y}_r) = y_r$ and $V_L(\hat{y}_r) = \sigma_r^2$ with uncorrelated $y_1, y_2, \ldots, y_n$. Then, DAS RAJ's (1968) proposed estimator for Y is

$$e_H = \frac{1}{n}\sum_{r=1}^{n}\frac{\hat{y}_r}{p_r}$$

for which the variance is

$$\begin{aligned} V(e_H) &= V_p(t_{HH}) + E_p\left[\frac{1}{n^2}\sum_{r=1}^{n}\frac{\sigma_r^2}{p_r^2}\right] \\ &= \frac{1}{n}\sum P_i\left(\frac{Y_i}{P_i} - Y\right)^2 + \frac{1}{n}\sum_{1}^{N}\frac{\sigma_i^2}{P_i} \\ &= V_H, \text{ say.} \end{aligned}$$

It follows that

$$v_H = \frac{1}{2n^2(n-1)}\sum_{r=1}^{n}\sum_{\substack{r'=1 \\ r\neq r'}}\left(\frac{\hat{y}_{r'}}{p_{r'}} - \frac{\hat{y}_r}{p_r}\right)^2$$

is an unbiased estimator for V_H because

$$E_l(v_H) = \frac{1}{2n^2(n-1)}\sum\sum_{r\neq r'}\left[\frac{y_r^2}{p_r^2} + \frac{y_{r'}^2}{p_{r'}^2} + \frac{\sigma_r^2}{p_r^2} + \frac{\sigma_{r'}^2}{p_{r'}^2} - 2\frac{y_r}{p_r}\frac{y_{r'}}{p_{r'}}\right]$$

$$\begin{aligned} E\,v_H = E_p\,E_L(v_H) &= \frac{1}{n}\left(\sum\frac{Y_i^2}{P_i} - Y^2\right) + \frac{1}{n}\sum\frac{\sigma_i^2}{P_i} \\ &= \frac{1}{n}\sum P_i\left(\frac{Y_i}{P_i} - Y\right)^2 + \frac{1}{n}\sum\frac{\sigma_i^2}{P_i} = V(e_H). \end{aligned}$$

Thus here an estimator for σ_r^2 is not required in estimating $V(e_H)$.

But it should be noted that

(a) sampling with replacement is not very desirable because it allows reappearance of the same unit leading to estimators that can be improved upon by Rao-Blackwellization, and

(b) resampling the same sampled cluster may be tedious and impracticable. So, even if a PPSWR sample (in n draws) of cluster may be selected, it may be considered prudent to subsample a chosen cluster only once irrespective of its frequency of appearance in the sample.

Thus one may consider the following alternative estimator for Y, namely,

$$e_A = \frac{1}{n}\sum_i \frac{\hat{Y}_i}{P_i} f_{si}.$$

Here f_{si} is the frequency of i in s, $\hat{Y}_i$ is an estimator for Y_i based on sampling at later stages of the cluster i in such a way that

$$E_L(\hat{Y}_i) = Y_i,\ V_L(\hat{Y}_i) = \sigma_i^2$$

and further, based on sampling of ith cluster at later stages $\hat{\sigma}_i^2$ is available as an estimator for σ_i^2 such that

$$E_L(\hat{\sigma_i^2}) = \sigma_i^2.$$

Then,

$$E_L(e_A) = \frac{1}{n}\sum_i \frac{Y_i}{P_i} f_{si} = t_A, \text{ say,}$$

and $E(e_A) = E_p(t_A) = Y$ because $E_p(f_{si}) = nP_i$. Furthermore

$$\begin{aligned} V(e_A) &= V_p(t_A) + E_p\left[V_L(e_A)\right] \\ &= \frac{1}{n}\left[\sum \frac{Y_i^2}{P_i} - Y^2\right] + \frac{1}{n}\sum \frac{\sigma_i^2}{P_i} + \frac{n-1}{n}\sum \sigma_i^2 \end{aligned}$$

noting that $V_p(f_{si}) = nP_i(1-P_i)$, $cov_p(f_{si}, f_{sj}) = -nP_iP_j$.

An unbiased estimator for $V(e_A)$ may be taken as

$$v_A = \frac{1}{(n-1)}\left[\frac{1}{n}\sum \frac{\hat{Y}_i^2}{p_i^2} f_{si} - e_A^2 + \frac{n-1}{n}\sum \frac{\hat{\sigma}_i^2}{P_i} f_{si}\right]$$

$$E_L(v_A) = \frac{1}{(n-1)}\left[\frac{1}{n}\sum \frac{\hat{Y}_i^2}{p_i^2} f_{si} + \frac{1}{n}\sum \frac{\sigma_i^2}{p_i^2} f_{si} - E_L(e_A^2)\right.$$
$$\left. + \frac{n-1}{n}\sum \frac{\sigma_i^2}{p_i^2} f_{si}\right]$$

$$E(v_A) = \frac{1}{(n-1)}\left[\sum \frac{\hat{Y}_i^2}{P_i} + \sum \frac{\sigma_i^2}{P_i} - V(e_A) - Y^2\right.$$
$$\left. + (n-1)\sum \sigma_i^2\right] = V(e_A)$$

Thus, this estimator of variance is not free of $\hat{\sigma}_i^2$ and, interestingly, the estimator e_A is less efficient than e_H. So, if repeated subsampling is feasible, then DES RAJ's (1968) procedure is better than this alternative. However, if repeated subsampling is to be eschewed from practical considerations, this alternative may be tried in case, again from practical considerations, it is considered desirable to choose a sample of fsus by PPSWR method.

8.1.3 Subsampling of Second-Stage Units to Simplify Variance Estimation

CHAUDHURI and ARNAB (1982) have shown that if the fsus are chosen according to any sampling scheme without replacement, or they are selected with replacement but an estimator is based on the distinct units that are each subsampled only once, then for any homogeneous linear function of estimated fsu totals used to estimate the population total, among all homogeneous quadratic functions of estimated fsu totals there does not exist one that is unbiased for the variance of the estimated population total. For the existence of an unbiased variance estimator one needs necessarily an unbiased estimator for the variance of the estimated fsu total for such strategies as noted above.

SRINATH and HIDIROGLOU (1980) contrived the following device to bypass the requirement of estimating $V_L(T_i)$. They consider choosing the fsus by SRSWOR scheme, choosing from each sampled fsu i in the sample s again an SRSWOR s_i, in independent manners cluster-wise of size m_i from M_i ssus in it, and using

$$e = \frac{N}{n} \sum_{i \in s} M_i \overline{y}_i$$

as an estimator for Y. Here $\overline{y}_i$ is the mean of the y values of the ssus in s_i for $i \in s$. Then they recommend taking a subsample $s_i^{'}$ of size $m_i^{'}$ out of s_i again by SRSWOR method, getting $\overline{y}_i^{'}$ as the mean of y based on the ssus in $s_i^{'}$. They show that an unbiased estimator for $V(e)$ is available exclusively in terms of $\overline{y}_i^{'}$ for $i \in s$ although not in terms of $\overline{y}_i$ as, ideally, one would like to have.

ARNAB (1988) argues that restriction to SRSWOR is neither necessary nor desirable and discarding the ssus in s_i or $s_i^{'}$ is neither desirable nor necessary, and gives further generalizations of this basic idea of SRINATH and HIDIROGLOU (1980). Following DES RAJ's (1968) general strategy, he suggests starting with the estimator

$$e_D = \sum_s b_{si} I_{si} T_i$$

with

$$V(e_D) = \sum Y_i^2(\alpha_i - 1) + \sum_{i \neq j} \sum Y_i Y_j(\alpha_{ij} - 1) + \sum \alpha_i \sigma_i^2$$

$$V_L(T_i) = \sigma_i^2$$

Let s_i be a sample of ssus chosen from the ith fsu chosen in the sample s selected such that ψ_i, based on $s_i^{'}$, is an unbiased estimator of Y_i, that is, $E_L(\psi_i) = Y_i$ with $V_L(\psi_i) = \phi_i^2$ so chosen that $(\alpha_i - 1)\phi_i^2 = \alpha_i \sigma_i^2$. He shows that the variance of

$$e_{AR} = \sum_s I_{si} T_i / \pi_i$$

then is unbiasedly estimated by

$$v_{AR} = \sum_s d_{si} T_i^2 + \sum_{i \neq j \in s} \sum d_{sij} \Psi_i \Psi_j$$

where

$$d_{si} = \frac{\alpha_i - 1}{\pi_i}, \alpha_i = \frac{1}{\pi_i}, d_{sij} = \frac{\alpha_{ij} - 1}{\pi_{ij}}.$$

He illustrates various schemes for which this approach is successful and also explains how a weighted combination based on a number of disjoint and exhaustive subsamples $s_i^{'}$ of s_i may also be derived for the same purpose, thereby avoiding loss of data available from the entire sample by discarding ssus in s_i or $s_i^{'}$.

8.1.4 Estimation of $\overline{Y}$

We have so far restricted ourselves to only unbiased estimators of Y. But suppose we want to estimate

$$\overline{Y} = \sum_1^N Y_i \Big/ \sum_1^N M_i$$

where $\sum_1^N M_i$ may also be unknown like $Y = \sum_1^N Y_i$ and we may know or ascertain only the values of M_i for the clusters actually selected. In that case, an unbiased estimator is unlikely to be available for $\overline{Y}$. Rather, a biased ratio estimator $t_R = \Sigma_s Y_i / \Sigma_s M_i$ may be based on an SRSWOR s of selected clusters if Y_i's are ascertainable. If not, one may employ

$$e_R = \frac{\sum_s T_i}{\sum_s M_i},$$

a biased estimator for $\overline{Y}$, using T_i's as unbiased estimators for Y_i based on samples taken at later stages of sampling from the fsu i such that $E_L(T_i) = Y_i$ with $V_L(T_i)$ equal to V_{si} or σ_i^2 admitting respectively unbiased estimators $\hat{V}_{si}$ or $\hat{\sigma}_i^2$ such that $E_L(\hat{V}_{si}) = V_{si}$ or $E_L(\hat{\sigma}_i^2) = \sigma_i^2$.

In general, following RAO and VIJAYAN (1977) and RAO (1979), let us start with

$$t = \sum_s b_{si} I_{si} Y_i$$

not necessarily unbiased for Y such that

$$M = E_p(t - Y)^2 = \sum \sum Y_i Y_j d_{ij}$$

with

$$E_p(b_{si}I_{si} - 1)(b_{sj}I_{sj} - 1) = d_{ij}.$$

Let us assume that there exist $W_i \neq 0$ such that if $Z_i = Y_i/W_i = c$ (a non-zero constant) for all i, then M equals zero. In that case, from chapter 2 we know that we may write

$$\begin{aligned} M &= -\sum\sum_{i<j} d_{ij} W_i W_j (Z_i - Z_j)^2 \\ &= -\sum\sum_{i<j} d_{ij} W_i W_j \left(\frac{Y_i}{W_i} - \frac{Y_j}{W_j}\right)^2. \end{aligned}$$

Assuming that we may find out d_{sij} such that

$$E_p(d_{sij}I_{sij}) = d_{ij},$$

then

$$m = -\sum\sum_{i<j} d_{sij} I_{sij} W_i W_j \left(\frac{Y_i}{W_i} - \frac{Y_j}{W_j}\right)^2$$

is unbiased for M, that is, $E_p(m) = M$.

Now, supposing Y_i's are unascertainable, we replace Y_i by T_i with $E_L T_i = Y_i$ so as to use $e = \Sigma b_{si} I_{si} T_i$ to estimate Y. Then

$$\begin{aligned} E_p E_L(e - Y)^2 &= E_p E_L[(e - t) + (t - Y)]^2 \\ &= E_p E_L\left[\sum_i b_{si} I_{si}(T_i - Y_i) + \sum_i Y_i(b_{si} I_{si} - 1)\right]^2 \\ &= E_p\left[\sum b_{si}^2 I_{si} \sigma_i^2\right] + M \\ &= \sum \sigma_i^2 E_p(b_{si}^2 I_{si}) - \sum\sum_{i<j} d_{ij} W_i W_j \left(\frac{Y_i}{W_i} - \frac{Y_j}{W_j}\right)^2 \\ &= \sum \alpha_i \sigma_i^2 - \sum\sum_{i<j} d_{ij} W_i W_j \left(\frac{Y_i}{W_i} - \frac{Y_j}{W_j}\right)^2. \end{aligned}$$

An unbiased estimator for $E_p E_L(e - Y)^2$ is then

$$\sum b_{si}^2 I_{si} \hat{\sigma}_i^2 - \sum\sum_{i<j} d_{sij} I_{sij} \left(\frac{T_i}{W_i} - \frac{T_j}{W_j}\right)^2 + \sum\sum_{i<j} d_{sij} I_{sij} \left(\frac{\hat{\sigma}_i^2}{W_i^2} + \frac{\hat{\sigma}_i^2}{W_j^2}\right).$$

If σ_i^2 is not applicable, but V_{si} must be used, then

$$E_p E_L(e - Y)^2 = E_p \sum b_{si}^2 V_{si} I_{si} - \sum\sum_{i<j} d_{ij} W_i W_j \left(\frac{Y_i}{W_i} - \frac{Y_j}{W_j}\right)^2$$

and an unbiased estimator for this is

$$\sum b_{si}^2 \hat{V}_{si} I_{si} - \sum\sum_{i<j} d_{sij} I_{sij} W_i W_j \left(\frac{T_i}{W_i} - \frac{T_j}{W_j}\right)^2 + \sum\sum_{i<j} d_{sij} I_{sij} \left(\frac{\hat{V}_{si}}{W_i^2} + \frac{\hat{V}_{sj}}{W_j^2}\right).$$

Finally, in order to estimate $\overline{Y} = \Sigma_1^N Y_i / \Sigma_1^N M_i$ when Y_i is not ascertainable and M_i is unknown for $i \notin s$ we may proceed as follows:

Take for an SRSWOR s of fsus

$$\bar{e} = \sum_s T_i / \sum_s M_i$$

$$E_p \left[\frac{\sum_s V_{si}}{\left(\sum_s M_i\right)^2}\right] + \frac{N^2(1-f)}{\left(\sum_1^N M_i\right)^2} \frac{1}{n} \frac{1}{(N-1)} \sum_1^N \left[Y_i - \frac{\sum_1^N Y_i}{\sum_1^N M_i} M_i\right]^2$$

and this may be reasonably estimated by

$$\sum_s \hat{V}_{si}/(\sum_s M_i)^2$$
$$+\frac{(1-f)}{(\sum_s M_i)^2}\frac{n}{(n-1)}\left[\sum_s\left(T_i-\frac{\sum_s T_i}{\sum M_i}M_i\right)^2\right.$$
$$\left.-\sum_s \hat{V}_{si}-\frac{\sum_s \hat{V}_{si}\sum_s M_i^2}{(\sum_s M_i)^2}+2\frac{\sum_s M_i\hat{V}_{si}}{\sum_s M_i}\right]$$

neglecting the error in replacing $\Sigma_1^N M_i$ throughout by its unbiased estimator

$$\frac{N}{n}\sum_s M_i.$$

For further discussion on multistage sampling, one may consult RAO (1988) and BELLHOUSE (1985).

8.2 DOUBLE SAMPLING WITH EQUAL AND VARYING PROBABILITIES: DESIGN-UNBIASED AND REGRESSION ESTIMATORS

Assume that positive size measures W_i with a total (mean) $W\,(\overline{W})$ are available for the units of a finite population $U = (1,\ldots,i,\ldots,N)$. Suppose that it is difficult and expensive to measure the values Y_i of the variable y of interest and that it is less expensive to ascertain the values X_i of an auxiliary variable x. Then it seems to be reasonable to take an initial sample s_1, of large size n_1, with a probability $p_1(s_1)$ according to a design p_1 that may depend on $\underline{W} = (W_1,\ldots,W_N)$ and to observe the values X_i for $i \in s_1$. Supposing that y is correlated with not only x but also with w for which the values are W_i, $i = 1,\ldots N$, one may now take a subsample s_1 of size n_2 ($< n_1$, possibly $n_2 << n_1$) with a conditional probability $p_2(s_2/s_1)$ from s_1. This conditional probability sampling design $p_2(./.)$ may utilize the values W_j and also X_j for $j \in s_1$. The overall sample may be denoted as $\overline{s} = (s_1, s_2) = [(i,j)|i \in s_i, j \in s_2]$ and the overall sampling design as p such that

$$p(\bar{s}) = p_1(s_1)\, p_2(s_2/s_1).$$

The ascertained survey data may be denoted as $d = [(i, j, X_i, Y_j)|i \in s_1, j \in s_2]$. This procedure is called **two-phase** or **double sampling** in the literature.

For the time being, we suppose that p_2 does not involve $\underline{X} = (X_1, \ldots, X_i, \ldots, X_N)'$ but may involve only $\underline{W} = (W_1, \ldots, W_i, \ldots, W_N)'$. In order to estimate $\overline{Y}$, RAO and BELLHOUSE (1978) considered the following class of nonhomogeneous linear estimators

$$t_b = b_{\bar{s}} + \sum_{j \in s_2} b_{\bar{s}j} Y_j + \sum_{i \in s_1} b_{\bar{s}i} X_i.$$

They assumed that X_j are ascertainable free of observational errors, but the Y_j's are observable as $\hat{Y}_j$'s with unknown random errors $(\hat{Y}_j - Y_j)$'s.

In the following, we specialize their model assuming error-free observation of the y values. Writing

$$R_j = \frac{Y_j}{W_j}, \overline{R} = \frac{1}{N}\sum_1^N R_j, T_j = \frac{X_j}{W_j}, \overline{T} = \frac{1}{N}\sum_1^N T_j,$$

they postulated a model:

$$\frac{Y_j}{W_j} = \overline{R} + \epsilon_j, \frac{X_j}{W_j} = \overline{T} + \epsilon_j^{'}, E_m(\overline{R}) = \overline{R}, E_m(\overline{T}) = \overline{T},$$

$$E_m(\epsilon_j) = E_m(\epsilon_j^{'}) = 0$$

$$E_m(\epsilon_j^2) = \delta_1 (> 0), E_m(\epsilon_j \epsilon_j^{'}) = \gamma_1, E_m\left(\epsilon_j^{'2}\right) = \eta_1 > 0$$

$$E_m(\epsilon_j \epsilon_k) = \delta_2 (j \neq k), E_m(\epsilon_j \epsilon_k^{'}) = \gamma_2, E_m(\epsilon_j^{'} \epsilon_k^{'}) = \eta_2 (j \neq k),$$

where E_m is the operator for expectation with respect to the joint probability distribution of the vectors $\underline{R} = (R_1, \ldots, R_N)'$ and $\underline{T} = (T_1, \ldots, T_N)'$. From the above, it is apparent that the pairs of random variables (R_j, T_j) have a joint exchangeable distribution. For example, this exchangeable distribution may be a permutation distribution that regards a particular realization $[(R_{i_1}, T_{i_1}), \ldots, (R_{i_N}, T_{i_N})]^{'}$ for a permutation $(i_1, \ldots, i_N)$ of $(1, \ldots, N)$ as one of the $N!$ possible vectors $[(R_{j_1}, T_{j_1}), \ldots, (R_{j_N}, T_{j_N})]^{'}$ chosen with a common probability $1/N!$, there being $N!$ such vectors corresponding to as many permutations $(j_1, \ldots, j_N)$ of the fixed vector $(1, \ldots, N)$. Such an assumption of a permutation model, or, more generally, an exchangeable model as postulated above, presuppose that the R_j's and T_j's are unrelated to the W_j's and especially that the labels $1, \ldots, N$ bear no information on $\underline{R}$ and $\underline{T}$. For permutation models, important references are KEMPTHORNE (1969), C. R. RAO (1971), THOMPSON (1971) and T. J. RAO (1984).

Under this model, they show that among all estimators of the form t_b above, subject to the model-design unbiasedness restriction $E_m E_p(t_b - \overline{Y}) = 0$,

$$t_b^* = \overline{W}\left[\frac{1}{n_2}\sum_{s_2}\frac{Y_i}{W_i} + \beta\left(\frac{1}{n_1}\sum_{s_1}\frac{X_i}{W_i} - \frac{1}{n_2}\sum_{s_2}\frac{X_i}{W_i}\right)\right],$$

where $\beta = \frac{\gamma_1 - \gamma_2}{\eta_1 - \eta_2}$ minimizes $E_m E_p(t_b - \overline{Y})^2$.

If the estimator t_b is restricted to be design-unbiased for $\overline{Y}$, then they show that the optimal strategy among (p, t_b) is $(p*, t_b*)$ where $p*$ is a double sampling design for which $\pi_{1i} = n_1 W_i / W$ and $\pi_{2i} = n_2/n_1$, $i = 1, \ldots, N$. Here by $\pi_{1i}(\pi_{2i})$ we mean the inclusion probability of a unit according to first-phase sampling design p_1 and second-phase conditional inclusion probability according to second-phase sampling design p_2 discussed above.

A shortcoming of t_b^* is that it contains an unknown parameter β and hence is not practicable as such. In practice one may employ the double sample regression estimator obtained by replacing β by $\hat{\beta}$ where

$$\hat{\beta} = \frac{\hat{\gamma}_1 - \hat{\gamma}_2}{\hat{\eta}_1 - \hat{\eta}_2}$$

where by $\hat{\gamma}_1$, $\hat{\gamma}_2$, $\hat{\eta}_1$ and $\hat{\eta}_2$ we mean sample-based estimators of the quantities of the form $E_p(u_j - E_p u_j)(v_k - E_p v_k)$ where u_j, v_k stand for $\frac{Y_j}{W_j}$, $\frac{X_k}{W_k}$, etc., taken in obvious manners. But the consequence of this replacement on t_b^* in respect of bias and efficiency is neither known nor studied.

Considering the same class of fixed-sample-size two-phase sampling designs p, as above, CHAUDHURI and ADHIKARI (1983, 1985) proposed the estimator for Y based on data d as

$\bar{t}_b = \sum_{s_1} \frac{X_j}{\pi_{ij}} + \sum_{s_2} \frac{(Y_j - X_j)}{\pi_{2j}}$, which is an extension of the Horvitz-Thompson (1952) method to the two-phase sampling. This estimator is free from unknown parameters, but its scope is limited because it does not include anything like the regression coefficient of y on x or on w or of y/w on x/w, etc. But following GODAMBE and JOSHI (1965), they proved many desirable and optimal properties of $\bar{t}_b$ and also proved optimality properties of the subclass of strategies $(\bar{p}, \bar{t}_b)$ with $\bar{p}$ as the class of two-phase sampling designs for which $\pi_{1i} = n_1 W_i / W$ and $\pi_{2i} = n_2 W_i / W$, $i = 1, \ldots, N$. Details may be found in CHAUDHURI and VOS (1988) and CHAUDHURI (1988), among others.

MUKERJEE and CHAUDHURI (1990) extended the design p to allow p_2 to involve X_i for $i \in s_1$ and proposed the regression estimator for Y as

$$t_r = \sum_{s_2} \frac{Y_i}{\pi_{1i}\pi_{2i}} - \hat{\beta}_1 \left[\sum_{s_2} \frac{X_i}{\pi_{1i}\pi_{2i}} \left\{ \sum_{s_1} \frac{X_i}{\pi_{1i}} - \hat{\beta}_3 \left(\sum_{s_1} \frac{W_i}{\pi_{1i}} - W \right) \right\} \right]$$
$$- \hat{\beta}_2 \left(\sum_{s_2} \frac{W_i}{\pi_{1i}\pi_{2i}} - W \right)$$

motivated by consideration of the model for which they postulate the following:

$$E_m(Y_i(X_i) = \beta_1 X_i + \beta_2 W_i, E_m(X_i) = \beta_3 W_i, i = 1, 2, \ldots$$

Another motivation to hit upon this regression form is the following: if X_i were known for every i in U, then one might employ the regression estimator

$$t_r^{'} = \sum_{s_2} \frac{Y_i}{\pi_{1i}\pi_{2i}} - \hat{\beta}_1 \left(\sum_{s_2} \frac{X_i}{\pi_{1i}\pi_{2i}} - X \right) - \hat{\beta}_2 \left(\sum_{s_2} \frac{W_i}{\pi_{1i}\pi_{2i}} - W \right)$$

noting that the unknown X in $t_r^{'}$ is just replaced in t_r by the sample-based quantity

$$\sum_{s_1} \frac{X_i}{\pi_{1i}} - \hat{\beta}_3 \left(\sum_{s_1} \frac{W_i}{\pi_{1i}} - W \right).$$

Here $\hat{\beta}_j, j = 1, 2, 3$ are suitable estimators for β_j, $j = 1, 2, 3$, respectively.

In order to find appropriate $\hat{\beta}_j$'s, choose appropriate classes of designs, and establish desirable properties for the resulting strategies involving t_r as the estimator for Y, they considered asymptotic design unbiasedness (ADU), asymptotic design consistency (ADC), and derived lower bounds for plim $E_m E_p(t_r - Y)^2$ following the approach of ROBINSON and SÄRNDAL (1983) who made a similar investigation to derive asymtotically desirable properties of regression estimators in case of single-phase sampling. The details are too technical and hence are omitted here, inviting the interested readers to see the original sources cited above.

8.3 SAMPLING ON SUCCESSIVE OCCASIONS WITH VARYING PROBALITIES

Suppose a finite population $U = (1, \ldots, N)$ is required to be surveyed to estimate the total or mean a number of times over which its composition remains intact. But a variable of interest should be supposed to undergo changes, though the values on close intervals apart should be highly correlated, the degree of correlation decreasing with time. For two occasions called, respectively, (1) the previous and (2) the current occasions, let us denote the values as X_i and Y_i $(i = 1, \ldots, N)$, regarding them, respectively, as values of a variable x denoting the previous and a variable y denoting the current values. Suppose on the first occasion a sample s_1 is chosen from U adopting a design p_1 with a fixed size n_1 for which the values X_i, $i \in s_1$, are ascertained. On the current occasion

(a) a subsample s_2 of size $n_2 (< n_1)$ is drawn from s_1 following a design p_2, and
(b) a subsample s_3 of size $n_3 (< N - n_1)$ is drawn from $U - s_1$ adopting a design p_3.

The designs p_2 and p_3 are both conditional probability sampling designs. In employing p_1, p_2, p_3, the known values $W_i (i = 1, \ldots, N)$ of some variable w correlated with x and y may be utilized, and, in case of p_2, the realized values X_i, $i \in s_1$ may further be utilized. We will refer to the overall design thus employed as p for which the total sample size is $n_1 + n_2 + n_3 = n$. The main interest here is to estimate $Y = \Sigma_1^N Y_i$ or $\overline{Y} = \Sigma_1^N Y_i/N$, but the problem is to exploit the information gathered on X_i, $i \in s_1$ and the association between x and y that may be assessed through the data on X_i, Y_i for $i \in s_2$. The overall data at hand may be summarized by the notation $d = [(i, j, X_i, Y_j)|i \in s_1, j \in s_2 \cup s_3]$ and the overall sample of size n by $s = (s_1, s_2, s_3)$. The main difference between the situation here and in double sampling is that here, in addition to the subsample s_2 (of s_1), which in this case is called the **matched subsample**, there is an additional **unmatched subsample** s_3 of $U - s_1$. RAO and BELLHOUSE (1978) postulated the same model connecting $\underline{X} = (X_1, \ldots, X_N)'$, $\underline{Y} = (Y_1, \ldots, Y_N)'$ and $\underline{W} = (W_1, \ldots, W_N)'$ as stated in section 8.2 and considered estimators of the term

$$t_{Rb} = b_s + \sum_{s_2} b_{sj} Y_j + \sum_{s_3} b_{\acute{s}j} Y_j + \sum_{s_2} b_{\acute{s}\acute{j}} X_j + \sum_{s_1 - s_2} b_{\acute{s}\acute{j}'} X_j$$

required to satisfy $E_m E_p(t_{Rb}) = \overline{R}\,\overline{W} = \mu$. They showed that an optimal estimator in this class is t_{Rb}^* for which

$$E_m E_p(t_{Rb} - \mu)^2 > E_m E_p(t_{Rb}^* - \mu)^2$$

and t_{Rb}^* is given by

$$t_{Rb}^* = \overline{W}\left[\psi t + (1 - \psi) t_1\right]$$

where

$$t = \frac{1}{n_2}\left(\sum_{s_2} \frac{Y_j}{W_j}\right) + \beta\left[\frac{1}{n_1}\left(\sum_{s_1} \frac{X_j}{W_j}\right) - \frac{1}{n_2}\left(\sum_{s_2} \frac{X_j}{W_j}\right)\right]$$

$$t_1 = \frac{1}{n_3}\left(\sum_{s_3} \frac{Y_j}{W_j}\right), \quad \beta = \frac{\gamma_1 - \gamma_2}{\eta_1 - \eta_2}, \quad \beta' = \frac{\gamma_1 - \gamma_2}{\delta_1 - \delta_2},$$

$$\xi^2 = \beta\beta', \quad \phi = 1 - \frac{n_2}{n_1}, \quad \psi = \frac{1 - \phi}{1 - \phi\xi^2}.$$

Requiring the class of estimators t_{Rb} above to be design-unbiased for $\overline{Y}$ and denoting by p^* the subclass of the above designs for which p_1, p_2, p_3 are restricted to have respective inclusion probabilities,

$$\pi_{1i} = \frac{n_1 W_i}{W}, \; i \in U,$$

$$\pi_{2i} = \frac{n_2}{n_1}, \; i \in s_1,$$

$$\pi_{3i} = \frac{n_3 W_i}{\sum_{U-s_1} W_i} \quad \text{for} \quad i \in U - s_1,$$

CHAUDHURI (1985) showed that

$$E_m E_p (t_{Rb} - \overline{Y})^2 > E_m E_{p*} (t^*_{Rb} - \overline{Y})^2.$$

He also showed how to implement sample selection so as to realize $p*$ by adapting FELLEGI's (1963) scheme of sampling.

GHOSH and LAHIRI (1987) have mentioned how their empirical Bayes estimators (EBE) can be used in the context of sampling on successive occasions. Their EBE procedure has been described by us briefly in section 4.2. But in actual large-scale surveys, this procedure is not yet known to have been put into practice, though we feel that projects deserve to be undertaken toward applications of EBE in this repetitive sampling context.

Numerous strategies for sampling on successive occasions are discussed in COCHRAN's (1977) standard text; CHAUDHURI and VOS (1988) have reviewed many more. They point out many amendments to our above designs p. For example, they differentiate between designs for which s_3 is to be subsampled from U itself, from $U - s_1$, or from $U - s_2$, and discuss corresponding advantages and disadvantages. They refer to various combinations of known sampling schemes to be adopted to realize p_1, p_2, and p_3, present various classes of estimators for Y or $\overline{Y}$, and refer to resulting consequences. An interested reader may be persuaded to look at the original references cited in CHAUDHURI and VOS (1988).

Chapter 9

Resampling and Variance Estimation in Complex Surveys

By a complex survey, we mean one in which any scheme of sampling other than simple random sampling (SRS) with replacement (WR) or without replacement (WOR) is employed; a common name for these two SRS schemes will be adopted as **epsem**, that is, **equal probability selection methods**. Estimating population totals or means involves weighting the sample observations using design parameters. Estimators for totals and means that are of practical uses are linear in observations on the values of the variables of interest. For such linear functions of single variables, variances or mean square errors (MSE) are quadratic forms, and suitable sample-based estimators for them are easily found, as we have discussed and illustrated in the preceding chapters. But the problem no longer remains so simple if we intend to estimate nonlinear functions of totals or means of more than one variable. In such cases, estimators that are linear functions of observations on more than one variable are not usually available, but nonlinear functions become indispensable. Their variances or MSEs, however, are difficult to express in simple exact forms, and estimators thereof with desirable properties and simple cosmetic forms are not easy to work out. To get over these situations, alternative techniques are needed, and the following sections give a brief account of them.

9.1 LINEARIZATION

Let us suppose that $\theta_1, \ldots, \theta_K$ are K population parameters and $f = f(\theta_1, \ldots, \theta_K)$ is a parametric function we intend to estimate. Let $t_1, \ldots, t_K$ be respective linear estimators based on a common sample s of size n, for $\theta_1, \ldots, \theta_K$. We assume that $f(t_1, \ldots, t_K)$ can be expanded in a TAYLOR series and well-approximated for large n by the linear function in t_i, $i = 1, 2, \ldots, K$:

$$f(\theta_1, \ldots, \theta_K) + \sum_1^K \lambda_i (t_i - \theta_i)$$

where

$$\lambda_i = \frac{\partial}{\partial t_i}(t_1, \ldots, t_K)]_{\underline{t}=\underline{\theta}}, i = 1, \ldots, k$$
$$\underline{t} = (t_1, \ldots, t_K), \underline{\theta} = (\theta_1, \ldots, \theta_K),$$

and of course we assume that n is large. Since θ_i's and λ_i's are constants, we approximate the variance of $f(\underline{t})$ by the variance of

$$\sum_1^K \lambda_i t_i$$

that is, we take

$$V[f(\underline{t})] = V\left[\sum_1^K \lambda_j t_j\right].$$

Let θ_j for $j = 1, \ldots, K$ denote the finite population total for a certain real variable ξ_j, $j = 1, \ldots, K$, that is, $\theta_j = \sum_1^N \xi_{ji}$,

$j = 1, \ldots, K$ and t_j's be of the form

$$t_j = \sum_{i \in s} b_{si} \xi_{ji}, (j = 1, \ldots, K)$$

using b_{si} as sample-based weights for the values $\xi_{ji}, i = 1, \ldots, N$ of the ξ_j's for a finite population $U = (1, \ldots, N)$ of size N.

So, we may write

$$V\left[f\left(\underset{\sim}{t}\right)\right] \simeq V\left[\sum_{i \in s}\left(\sum_{j=1}^{K} \lambda_j b_{si} \xi_{ji}\right)\right] = V\left[\sum_{i \in s} b_{si} \phi_i\right]$$

where

$$\phi_i = \sum_{j=1}^{K} \lambda_j \xi_{ji}.$$

This ϕ_i, which is obtained by aggregating over all the K variables, may be described as a **synthetic variable**. Now,

$$\sum_{i \in S} b_{si} \phi_i$$

is a linear function, and so, applying usual methods of finding variances or approximate variances of linear functions, one may proceed to work out formulae for exact or approximate unbiased estimators for

$$V\left[\sum_{i \in S} b_{si} \phi_i\right]$$

and treat them as approximately unbiased estimators of variances or MSEs of the original estimator $f\left(\underset{\sim}{t}\right)$.

The only conditions for applicability of this procedure are (a) large sample size n and (b) conformability of f to its Taylor expansion. A detailed exposition of this topic is given by RAO (1975b).

Let us illustrate an application of this procedure. This form of the procedure is due to WOODRUFF (1971). Suppose $K = 2, \xi_1 = y, \theta_1 = Y = \sum_1^N Y_i, \xi_2 = x, \theta_2 = X = \sum_1^N X_i$,

$$f(\theta_1, \theta_2) = \frac{\theta_1}{\theta_2} = \frac{Y}{X} = R.$$

Let an SRSWOR of size n be taken, yielding

$$t_1 = \frac{N}{n}\sum_s Y_i,\ t_2 = \frac{N}{n}\sum_s X_i,$$

$$f(t_1, t_2) = \frac{\sum_s Y_i}{\sum_s X_i} = \frac{\overline{y}_s}{\overline{x}_s}$$

$$\lambda_1 = (1/\overline{X}), \lambda_2 = (-\overline{Y}/\overline{X}^2) = -R/\overline{X}.$$

Then,

$$\begin{aligned} V\left(\frac{\overline{y}}{\overline{x}}\right) &\simeq V\left[\frac{N}{n}\sum_s\left(\frac{1}{\overline{X}}Y_i - \frac{R}{\overline{X}}X_i\right)\right] \\ &= \frac{N^2}{n^2}\left(\frac{1}{\overline{X}}\right)^2 V\left[\sum_s (Y_i - RX_i)\right] \\ &= \frac{N^2}{\overline{X}^2}\frac{1-f}{n}\frac{1}{N-1}\sum_1^N (Y_i - RX_i)^2 \end{aligned}$$

and this has the usual estimator

$$\frac{N^2}{\overline{x}^2}\frac{(1-f)}{n}\frac{1}{(n-1)}\sum_s (Y_i - \widehat{R}X_i)^2$$

where $\widehat{R} = \overline{y}/\overline{x}$.

As another example let us consider $K = 6, \xi_1 = 1, \xi_2 = y, \xi_3 = x, \xi_4 = y^2, \xi_5 = x^2$ and $\xi_6 = xy$. Let $\theta_1 = \sum_1^N \xi_{1i} = N, \theta_2 = \sum_1^N Y_i, \theta_3 = \sum X_i, \theta_4 = \sum_1^N Y_i^2, \theta_5 = \sum_1^N X_i^2, \theta_6 = \sum_1^N X_i Y_i$ and

$$f(\theta_1, \ldots, \theta_6) = \frac{\theta_1\theta_6 - \theta_2\theta_3}{\left[(\theta_1\theta_4 - \theta_2^2)(\theta_1\theta_5 - \theta_3^2)\right]^{1/2}}$$

which is obviously the finite population correlation coefficient

$$\rho_N = \frac{N\sum X_i Y_i - (\sum Y_i)(\sum X_i)}{\left[N\sum Y_i^2 - (\sum Y_i)^2\right]^{1/2}\left[N\sum X_i^2 - (\sum X_i)^2\right]^{1/2}}.$$

Let p be any sampling design with $\pi_i > 0$

$$t_j = \sum_s \frac{\xi_{ji}}{\pi_i}, \quad \text{for} \quad j = 1, \ldots, 6.$$

Then, $f(t_1, \ldots, t_6)$ takes the form, say,

$$\widehat{\rho}_s = \frac{\left(\sum_s \frac{1}{\pi_i}\right)\left(\sum_s \frac{Y_i X_i}{\pi_i}\right) - \left(\sum_s \frac{Y_i}{\pi_i}\right)\left(\sum_s \frac{X_i}{\pi_i}\right)}{\left[\left(\sum_s \frac{1}{\pi_i}\right)\left(\sum_s \frac{Y_i^2}{\pi_i}\right) - \left(\sum_s \frac{Y_i}{\pi_i}\right)^2\right]^{1/2}\left[\left(\sum_s \frac{1}{\pi_i}\right)\left(\sum_s \frac{X_i^2}{\pi_i}\right) - \left(\sum_s \frac{X_i^2}{\pi_i}\right)^2\right]^{1/2}}.$$

Here $b_{si} = \frac{1}{\pi_i}$, for every $j = 1, \ldots, 6$ and every $s \ni i$.

$$\lambda_j = \frac{\partial}{\partial t_j} f(t_1, \ldots, t_6)|_{\underline{t}=\underline{\theta}} = \psi_j(\underline{\theta})$$

is not difficult to work out. So, $\sum_{i \in S} \phi_i$ takes the form

$$\sum_{i \in s} \left\{ \sum_{j=1}^{6} \psi_j(\underline{\theta}) \xi_{ji} \right\} / \pi_i = \sum_s \frac{Z_i}{\pi_i}, \text{ say,}$$

which has the HORVITZ–THOMPSON (1952) estimator form. This immediately yields a known variance form and well-known estimators.

To consider another example, let us turn to HÁJEK's (1971) estimator

$$t_H = \frac{\sum_s Y_i / \pi_i}{\sum_s 1/\pi_i}$$

of the population mean $\overline{Y}$ based on an arbitrary design with $\pi_i > 0$, $i = 1, \ldots, N$. Then, let $\xi_1 = 1$, $\sum \xi_{1i} = N = \theta_1$, $\xi_2 = y$, $\sum \xi_{2i} = Y = \theta_2$,

$$f(\theta_1, \theta_2) = \frac{\theta_2}{\theta_1},$$

$$t_1 = \sum_s 1/\pi_i, \; t_2 = \sum_s Y_i/\pi_i.$$

Then the variance of

$$f(t_1, t_2) = \frac{\sum_s Y_i/\pi_i}{\sum_s 1/\pi_i}$$

is approximately equal to

$$V\left[\sum_s \left(\frac{\lambda_1 + \lambda_2 Y_i}{\pi_i}\right)\right] = \frac{1}{N^2} V \sum_s \left(\frac{Y_i - \overline{Y}}{\pi_i}\right).$$

9.2 JACKKNIFE

Let θ be a parameter required to be estimated from a sample s of size n and $t = t(n)$ be an estimator for θ based on s. Let t be a biased estimator of θ with a bias $B(t) = B_n(\theta) = E(t(n) - \theta)$ expressible in the form

$$B_n(\theta) = \frac{b_1(\theta)}{n} + \frac{b_2(\theta)}{n^2} + \frac{b_3(\theta)}{n^3} + \dots$$

where $b_j(\theta), j = 1, 2, \dots$ are unknown functions of θ and $b_1(\theta) \neq 0$. Then, in the following way, we can derive another estimator for θ with a bias of order $1/n^2$, that is, of the form

$$\frac{b_2'(\theta)}{n^2} + \frac{b_3'(\theta)}{n^3} + \dots$$

Let the sample s be split up into $g(\geq 1)$ disjoint groups, each of a size $m(= \frac{n}{g})$. Let the groups be marked $1, \dots, g$ and the statistic t be now calculated on the basis of the values in s excluding those in the ith group. The new statistic may be denoted as $t_i = t_i(n - m)$ as it is based on $n - m$ units, omitting from s of size n the m units in the ith group. Let us now consider a new statistic

$$e_i = gt(n) - (g - 1)t_i(n - m)$$

called the ith **pseudo-value**. Then we have the expectation as

$$\begin{aligned}
E(e_i) &= gEt(n) - (g - 1)E(t_i(n - m)) \\
&= \left[\theta + \frac{b_1(\theta)}{n} + \frac{b_2(\theta)}{n^2} + \dots\right] \\
&\quad - (g - 1)\left[\theta + \frac{b_1(\theta)}{n - m} + \frac{b_2(\theta)}{(n - m)^2} + \dots\right] \\
&= \theta + b_1(\theta)\left(\frac{g}{n} - \frac{g - 1}{n - m}\right) \\
&\quad + b_2(\theta)\left\{\frac{g}{n^2} - \frac{g - 1}{(n - m)^2}\right\} + \dots \\
&= \theta - \frac{g}{g - 1}\frac{b_2(\theta)}{n^2} + \dots
\end{aligned}$$

Repeating this process we may derive g such pseudo-values e_i, $i = 1, \dots, g$, each with a bias of order $1/n^2$. Now using these

e_i's we may construct a new statistic, viz.,

$$t_J = \frac{1}{g}\sum_{i=1}^{g} e_i = gt(n) - \frac{g-1}{g}\sum_{i=1}^{g} t_i(n-m)$$
$$= gt(n) - (g-1)\bar{t}, \text{ say.}$$

Obviously, this new statistic t_J has also a bias of order $1/n^2$ as an estimator for θ. Starting with t_J and applying this technique, we may get another estimator with a bias of order $1/n^3$.

The statistic t_J is called a **jackknife** statistic. It was introduced by QUENOUILLE (1949) as a bias reduction technique (seen above). But later TUKEY (1958) started using the jackknife statistics in estimating mean square errors of biased estimators for parameters.

In order to estimate the mean square error (MSE) of the jackknife statistic

$$t_J = \frac{1}{g}\sum_{i=1}^{g} e_i$$

one may consider the estimator

$$v_J = \frac{1}{g(g-1)}\sum_{i=1}^{g}\left(e_i - \frac{1}{g}\sum_{1}^{g} e_i\right)^2$$
$$= \frac{1}{g(g-1)}\sum_{1}^{g}(e_i - t_J)^2$$
$$= \frac{(g-1)}{g}\sum_{1}^{g}(t_i - \bar{t})^2.$$

The **pivotal**

$$\frac{(t_J - \theta)}{\sqrt{v_J}},$$

for large n and moderate g is supposed to have approximately STUDENT's t distribution with $(g-1)$ degrees of freedom (df), and for very large g its distribution may be approximated by that of the standardized normal deviate τ. Then $t_J \pm t_{g-1,\alpha/2}\sqrt{v_J}$

or $t_J \pm \tau_{\alpha/2}\sqrt{v}_J$ is used to construct $100(1-\alpha)\%$ confidence intervals for θ for large n, writing $t_{g-1,\alpha/2}$ $(\tau_{\alpha/2})$ for the $100\alpha/2\%$ point in the right tail area of the distribution of STUDENT's statistic with $(g-1)$ df (standardized normal deviate τ).

9.3 INTERPENETRATING NETWORK OF SUBSAMPLING AND REPLICATED SAMPLING

MAHALANOBIS (1946) introduced the technique of **interpenetrating network of subsampling** (IPNS) (1) to improve the accuracy of data collection and (2) to throw interim measures of error in estimation even before the completion of the entire fieldwork in surveys and processing-cum-tabulation. The method consists in dividing a sample into two or more parts, entrusting each part to a separate batch of field workers. Since each part is supposed to provide an estimate of the same parameter, any awkward divergences among the estimates emerging from the various parts are likely to create suspicion about the quality of field work carried out by the various teams. This realization should induce vigilance on their functions, engendering higher qualities of work. Moreover, with the completion of each part, a separate estimate is produced, and with two or more parts of data at hand using the separate comparable estimates, a measure of error is available as soon as at least two estimates are obtained. DEMING (1956) applied essentially the same technique, but mainly with the intention of getting an easy and simple estimate of the variance of an estimator of any parameter, no matter how complicated the sampling scheme. He called this the method of **replicated sampling**, which is equivalent to IPNS. Let us see how it works.

Let K independent samples be selected from a given finite population each following the same scheme of sampling. Let each sample throw up an estimator that is unbiased for a parameter θ of interest relating to the population. Let $t_1, \ldots, t_i, \ldots, t_K$ be K such independent estimators for θ. Then, $E(t_i) = \theta$ for every $i = 1, \ldots, K$. Also each t_i has the same variance because each is based on a design that is identical in all respects.

Thus, $V(t_i) = V$, for every $i = 1, \ldots, K$. Then, for

$$\bar{t} = \frac{1}{K}\sum_{1}^{K} t_i$$

we have

$$E(\bar{t}) = \theta, \ V(\bar{t}) = \frac{1}{K^2}\sum V(t_i) = \frac{V}{K}.$$

It follows that

$$v = \frac{1}{K(K-1)}\sum_{1}^{K}(t_i - \bar{t})^2$$

is an unbiased estimator for $V(\bar{t})$.

In case $K = 2$, $V(\bar{t}) = \frac{V}{2}$ and

$$v = \frac{1}{2}\left[\left(t_1 - \frac{t_1 + t_2}{2}\right)^2 + \left(t_2 - \frac{t_1 + t_2}{2}\right)^2\right] = \frac{1}{4}(t_1 - t_2)^2$$

and $\frac{1}{2}|t_1 - t_2|$ is taken as a measure of the standard error of the estimator $\bar{t} = \frac{1}{2}(t_1 + t_2)$ for θ. For the case $K = 2$, the IPNS is called **half-sampling**.

If the samples are independently chosen, this procedure, of course, is useful in estimating any finite population parameter no matter how complicated, and also it is immaterial how complicated is the sampling scheme, provided an unbiased estimator is available. But in practice, for complicated parameters like population multiple correlation coefficient, ratio of two means based on stratified two-stage sampling, etc., unbiased estimators cannot be found. Moreover MAHALANOBIS's IPNS does not ensure independent sampling and hence the estimators t_i for θ are not independent but correlated. In IPNS a realized sample s of size n is usually split up at random into two or more groups usually of a common size. The manner of forming the groups required to turn out mutually exclusive results cannot but lead to estimates that are correlated. So, it is necessary to examine both the bias of an estimator $\bar{t} = \frac{1}{K}\sum_1^K t_i$ for θ when θ is a complex parameter for which t_i's are each biased estimators and also of

$$\frac{1}{K(K-1)}\sum_{1}^{K}(t_i - \bar{t})^2$$

as an estimator for the variance or the mean square error of $\bar{t}$ as an estimator for θ. WOLTER (1985) has made detailed investigation of IPNS and **random group methods** in tackling the advantages and shortcomings of this method of replication. These may really be called **pseudo-replication** or **sample re-use techniques** because here essentially we have a single sample from which an estimator t for a parameter might be obtained, but since it is difficult to estimate its variance, the sample is artificially split up into components leading to several estimators for the same parameter, and from the variations among these estimators a measure of error for an overall combined estimator is derived. There is a considerable literature on this topic, but WOLTER's (1985) text seems to provide an adequate coverage. KOOP (1967) demonstrated certain merits in dividing a sample into unequal rather than equal groups, ROY and SINGH (1973) showed advantages in forming the groups on taking the units from the chosen sample by SRS without replacement rather than with replacement. CHAUDHURI and ADHIKARI (1987) derive further results as followups to them.

9.4 BALANCED REPEATED REPLICATION

Suppose a finite population of N units is divided into L strata of $N_1, N_2, \ldots, N_L$ units, respectively. From each stratum let SRSWORs be independently selected, making n_h draws from the hth, $h = 1, \ldots, L$. Let L be sufficiently large and n_h be taken as 2 for each $h = 1, \ldots, L$. Let us write (y_{h1}, y_{h2}) as the vector of variable values on the variable of interest y observed for the sample from the hth stratum. Then, with $W_h = N_h/N$,

$$\frac{1}{N}\sum N_h\left(\frac{y_{h1}+y_{h2}}{2}\right) = \sum W_h\bar{y}_h = \bar{y}_{st}, \text{ say}$$

is taken as the usual unbiased estimator for $\overline{Y} = \sum W_h\overline{Y}_h$, the population mean. Neglecting $n_h/N_h = f_h$, that is, ignoring the finite population correction $1 - f_h$ for every h, we have the variance of $\bar{y}_{st}$ as

$$V(\bar{y}_{st}) = \sum W_h^2 S_h^2/2$$

where

$$S_h^2 = \frac{1}{N_h - 1}\sum_1^{N_h}(Y_{hi} - \overline{Y}_h)^2,$$

writing Y_{hi} as the value of ith unit of hth stratum and $\overline{Y}_h$ for their mean. This $V(\overline{y}_{st})$ is unbiasedly estimated by

$$v = \frac{1}{4}\sum W_h^2 d_h^2,$$

where $d_h = (y_{h1} - y_{h2})$. Let us now form two half-samples by taking into the first half-sample one of y_{h1} and y_{h2} for every $h = 1, \ldots, L$ leaving the other ones, which together, over $h = 1, \ldots, L$, form the second half-sample. We denote the first half-sample by I and the second by II. There are, in all, 2^L possible ways of forming these half-samples. For the jth ($j = 1, \ldots, 2^L$) such formation, let $\delta_{hj} = 1(0)$ if y_{h1} appears in I (II). Then,

$$t_{h1} = \sum W_h \left[\delta_{hj} y_{h1} + (1 - \delta_{hj}) y_{h2}\right]$$
$$t_{h2} = \sum W_h \left[(1 - \delta_{hj}) y_{h1} + \delta_{hj} y_{h2}\right]$$

form two unbiased estimators of $\overline{Y}$ based respectively on I and II. Then, $\overline{t}_j = \frac{1}{2}(t_{j1} + t_{j2}) = \sum W_h \overline{y}_h$ for every $j = 1, \ldots, 2^L$. Also

$$v_j = (t_{j1} - t_{j2})^2/4$$

may be taken as an estimator for

$$V(\overline{t}_j) = V\left(\sum W_h \overline{y}_h\right) = V(\overline{y}_{st}).$$

We may note that

$$\frac{1}{4}(t_{j1} - t_{j2})^2 = \frac{1}{4}\left(\sum_h W_h \psi_{hj} d_h\right)^2,$$

writing $\psi_{hj} = 2\delta_{hj} - 1 = \pm 1$ for every $j = 1, \ldots, 2^L$. Thus,

$$v_j = \frac{1}{4}\sum_h W_h^2 d_h^2 + \frac{1}{4}\sum\sum_{h \neq h'} W_h W_{h'} d_h d_{h'} \psi_{hj} \psi_{hj'}$$

and

$$\overline{v} = \frac{1}{2^L}\sum_{j=1}^{2^L} v_j = \frac{1}{4}\sum_h W_h^2 d_h^2 = v$$

because $\sum_j \psi_{hj}\psi_{h'j} = 0$, the sum being over $j = 1, \ldots, 2^L$. But even for $L = 10$, $2^L = 1024$ so that numerous v_j's must be calculated to produce $\bar{v}$ that equals the standard or customary variance estimator v. So, it is desirable to form a small subset of a moderate number, K, of replicates of I and II so that the average of v_j's over that small subset may also equal v. In order to do so, we are to form K half-samples I and II such that $\Sigma'\psi_{hj}\psi_{h'j} = 0$, writing Σ' for the sum over this small subset of half-sample formations. Using **Hadamard matrices** with entries ± 1, which are square matrices of orders that are multiples of 4, it is easy to construct such half-sample replicates and the number of such replicates, namely K, is a multiple of 4 and is within the range $(L, L+3)$. Thus, for $L = 10$ strata, $K = 12$ replicates are enough to yield $\Sigma'\psi_{hj}\psi_{h'j} = 0$ giving

$$\frac{1}{K}\Sigma' v_j = v.$$

Let us illustrate below the choice of the values of ψ_{hj} (writing + for +1 and – for –1) for $L = 5$ or 6 and $K = 8$.

Values of $\psi_{hj}(\pm)$

Replicate number j	Stratum number h					
	1	2	3	4	5	6
1	+	+	–	–	–	+
2	+	+	–	+	–	–
3	–	+	+	–	–	–
4	–	+	+	+	–	+
5	+	–	+	–	–	+
6	+	–	+	+	–	–
7	–	–	–	+	–	+
8	–	–	–	–	–	–

It should be noted that if the parameter of interest is the simple linear parameter, namely the population mean, and the estimator is the standard linear unbiased estimator $\bar{y}_{st} = \sum W_h\bar{y}_h$, then a standard unbiased estimator ignoring *fpc*, namely $v = \frac{1}{4}\sum W_h^2 d_h^2$, is available, and the above exercise of forming replicates of half-samples in a balanced manner ensuring the condition $\Sigma'_j\psi_{hj}\psi_{h'j} = 0$ of **orthogonality** to achieve $\Sigma'_j v_j/K$ equal to v seems redundant. Actually, this procedure of forming **balanced replications** is considered useful to apply to alternative variance estimator formation when, in a more complicated and nonlinear case, a standard estimator is not available. For example, in estimating the finite population correlation coefficient ρ_N between two variables y and x, one may calculate the sample correlation coefficient based on the first half-sample values

$$[\delta_{hj}y_{h1} + (1 - \delta_{hj})y_{h2}, \delta_{hj}x_{h1} + (1 - \delta_{hj})x_{h2}]$$

for $h = 1, \ldots, L$, call it r_{1j}, and the same based on the second half-sample values

$$[(1 - \delta_{hj})y_{h1} + \delta_{hj}y_{h2}, (1 - \delta_{hj})x_{h1} + \delta_{hj}x_{h2}]$$

over all the strata $h = 1, \ldots, L$ and call it r_{2j}. Then, $\bar{r} = \frac{1}{2K}\Sigma'(r_{1j} + r_{2j})$ may be taken as an overall estimator for ρ_N and $\frac{1}{4K}\Sigma'(r_{1j} - r_{2j})^2$ as an estimator for the variance of $\bar{r}$, Σ' denoting the sum over a balanced set of K replicates for which $\Sigma'\psi_{hj}\psi_{h'j} = 0$. In this case, a standard variance estimator is not available, and hence the utility of the procedure.

KEYFITZ (1957) earlier considered estimation of variances of estimators when only two sample observations are recorded from each of several strata. But the above repeated orthogonal replication method (or balanced repeated replication method or balanced half-sampling method) was introduced and studied by MCCARTHY (1966, 1969) to consider variance estimation for nonlinear statistics like correlation and regression estimates, in particular when only two observations on each variable are available from several strata. To ensure orthogonality, or balancing, and keep the number of replicates down, HADAMARD matrices are utilized. GURNEY and JEWETT (1975) extended this to cover the case of exactly $p(>2)$ observations per stratum, with p as any prime positive integer. GUPTA and NIGAM (1987) extended it to cover the case of any arbitrary number of observations per stratum. They showed that balanced subsamples strata-wise may be derived for useful variance estimation using mixed orthogonal arrays of strength two or equivalently equal frequency orthogonal main effects plans for asymmetrical factorials. WU (1991) pointed out that an easy way to cover arbitrary number of units per stratum is to divide the units in each stratum separately and independently into two groups of a common number of units, or closely as far as practicable, and then apply the balanced half-sampling method to the two groups.

He also notes that neither this method nor GUPTA and NIGAM's (1987) method is efficient enough and recommends a revised method of balanced repeated replications based on mixed orthogonal arrays. SITTER (1993) points out the difficulty with the mixed orthogonal arrays to keep the number of replicates in check while constructing the orthogonal arrays. As a remedy, he prescribes the use of orthogonal multi-arrays to produce balanced repeated replications.

In the linear case we have seen that $\frac{1}{2}(t_{1j} + t_{2j})$ equals the standard estimator $\sum_h W_h \overline{y}_h$ for every j. But $\overline{r}$ does not equal the sample correlation coefficient that might be calculated from the entire sample. If in nonlinear cases, in specific situations, there is such a match of the half-sample estimates when averaged over the replicates satisfying the balancing condition, then we say that we have **double balancing**.

9.5 BOOTSTRAP

Consider a population $U = (1, 2, \ldots, N)$ and unknown values $Y_1, Y_2, \ldots, Y_N$ associated with the units $1, 2, \ldots, N$. Let $\theta = \theta(\underline{Y})$ be a population parameter, for example, the population mean $\overline{Y}$, or some not necessarily linear function $f(\overline{Y})$ of $\overline{Y}$, or the median of the values $Y_1, \ldots, Y_N$, etc. Suppose a sample $s = (i_1, \ldots, i_n)$ is drawn by SRSWR, write for $j = 1, 2, \ldots, n$

$$y_j = Y_{i_j}$$

and define

$$\underline{y} = (y_1, y_2, \ldots, y_n)'$$

Let $\widehat{\theta} = \widehat{\theta}(\underline{y})$ be an estimator of θ; in the special case $\theta = f(\overline{Y})$, for example, it suggests itself to choose $\widehat{\theta} = f(\overline{y})$, where $\overline{y}$ is the sample mean. To calculate confidence intervals for θ we need some information on the distribution of $\widehat{\theta}$ relative to θ.

Now, choose a sample s^* of size n from s by SRSWR, denote the observed values by

$$\overset{*}{y}_{11}, \overset{*}{y}_{21}, \ldots, \overset{*}{y}_{n1}$$

and define

$$\underset{-}{\overset{*}{y}}_1 = (\overset{*}{y}_{11}, \overset{*}{y}_{21}, \ldots, \overset{*}{y}_{n1})'$$

and $\overset{*}{s}$ is called a **bootstrap sample**. If, for example, $s = (4, 2, 4, 5)$, then $\overset{*}{s} = (2, 2, 4, 2)$ would be possible, and in this case $\overset{*}{y}_1 = (y_2, y_2, y_4, y_2)$.

Repeat the selection of a bootstrap sample independently to obtain

$$\underset{-}{\overset{*}{y}}_2, \underset{-}{\overset{*}{y}}_3, \ldots, \underset{-}{\overset{*}{y}}_B$$

where $B = 500, 1000$, or even larger, and calculate

$$\widehat{\theta}_0 = \frac{1}{B} \sum_{b=1}^{B} \widehat{\theta}(\underset{-}{\overset{*}{y}}_b)$$

$$v_B = \frac{1}{B-1} \sum_{b=1}^{B} [\widehat{\theta}(\underset{-}{\overset{*}{y}}_b - \widehat{\theta}_0]^2$$

It may be shown that the empirical distribution of

$$\widehat{\theta}(\underset{-}{\overset{*}{y}}_b) - \widehat{\theta}(\underset{-}{y}), b = 1, 2, \ldots, B$$

for large n and B approximates closely the distribution of

$$\widehat{\theta}(\underset{-}{y}) - \theta(\underset{-}{Y})$$

and that v_B approximates the variance of $\widehat{\theta}(\underset{-}{y})$. For details, good references are RAO and WU (1985, 1988).

Since B is usually taken as a very large number, it is useful to construct a histogram based on the values $\widehat{\theta}(\underset{-}{y}_b), b = 1, \ldots, B$. This **bootstrap histogram** is a close approximation to the true distribution of the statistic $\widehat{\theta}(\underset{-}{y})$. Let $100\alpha/2\%$ of the histogram area be below $\theta_{\alpha/2,l}$ and above $\theta_{\alpha/2,u}$. Then

$$[\widehat{\theta}_{\alpha/2,l}, \widehat{\theta}_{\alpha/2,u}]$$

is taken as a $100(1-\alpha)\%$ confidence interval for θ. This procedure is called the **percentile method** of confidence interval estimation.

An alternative procedure is the following. The statistic of the form of STUDENT's t, namely

$$[\widehat{\theta}(\underline{y}_b) - \widehat{\theta}(\underline{y})]/\sqrt{v_B} = t_b$$

is considered and the bootstrap histogram of the values $t_b, b = 1, 2, \ldots, B$ is constructed. Then, values $t_{\alpha/2,l}$ and $t_{\alpha/2,u}$ are found such that the proportions of the areas under this bootstrap histogram, respectively below and above these two values, are both $\alpha/2, (0 < \alpha < 1)$. Then the interval

$$(\widehat{\theta}(\underline{y}) - t_{\alpha/2,l}\sqrt{v_B}, \widehat{\theta}(\underline{y}) + t_{\alpha/2,u}\sqrt{v_B})$$

is a $100(1-\alpha)\%$ confidence interval because this bootstrap histogram is supposed to closely approximate the distribution of

$$\frac{\widehat{\theta}(\underline{y}) - \theta}{\sqrt{v(\widehat{\theta}(\underline{y}))}}$$

and $v(\widehat{\theta}(\underline{y}))$ is approximated by v_B.

So far only SRSWR has been considered. Now, samples are often taken without replacement and selections are from highly clustered groups of individuals. In addition, numerous strata are often formed, but the numbers of units selected from within each stratum are quite small, say, 2, 3, 4. So, within each stratum, separate application of the bootstrap method may not be reasonable. However, modifications are now available in the literature to effectively bypass these problems, and successful applications of bootstrap in complex sample surveys are reported. An interested reader may consult RAO and WU (1988).

It is necessary and important to compare the relative performances of the techniques of (a) linearization, (b) jackknife, (c) BRR (balanced repeated replication), (d) IPNS, and (e) bootstrap in yielding variance estimators in respect of bias, stability, and coverage probabilities for confidence intervals they lead to. J. N. K. RAO (1988) is an important reference for this.

A few methods of drawing bootstrap samples in the context of finite survey populations that are available in the current literature are briefly recounted below.

(1) Naive bootstrap

Let $\overline{Y}_j = \frac{1}{N}\sum_{i=1}^{N} y_{ji}$, $j = 1, \ldots, T$ and $\underline{\overline{Y}} = (\overline{Y}_1, \ldots, \overline{Y}_T)$, a vector of T finite population means of T variables $y_j (j = 1, \ldots, T)$ with values y_{ji} for the ith unit, $i \in U = (1, \ldots, N)$. Let $\theta = g(\underline{\overline{Y}})$ be a nonlinear function of $\underline{\overline{Y}}$. For example, the generalized regression estimator for $\overline{Y}$, namely

$$t_g = \frac{1}{N}\sum_{i\in s}\frac{y_i}{\pi_i} + \left(\overline{X} - \frac{1}{N}\sum_{i\in s}\frac{x_i}{\pi_i}\right)\frac{\sum_{i\in s} y_i x_i Q_i}{\sum_{i\in s} x_i^2 Q_i}, \quad Q_i(>0)$$
$$= t_g(., ., ., .)$$

is a nonlinear function of four statistics that are unbiased estimators of 4 population means, namely $\overline{Y} = \frac{1}{N}\sum y_i$, $\overline{X} = \frac{1}{N}\sum x_i$, $\frac{1}{N}\sum y_i x_i Q_i \pi_i = \overline{W}$, and $\frac{1}{N}\sum x_i^2 Q_i \pi_i = \overline{Z}$. So, θ may be written as $\theta = g(\overline{Y}, \overline{X}, \overline{W}, \overline{Z})$, which in this case reduces to $\theta = \overline{Y}$. Also, t_g may be written as an estimator $\hat{\theta}$ for θ.

Suppose U is split up into H strata of sizes N_h, with means $\overline{Y}_h$ $(h = 1, \ldots, H)$. Then, $\overline{Y} = \sum W_h \overline{Y}_h$, $W_h = \frac{N_h}{N}$. Let $\overline{y}_h$ be the mean based on an SRSWR from the hth stratum. Letting $\overline{y}_{st} = \Sigma W_h \overline{y}_h$, $\hat{\theta} = g(\overline{y}_{1st}, \ldots, \overline{y}_{Tst})$ may be taken as an estimator for $\theta = g(\overline{Y}_1, \ldots, \overline{Y}_T)$.

Let from the SRSWR $(y_{h1}, \ldots, y_{hn_h})$ coming from the hth stratum, $(y_{h1}^*, \ldots, y_{hn_h}^*)$ be an SRSWR in n_h draws called a bootstrap sample, $\overline{y}_h^* = \frac{1}{n_h}\sum_1^{n_h} y_{h_i}^*$, $\overline{y}_{st}^* = \sum W_h \overline{y}_h^*$, $\hat{\theta}^* = g(\overline{y}_h^*)$, writing $\underline{y}_h^* = (\overline{y}_{1h}^*, \cdots, \overline{y}_{Th}^*)$, the sample mean vector. Let this be repeated a large number of times B, and for the bth replicate $\hat{\theta}_b^*$ be calculated by the above formula $(b = 1, \ldots, B)$. Letting $\hat{\theta}^*(.) = \hat{\theta}_B^*(.) = \frac{1}{B}\sum_{b=1}^{B}\hat{\theta}_b^*$ be the bootstrap estimator for θ,

$$v_B = \frac{1}{B-1}\sum_{b=1}^{B}(\hat{\theta}_b^* - \hat{\theta}_B^*(.))^2$$

is taken as the bootstrap variance estimator for the estimator $\hat{\theta}^*(.)$ and also forms $\hat{\theta} = g(., \ldots, .)$.

If we write E_*, V_* the expectation and variance operators with respect to the above bootstrap sampling continued indefinitely, then $\hat{\theta}^*(.)$ is an approximation for $E_*(\hat{\theta}^*)$ and v_B is an approximation for $V_*(\hat{\theta}^*)$. For the case $T=1$ it follows that $\hat{\theta}^* = \sum W_h \overline{y}_h^*$ and also writing $\overline{y}_h$ the mean for the original sample, $v_B = \sum W_h^2 \frac{n_h-1}{n_h}\frac{s_h^2}{n_h}$, $s_h^2 = \frac{1}{n_h-1}\sum_1^{n_h}(y_{hi}-\overline{y}_h)^2$. But for $\overline{y}_{st} = \sum W_h \overline{y}_h$ we have $V(\overline{y}_{st}) = \sum W_h^2 \frac{s_h^2}{n_h}$.

So, unless n_h is very large

$$V_*(\hat{\theta}^*) \neq V(\overline{y}_{st}).$$

So, $\hat{\theta}_B^*(.)$ is not a fair estimator of θ because $v_B(\overline{y}^*)$ is not a consistent estimator of $V(\overline{y}_{st})$.

If $n_h = k$ for every $h = 1, \ldots, H$, then, $\frac{k}{k-1}V_*(\hat{\theta}^*) = V(\overline{y}_{st})$ and there is consistency only in this special case.

EFRON (1982) calls it a scaling problem for this naive bootstrap procedure, and his remedy is to take the bootstrap sample of size $(n_h - 1)$ instead of n_h and thus take care of the scaling problem. Obviously, with this amendment $V_*(\hat{\theta}^*)$ would equal $V(\overline{y}_{st})$.

(2) RAO and WU's (1988) rescaling bootstrap

This is a modification of the naive bootstrap method. From the original SRSWR taken from the hth stratum in n_h draws, let an SRSWR bootstrap sample be drawn in $n_h^*(\geq 1)$ draws and repeated independently across $h = 1, \ldots, H$. Let

$$f_h = \frac{n_h}{N_h},$$

$$C_h = \sqrt{\frac{n_h^*}{n_h - 1}(1 - f_h)},$$

$$\tilde{y}_h^* = \overline{y}_h + C_h(\overline{y}_h^* - \overline{y}_h),$$

with $\overline{y}_h^*$ as the mean of the bootstrap SRSWR of size n_h^*,

$$\underline{\tilde{y}}^* = \sum_{h=1}^{H} \underline{\tilde{y}}_h^*, \quad \tilde{\theta}^* = g(\underline{\tilde{y}}^*)$$

using a lower bar to denote the $T-$ vector of the obvious entities.

Let the bootstrap sampling above be repeated a large number of times B and let $\widetilde{\theta}_b^*$ denote the above $\widetilde{\theta}^*$ for the bth bootstrap sample ($b = 1, \ldots, B$). Then $\widetilde{\theta}_B^*(.) = \frac{1}{B}\sum_{b=1}^{B}\widetilde{\theta}_b^*$ is taken as the final estimator for θ and $\widetilde{v}_B = \frac{1}{B-1}\sum_{b=1}^{B}(\widetilde{\theta}_b^* - \widetilde{\theta}_B^*(.))^2$ as the variance estimator for $\widetilde{\theta}_B^*(.)$.

This procedure eliminates the scaling problem of the naive bootstrap method and ensures consistency of $\widetilde{v}_B$.

(3) RAO and WU's (1988) general with replacement bootstrap

For the $T-$ vector of totals $Y_t (t = 1, \ldots, T)$ if one defines $\theta = g(\underline{Y})$, $\underline{Y} = (\overline{Y}_1, \ldots, \overline{Y}_t, \ldots \overline{Y}_T)$ and employs the homogeneous linear estimator, $\hat{Y}_t = \sum_{i\in s} b_{si} y_{ti}$ for Y_t such that the mean square error MSE of $\hat{Y}_t$ is zero if $\frac{y_{ti}}{w_{ti}} =$ constant for every $i \in U = (1, \ldots, N)$, with $w_{ti} (\neq 0)$ as known non-zero constants, then from RAO (1979) it is known that

$$m(\hat{Y}_t) = -\sum_{i<j} I_{sij} d_{sij} w_{ti} w_{tj} \left(\frac{y_{ti}}{W_{ti}} - \frac{y_{tj}}{W_{tj}}\right)^2$$

with

$$E(d_{sij} I_{sij}) = d_{ij} = E_p(b_{si} I_{si} - 1)(b_{sj} I_{sj} - 1).$$

Then, in order to estimate $\theta = g(\underline{Y})$ and its variance, rather MSE estimator, RAO and WU (1988) recommend the following bootstrap procedure.

Let for any sample s the selection probability $p(s)$ be positive only for every s with n as the number of units in it all distinct. A bootstrap sample from s is chosen in the following way. First $n(n-1)$ ordered pairs of units $i, j (i \neq j)$ in s are formed. From them, m pairs (i^*, j^*) are chosen with replacement (WR) with probabilities $\lambda_{ij} (= \lambda_{ji})$ with their values as specified below. The sample drawn is denoted s^*. For simplicity of notation we drop the subscript t throughout the symbols used above.

Let us define

$$\widetilde{Y} = \hat{Y} + \frac{1}{m} \sum_{i^*, j^* \in s^*} k_{i^* j^*} \left(\frac{y_{i^*}}{w_{i^*}} - \frac{y_{j^*}}{w_{j^*}} \right)$$

with k_{ij}'s to be specified as below.

Let

$$\widetilde{\widetilde{Y}}_t = \frac{\hat{Y}_t}{N}, \underline{\widetilde{\widetilde{Y}}} = (\widetilde{\widetilde{Y}}_1, \ldots, \widetilde{\widetilde{Y}}_t, \ldots, \widetilde{\widetilde{Y}}_T),$$

$$\widetilde{\theta} = g(\underline{\widetilde{\widetilde{Y}}}).$$

Let the bootstrap sampling as above be independently repeated a large number of times B. Let for the bth bootstrap sample the above statistics be denoted as $\widetilde{Y}_b, \underline{\widetilde{\widetilde{Y}}}_b, \widetilde{\theta}_b = g(\underline{\widetilde{\widetilde{Y}}}_b)$. In case $T = 1$ and $\theta = \overline{Y}$, it will follow that $E_*(\widetilde{\widetilde{Y}}) = \frac{\hat{Y}}{N}$ because

$$E_*(\widetilde{Y}) = \hat{Y} + E_* \left\{ k_{i^* j^*} \left(\frac{y_{i^*}}{w_{i^*}} - \frac{y_{j^*}}{w_{j^*}} \right) \right\}$$

$$= \hat{Y} + \sum_{i \neq j \in s} k_{ij} \lambda_{ij} \left(\frac{y_i}{w_i} - \frac{y_j}{w_j} \right) = \hat{Y}$$

because $k_{ij} \lambda_{ij} = k_{ji} \lambda_{ji}$. Also

$$V_*(\widetilde{Y}) = \frac{1}{m} E_* \left\{ k_{i^* j^*} \left(\frac{y_{i^*}}{w_{i^*}} - \frac{y_{j^*}}{w_{j^*}} \right)^2 \right\}$$

$$= \frac{1}{m} \sum \sum_{i \neq j \in s} k_{ij}^2 \lambda_{ij} \left(\frac{y_i}{w_i} - \frac{y_j}{w_j} \right)^2$$

Then $k_{ij} \lambda_{ij}$ and m are to be so chosen that

$$k_{ij}^2 \frac{\lambda_{ij}}{m} = -\frac{1}{2} d_{ij}(s) w_i w_j.$$

In that case $V_*(\widetilde{Y})$ would match the estimate $m(\hat{Y})$ of MSE $(\hat{Y})$.

RAO and WU (1988) recommend that in the linear case, that is, when $T = 1$ and the initial estimator e_b is linear in $y_i, i \in s$, if its variance or MSE can be matched by an estimator based on a bootstrap sample for which the bootstrap variance equals it, then in the nonlinear case $\theta = g(\underline{Y})$ should be estimated by the bootstrap estimator, which is

$$\widetilde{\theta}_B = \frac{1}{B}\sum_{b=1}^{B}\widetilde{\theta}_b$$

writing $\widetilde{\theta}_b$ for the statistic defined as $\underline{\widetilde{\theta}} = g(\widetilde{\underline{\overline{Y}}})$ for the bth bootstrap sample. Then, the bootstrap variance estimator for $\widetilde{\theta}_B$ is

$$v_B = \frac{1}{B}\sum_{b=1}^{B}(\widetilde{Y}_b - \widetilde{\theta}_B)^2$$

In case RAO's (1979) approach is modified (a) eliminating the condition that MSE $(\hat{Y})$ equals zero when $y_i \propto w_i$ and (b) consequently adding a term $\sum \frac{y_i^2}{w_i}\beta_i$ to MSE $(\hat{Y})$ and a term $\sum \frac{y_i^2}{w_i}\beta_i \frac{I_{si}}{\pi_i}$ to $m(\hat{Y})$, then certain modifications in the above bootstrap are necessary because (a) the sample size may now vary with samples and (b) non-negativity of an estimator for the MSE $(\hat{Y})$ consequently can be ensured only under additional conditions. PAL (2002) has provided some solutions in this regard in her unpublished Ph.D. thesis.

(4) SITTER's (1992) mirror-match bootstrap

Here the original sample is a stratified SRSWOR with n_h units drawn from hth stratum with $\overline{y}_h$ as the sample mean. For the case $T = 1$, the unbiased traditional estimator for $\overline{Y}$ is $\overline{y}_{st} = \sum W_h \overline{y}_h$ with

$$\hat{Var}(\overline{y}_{st}) = \sum W_h^2 \frac{1-f_h}{n_h} s_h^2, \quad f_h = \frac{n_h}{N_h}, \quad h = 1, \ldots, H.$$

For bootstrap sampling the recommended steps are:

(a) Choose an integer $n_h'(1 < n_h' < n_h)$ and take SRSWOR of size n_h' from the initial SRSWOR of size n_h from the hth stratum to get $y_{h1}^*, \ldots, y_{hn_h'}^*$.

(b) Return this SRSWOR of size n'_h to the SRSWOR of size n_h and repeat step (a) a number of times equal to $k_h = \frac{n_h(1-f^*_h)}{n'_h(1-f_h)}$, $f^*_h = \frac{n'_h}{n_h}$. Then we have a total number of y values in this bootstrap sample given by

$$n'_h k_h = \frac{n_h(1 - f^*_h)}{(1 - f_h)} = n^*_h, \text{ say.}$$

If k_h is not an integer, take it as $[k_h]$ with probability q and as $[k_h] + 1$ with probability $1 - q_h$ with a suitable choice of q_h $(0 < q_h < 1)$.

(c) After realizing the sample observations

$$s^* = (y^*_{h1}, \ldots, y^*_{hn^*_h},\ h = 1, \ldots, H)$$

calculate $\hat{\theta}^* = \hat{\theta}(s^*)$.

(d) Repeat steps a large number of times B.

Denoting by $\hat{\theta}^*_b$ the $\hat{\theta}^*$ for the bth bootstrap sample ($b = 1, \ldots, B$) and writing $\hat{\theta}^*_B = \frac{1}{B}\sum_{b=1}^{B} \hat{\theta}^*_b$, take $\hat{\theta}^*_B$ as the bootstrap estimate of θ and take $v_B = \frac{1}{B-1}\sum_{i=1}^{B}(\hat{\theta}^*_b - \hat{\theta}^*_B)^2$ as the variance estimate of $\hat{\theta}^*_B$ and of $\hat{\theta}$.

If $T = 1$, then $E_*(\hat{\theta}^*_b - E\hat{\theta}^*_b)^2$ equals $V(\overline{y}_{st})$.

If $f_h \geq \frac{1}{n_h}$, that is, $n^2_h \geq N_h$, then the choice $n'_h = f_h n_h$ ensures $f^*_h = f_h$, implying that the bootstrap at the initial step mirrors the original sampling. The matching indeed is about the $Var(\overline{y}_{st})$ and the estimate of variance v_B.

(5) BWR bootstrap of McCarthy and Snowden (1985)

This is a modification of the naive bootstrap method by taking the sample size m_h for the bootstrap sample to be drawn by SRSWR method from the initial sample, which is drawn either by SRSWR or SRSWOR independently from each stratum in such a way that the bootstrap variance estimator

$$v_B = \sum_{h=1}^{H} \frac{W^2_h}{n_h} \frac{(n_h - 1)}{m_h} s^2_h$$

may match $\hat{V}(\overline{y}_{st}) = \sum W_h^2 \frac{s_h^2}{n_h}$ for SRSWR or

$$\hat{V}(\overline{y}_{st}) = \sum W_h^2 (1 - f_h) \frac{s_h^2}{n_h}.$$

Thus, either $m_h = (n_h - 1)$ or

$$m_h = \frac{n_h - 1}{1 - f_h}$$

(6) BWO boostrap of GROSS (1980)

For this method let the initial sample be an SRSWOR of size n. Let k be an integer such that $N = kn$. Then the following are the steps.

(a) Independently replicate the initial sample k times.

(b) Draw an SRSWOR of size n from the pseudo-population generated in step (a). Let the sample observations be

$$y_1^*, \ldots, y_n^*$$

and calculate

$$\hat{\theta}^* = g(y^*) = \hat{\theta}(y_1^*, \ldots, y_n^*)$$

(c) Repeat step (b) a large number of times B. Calculate θ_b^*, which is $\hat{\theta}^*$ for the bth bootstrap sample above ($b = 1, \ldots, B$). Writing

$$\theta_B^* = \frac{1}{B} \sum (\theta_b^*)$$

take

$$v_B = \frac{1}{B - 1} \sum_1^B (\theta_b^* - \theta_B^*)^2$$

as the variance estimator for θ_B^* and for $\hat{\theta}$.

BICKEL and FREEDMAN (1981) extended this to stratified SRSWOR, which was also discussed by MCCARTHY and SNOWDEN (1985).

(7) SITTER's (1992) extended BWO bootstrap method

Bickel–Freedman's BWO method is extended to stratified SRSWOR in the following way by SITTER (1992).

Ignoring the fractional parts in

$$n'_h = n_h - (1 - f_h)$$

and

$$k_h = \frac{N_h}{n_h}\left(1 - \frac{1 - f_h}{n_h}\right)$$

the following are the bootstrap sampling steps:

(a) Replicate $(y_{h1}, \ldots, y_{hn_h})$, separately and independently k_h times, $h = 1, \ldots, H$ to create H different pseudo-strata.

(b) Draw an SRSWOR of size n'_h from the hth pseudo-stratum, and repeat this independently for each $h = 1, \ldots, H$, thus generating bootstrap sample observations

$$s^* = \{(y^*_{h1}, \ldots, y^*_{hn'_h}),\ h = 1, \ldots, H\}$$

and let $\hat{\theta}^* = \hat{\theta}(s^*)$.

(c) Repeat steps (b) and (a) a large number of times B, and calculate for the bth bootstrap sample the statistics

$$\hat{\theta}^*_b,\ b = 1, \ldots, B,$$

and let

$$\hat{\theta}^*_B = \frac{1}{B}\sum_{b=1}^{B} \hat{\theta}^*_b$$

and

$$v_{BWO} = \frac{1}{B-1}\sum_{b=1}^{B}(\hat{\theta}^*_b - \hat{\theta}^*_B)^2$$

be taken as the variance estimator for θ^*_B as well as for $\hat{\theta}$, based on the original sample.

For $T = 1$ and $\hat{\theta} = \overline{y}_{st}$ it may be checked that

$$E_*(\hat{\theta}^* - E_*\hat{\theta}^*)^2 = V(\overline{y}_{st}).$$

Unlike Bickel–Freedman's extension of BWO to stratified SRSWOR, where it is necessary that

$N_h = k_h n_h$ with n_h as the re-sample size as well, in the present case n'_h and k_h are chosen to satisfy

$$f_h^* = f_h \quad \text{where} \quad f_h^* = \frac{n'_h}{(k_h n_h)}$$

and

$$V_*(\bar{y}_h^*) = \frac{1 - f_h}{n_h} s_h^2, \; h = 1, \ldots, H,$$

fractional parts whenever necessary being ignored. SITTER (1992) may be consulted for further details.

(8) SITTER's (1992) bootstrap for RHC initial samples

Suppose from the population $U = (1, \ldots, i, \ldots, N)$ on which $\underline{Y} = (y_1, \ldots, y_i, \ldots, y_N)$ and $\underline{p} = (p_1, \ldots, p_i, \ldots, p_N)$ are defined as the vectors of real values y_i and normed size measures $p_i (0 < p_i < 1, \sum p_i = 1)$ a sample s of n units is drawn by the RHC scheme. For this method integers N_i are chosen with their sum over $i = 1, \ldots, n$, namely $\Sigma_n N_i$ equal to N. Then n groups are formed taking N_i units chosen by SRSWOR from U into the ith group. Writing Q_i as the sum of the p_i values for the N_i units in the ith group, one unit from the ith group is chosen with a probability equal to its p_i value divided by Q_i and this is repeated independently for the n groups formed. Then RHC's unbiased estimator for Y is

$$t_{RHC} = \Sigma_n y_i \frac{Q_i}{p_i},$$

writing, (y_i, p_i) as the y_i and p_i value for the unit chosen from the ith group. Its variance is

$$V(t_{RHC}) = \frac{\Sigma_n N_i^2 - N}{N(N-1)} \left[\sum \frac{y_i^2}{p_i} - Y^2 \right]$$

and RHC's unbiased estimator for $V(t_{RHC})$ is

$$v(t_{RHC}) = \left(\frac{\sum_n N_i^2 - N}{N^2 - \sum_n N_i^2} \right) \left[\Sigma_n Q_i \left(\frac{y_i}{p_i} \right)^2 - t_{RHC}^2 \right]$$

The following are the steps for bootstrap sampling given by SITTER (1992) in this case.

(a) Choose an integer n^* such that $1 < n^* < n$. Divide the initially chosen RHC sample s of size n into n^* nonoverlapping groups, taking into the ith group $(i = 1, \ldots, n^*)$, n_i units of s such that the sum of n_i's over the n^* groups, namely $\Sigma_{n^*} n_i$, equals n. Treat the Q_i's, for which $\Sigma_n Q_i = 1$, as the normed size measures of the units in s. Calculate the sum R_i^* of the Q_i values for the n_i units in the ith group into which s is split up. Then from the ith group choose one unit with a probability proportional to the ratio of its Q_i value to R_i^* and repeat this independently for all the n^* groups. Thus, a sample s^* of size n^* is generated out of the original s.

(b) Repeat step (a) a total of times equal to

$$k = \left[\frac{\Sigma_{n^*} n_i^2 - n}{n(n-1)}\right] \frac{(N^2 - \Sigma_n N_i^2)}{(\Sigma_n N_i^2 - N)}$$

each time keeping s intact but replacing s^* each time.

(c) Let

$$y_1^* \frac{R_1^*}{Q_1^*}, \ldots, y_{n^*}^* \frac{R_{n^*}^*}{Q_n^*}$$

denote values respectively for the 1st, …, n^*th group from which one unit each is selected and pooling together the corresponding k replicates the values written as

$$y_1^* \frac{R_1^*}{Q_1^*}, \ldots, y_{n^*}^* \frac{R_{n^*}^*}{Q_{n^*}^*}, \ldots, y_{kn^*}^* \frac{R_{kn^*}^*}{Q_{kn^*}^*}$$

Then, calculate θ^* based on the kn^* samples, values.

(d) Repeat independently steps (a) to (c) a large number of times B. For the bth replicate, let θ_b^*

be the θ^* value and

$$\theta_B^* = \frac{1}{B} \sum_{b=1}^{B} \theta_b^*$$

Then,

$$v_b = \frac{1}{B-1} \sum_{b=1}^{B} (\theta_b^* - \theta_B^*)^2$$

is the variance estimator for θ^*.

SITTER (1992b) has shown that, in the linear case for the RHC estimator based on

$$\hat{\overline{Y}}^* = \frac{1}{kn^*} \left[y_1^* \frac{R_1^*}{Q_1^*} + \ldots + y_{kn^*}^* \frac{R_{kn^*}^*}{Q_{kn^*}^*} \right]$$

one has $E_*(\hat{\overline{Y}}^*) = \overline{Y}$ and $V_*(\hat{\overline{Y}}^*) = v(t_{RHC})$.

Finally, let us add one point, that, besides the percentile method of constructing the confidence interval discussed earlier, the following double bootstrap method is also often practicable.

Let $\hat{\theta}$ be a point estimator for a parameter θ with v as an estimator for the variance of $\hat{\theta}$.

Corresponding to the standardized pivotal quantity

$$\frac{\hat{\theta} - \theta}{\sqrt{v}},$$

let us consider $\delta_b = \frac{\hat{\theta}_b - \hat{\theta}}{\sqrt{v_b}}$, where $\hat{\theta}_b$ is a bootstrap estimator for θ based on the bth bootstrap sample when a large number of bootstrap samples are drawn by one of the bootstrap procedures. Let another set of B bootstrap samples by the same method be drawn from this bth bootstrap sample on which basis v_b is the variance estimator for $\hat{\theta}$.

Now, constructing the histogram based on the values of δ_b above, let l and u be the lower and upper $100\alpha/2\%$ points respectively of this histogram. Then, approximately,

$$1-\alpha = Prob\left[l < \frac{\hat{\theta}_b - \hat{\theta}}{\sqrt{v_b}} < u\right]$$

$$= Prob[\hat{\theta}_b - u\sqrt{v_b} < \hat{\theta} < \hat{\theta}_b - l\sqrt{v_b}]$$

Now replacing $\hat{\theta}$ by θ and $\hat{\theta}_b$ by $\hat{\theta}$ in this one may write

$$1-\alpha = Pr[\hat{\theta} - u\sqrt{v_b} < \theta < \hat{\theta} - l\sqrt{v_b}]$$

So $(\hat{\theta} - u\sqrt{v_b}, \hat{\theta} + l\sqrt{v_b})$ provides the $100(1-\alpha)\%$ double bootstrap confidence interval for θ.

Chapter 10

Sampling from Inadequate Frames

Suppose a finite population of N units is divisible into a number of groups. If the groups are mutually exclusive and, together, they exhaust the population, the number of units belonging to each group is known and it is also possible to identify at the start of the survey which individual univocally belongs to which group, then one may undertake standard procedures of sample selection and estimation of parameters of interest. For example, one may have stratified sampling if from each group with a known composition a predetermined number $t(\geq 1)$ of units is sampled. If instead, only some, but not all, the groups are decided to be sampled with preassigned selection probabilities, we have cluster sampling. The groups are called **strata** in case of stratified sampling where each stratum is represented in the sample with probability 1. The same groups are called clusters in case of cluster sampling when the groups are given positive selection probabilities less than 1. If the selected clusters are not fully surveyed, but only samples of individuals of the selected clusters are surveyed, then we have two-stage sampling and the clusters are called the first-stage units or primary sampling units (fsu or psu).

If instead, before sample selection it is not known as to which group an individual belongs to, but the groups are identifiable and distinguishable with respect to known characteristics like, for example, racial, educational, economic, occupational levels of distinction, etc., so that an individual after selection and interrogation is assignable unequivocally to one of the distinct groups, then the groups are called **domains**. Neither the compositions nor the sizes of the domains are known prior to at least the initial part of the survey.

But if, at the start of the survey, the sizes, that is, the number of units contained in the respective groups, are known, say, from recent censuses, but their compositions are not known so that one cannot utilize a **frame** to select members of the respective groups with predetermined probabilities, then the groups are called **post-strata**, provided that after the selection and survey the individuals are assignable to respective groups and data analysis takes account of the assignment to groups.

In the former case we are interested in inferring the characteristics of population members of one or more domains. In the second case the population is one of inferring parameters relating to the entire population, but we intend to make use of the knowledge of post-strata sizes and, if available, other post-strata characteristics, even though we fail to choose samples from the respective groups in adequate proportions.

In some cases we may have two or more overlapping frames. In that case one may choose the samples separately using several frames and face and work out associated additional problems of inference and interpretation. This is the problem of multiple-frame estimation.

Sometimes the domains of interest may be so numerous, while the total sample size one can afford is meager, that it is impossible to have adequate representations of all domains of interest in a sample. In that case, similar domains are conceptually pooled together and samples are amalgamated across the similar domains to borrow strength from the ensembles in order to derive improved estimators for the respective domain parameters. This is the problem of small area estimation.

In many of these cases, the sample sizes representing various domains or post-strata become random variables. Hence the problem of inferences conditional on certain sample configurations, as opposed to unconditional inferences where sample configurations are averaged over conceptually repeated realizations of samples, arises. In what follows, we give short descriptions of these issues.

10.1 DOMAIN ESTIMATION

Let D be a domain of interest within a population $U = (1, \ldots, i, \ldots, N)$. Let N_D be the unknown size of D. Let a sample s of size n be drawn from U with a probability $p(s)$ according to a design p admitting positive inclusion probabilities π_i, π_{ij}. Let for $i = 1, 2, \ldots, N$

$$\begin{aligned} I_{Di} &= 1(0) \quad \text{if} \quad i \in D\,(i \notin D) \\ Y_{Di} &= Y_i(0) \quad \text{if} \quad i \in D\,(i \notin D). \end{aligned}$$

Then the unknown domain size, total, and mean are, respectively,

$$N_D = \sum_1^N I_{Di}, T_D = \sum_1^N Y_{Di} \quad \text{and} \quad \overline{T}_D = \frac{T_D}{N_D}$$

In analogy to $\underline{Y} = (Y_1, \ldots, Y_i, \ldots, Y_N)'$ we write $\underline{I}_D = (I_{D1}, \ldots, I_{Di}, \ldots, I_{DN})'$ and $\underline{Y}_D = (Y_{D1}, \ldots, Y_{Di}, \ldots, Y_{DN})'$. Then, corresponding to any estimator $t = t(s, \underline{Y}) = \widehat{Y}$, for $Y = \Sigma_1^N Y_i$ we may immediately choose estimators for N_D and T_D, respectively,

$$\widehat{N}_D = t(s, \underline{I}_D) \quad \text{and} \quad \widehat{T}_D = t(s, \underline{Y}_D).$$

It may then be a natural step to take the estimator $\widehat{\overline{T}}_D$ for $\overline{T}_D$ as

$$\widehat{\overline{T}}_D = \frac{\widehat{T}_D}{\widehat{N}_D}.$$

If t is taken as a homogeneous linear unbiased estimator (HLUE), that is, if it is of the form

$$t = t(s, \underline{Y}) = \sum_{i \in s} b_{si} Y_i \quad \text{with} \quad \sum_{s \ni i} b_{si}\, p(s) = 1 \text{ for all } i,$$

then it has a variance

$$V_p(t) = \sum_i d_i Y_i^2 + \sum\sum_{i<j} d_{ij} Y_i Y_j$$

where

$$d_i = \sum_{s \ni i} b_{si}^2\, p(s) - 1, \; d_{ij} = \sum_{s \ni i,j} b_{si} b_{sj}\, p(s) - 1$$

and an unbiased estimator for $V_p(t)$ is

$$v_p(t) = \sum d_{si} I_{si} Y_i^2 + \sum\sum_{i \neq j} d_{sij} I_{sij} Y_i Y_j$$

if d_{si}, d_{sij}'s are available subject to

$$E_p(d_{si} I_{si}) = d_i, \; E_p(d_{sij} I_{sij}) = d_{ij},$$

writing as earlier

$$I_{si} = 1(0) \quad \text{if} \quad i \in s (i \notin s), I_{sij} = 1(0) \quad \text{if} \quad i, j \in s(i, j \notin s).$$

It follows then that

$$V_p(\hat{T}_D) = V_p(t)\,]_{\underline{Y} \to \underline{Y}_D}, v_p(\hat{T}_D) = v_p(t)\,]_{\underline{Y} \to \underline{Y}_D}$$
$$V_p(\widehat{N}_D) = V_p(t)\,]_{\underline{Y} \to \underline{I}_D}, v_p(\widehat{N}_D) = v_p(t)\,]_{\underline{Y} \to \underline{I}_D}$$

where

$$V_p(t)\,]_{\underline{Y} \to \underline{Y}_D}$$

means that $\underline{Y}$ in $V_P(t)$ is replaced by $\underline{Y}_D$ with a corresponding interpretation of the other expressions.

Next, if we may assume that the sample s_D consisting of the units of s contained in D, that is, $s_D = s \cap D$, has a size $n_D (\leq n)$ that is large enough so that we may apply the linearization technique of section 9.1, then we may have the following approximate formulae for the variance of $\widehat{\overline{T}}_D = \frac{\hat{T}_D}{\widehat{N}_D}$

and for an approximately unbiased estimator for that variance:

$$V_p(\widehat{T_D}) \simeq \frac{1}{N_D^2} V_p\left(\sum_s b_{si} Z_{Di}\right)$$

$$= \frac{1}{N_D^2}\left[\sum_1 d_i Z_{Di}^2 + \sum\sum_{i \neq j} d_{ij} Z_{Di} Z_{Dj}\right]$$

$$v_p(\widehat{T}) \simeq \frac{1}{\widehat{N}_D)^2}\left[\sum_i d_{si} I_{si} \widehat{Z}_i^2 + \sum\sum_{i \neq j} d_{sij} I_{sij} \widehat{Z}_i \widehat{Z}_j\right]$$

where

$$Z_{Di} = Y_{Di} - \frac{T_D}{N_D} I_{Di}$$

$$\widehat{Z}_i = Y_{Di} - \frac{\widehat{T}_D}{\widehat{N}_D} I_{Di},\ i = 1, \ldots, N.$$

10.2 POSTSTRATIFICATION

Suppose a finite population $U = (1, \ldots, i, \ldots, N)$ of N units consists of L post-strata of known sizes $N_h, h = 1, \ldots, L$ but unknown compositions with respective post-strata totals $Y_h = \sum_i^{N_h} Y_{hi}$ and means $\overline{Y}_h = Y_h/N_h$, $h = 1, \ldots, L$. Let a simple random sample s of size n have been drawn from U yielding the sample configuration $\underline{n} = (n_1, \ldots, n_h, \ldots, n_L)$ where $n_h(\geq 0)$ is the number of units of s coming from the hth post-stratum, $h = 1, \ldots, L$, $\sum_{h=1}^{L} n_h = n$. In order to estimate $\overline{Y} = \Sigma W_h \overline{Y}_h$, writing $W_h = \frac{N_h}{N}$, $h = 1, \ldots, L$ we proceed as follows.

Let $I_h = 1(0)$ if $n_h > 0$ $(n_h = 0)$. Then,

$$E(I_h) = \text{Prob}(I_h = 1) = 1 - \binom{N - N_h}{n} \bigg/ \binom{N}{n},\quad h = 1, \ldots, L.$$

For $\overline{Y}$ a reasonable estimator may be taken as

$$t_{pst} = t_{pst}(\underline{Y}) = \frac{\sum W_h \overline{y}_h I_h / E(I_h)}{\sum W_h I_h / E(I_h)}$$

writing $\overline{y}_h$ as the mean of the n_h units in the sample consisting of members of the hth post-stratum, if $n_h > 0$; if $n_h = 0$, then $\overline{y}_h$ is taken as $\overline{Y}_h$. It follows that $x = \sum W_h \overline{y}_h I_h / E(I_h)$ is an unbiased estimator for $\overline{Y}$ and $b = \sum W_h I_h / E(I_h)$ an unbiased estimator for 1. Yet, instead of taking just a as an unbiased estimator for $\overline{Y}$, this biased estimator of the ratio form $\frac{x}{b}$ is proposed by DOSS, HARTLEY and SOMAYAJULU (1979) because it has the following **linear invariance property** not shared by itself:

Assume $Y_i = \alpha + \beta Z_i$; then $\overline{y}_h = \alpha + \beta \overline{z}_h$ and $t_{pst}(\underline{Y}) = \alpha + \beta t_{pst}(\underline{Z})$, with obvious notations. Further properties of t_{pst} have been investigated by DOSS et al. (1979) but are too complicated to merit further discussion here.

10.3 ESTIMATION FROM MULTIPLE FRAMES

Suppose a finite population U of size N is covered exactly by the union of two overlapping frames A and B of sizes N_A and N_B. Let E_A denote the set of units of A that are not in B, E_{AB} denote those that are in both A and B, and E_B denote the units of B that are not in A; N_{EA}, N_{AB}, N_{EB} respectively denote the sizes of these three mutually exclusive sets. Let two samples of sizes n_A, n_B be drawn by SRSWOR from the two lists A and B respectively in independent manners. Let n_a, n_{ab}, n_{ba}, n_b denote respectively the sampled units of A that are in E_A, E_{AB} and of B that are in E_{AB}, E_B. Let us denote the corresponding sample means by $\overline{y}_a, \overline{y}_{ab}, \overline{y}_{ba}$, and $\overline{y}_b$. Then for the population total $Y = \sum_1^N Y_i$ one may employ the following estimators

$$\widehat{Y}_1 = (N_{EA}\overline{y}_a + N_{EB}\overline{y}_b) + N_{AB}(p\overline{y}_{ab} + q\overline{y}_{ba})$$

if N_{EA}, N_{EB}, and N_{AB} are known, or, without this assumption,

$$\widehat{Y}_2 = \frac{N_A}{n_A}(\overline{y}_a + p\overline{y}_{ab}) + \frac{N_B}{n_B}(\overline{y}_b + q\overline{y}_{ba}).$$

In $\widehat{Y}_1$ and $\widehat{Y}_2$, p is a suitable number, $0 < p < 1$ and $p + q = 1$. This procedure has been given by HARTLEY (1962, 1974). Supposing first that the variance of the variable of interest y for the respective sets E_A, E_{AB}, E_B are known quantities $\sigma_A^2, \sigma_{AB}^2, \sigma_B^2$ and choosing a simple cost function, he gave rules for optimal choices of n_A, n_B subject to a given value of $n = n_A + n_B$ and of p.

SAXENA, NARAIN and SRIVASTAVA (1984) consider the following extension of HARTLEY's (1962, 1974) technique to the case of two-stage sampling. Suppose that whatever has been stated above applied to the population of first-stage units (fsu). For each sampled fsu i, the total value Y_i over its second-stage units (ssu) is unavailable, but is estimated on taking samples of ssus independently. Then $\widehat{Y}_1, \widehat{Y}_2$ cannot be used and the following modifications are needed. Suppose for the ith fsu ($i = 1, \ldots, N$) two frames A_i, B_i are available that overlap but together coincide with the set of M_i ssus in the ith fsu. Let EA_i, EB_i, AB_i denote sets of ssus in ith fsu contained exclusively in A_i, B_i and both in A_i and B_i, respectively; let their sizes and variances be, respectively, $M_{A_i}, M_{B_i}, M_{AB_i}$, $\sigma^2_{A_i}, \sigma^2_{B_i}, \sigma^2_{AB_i}$. Let independent SRSWORs of sizes m_{A_i}, m_{B_i} be respectively drawn independently from A_i, B_i. Let m_{EA_i}, m_{AB_i}, m_{BA_i}, m_{EB_i} denote respectively the units out of m_{A_i} that are in EA_i, AB_i and of m_{B_i} that are in AB_i and EB_i. Let $\overline{y}_{a_i}, \overline{y}_{ab_i}, \overline{y}_{ba_i}$, $\overline{y}_{b_i}$ denote the corresponding sample means. Let $r_i (0 < r_i < 1)$ and s_i such that $r_i + s_i = 1$ be numbers suitably chosen. Then,

$$\widehat{Y}_i = M_{A_i}\overline{y}_{a_i} + M_{AB_i}(r_i\overline{y}_{ab_i} + s_i\overline{y}_{ba_i}) + M_{B_i}\overline{y}_{b_i}$$

is taken as an unbiased estimator for Y_i. Writing, with obvious notations,

$$\widehat{\overline{y}}_a = \frac{1}{n_a}\sum_1^{n_a}\widehat{Y}_a, \widehat{\overline{y}_b} = \frac{1}{n_b}\sum_1^{n_b}\widehat{Y}_{b_i}, \ \widehat{\overline{y}_{ab}} = \frac{1}{n_{ab}}\sum_1^{n_{ab}}\widehat{Y}_{ab_i},$$

$$\widehat{\overline{y}_{ba}} = \frac{1}{n_{ba}}\sum_1^{n_{ba}}\widehat{Y}_{ba_i}$$

an unbiased estimator for Y is taken as

$$\widehat{Y_1} = N_{EA}\widehat{\overline{y}_a} + N_{AB}(p\widehat{\overline{y}_{ab}} + q\widehat{\overline{y}_{ba}}) + N_{EB}\widehat{\overline{y}_b}$$

if N_{EA}, N_{AB}, N_{EB} are known, or as

$$\widehat{Y_2} = \frac{N_A}{n_A}(\widehat{\overline{y}_a} + p\widehat{\overline{y}_{ab}}) + \frac{N_B}{n_B}(\widehat{\overline{y}_b} + q\widehat{\overline{y}_{ba}}).$$

SAXENA et al. (1984) have worked out optimal choices of r_i, s_i, p, q, n_A, n_B, m_{A_i}, m_{B_i} considering suitable cost functions following HARTLEY's (1962, 1974) procedure of multiple frame estimation and recommended replacement of unknown parameters occurring in the optimal solutions by sample analogues, and have considered various special cases giving simpler solutions.

10.4 SMALL AREA ESTIMATION

10.4.1 Small Domains and Poststratification

Suppose a finite population U of N units labeled $1, \ldots, i, \ldots, N$ consists of a very large, say several thousand, domains of interest, like the households of people of various racial groups of different predominant occupational groups of their principal earning members located in various counties across various states like those in U.S.A. For certain overall general purposes a sample s of a size n, which may also be quite large, say a few thousand, may be supposed to have been chosen according to a design p admitting $\pi_i > 0$. Then the total $T_d = \sum_{U_d} Y_i$ for a variable of interest y relating to the members of a particular domain U_d of size N_d of interest may be estimated using the direct estimators

$$t_d = \left(\sum_{s_d} Y_i/\pi_i\right)$$

or

$$t'_d = N_d \left(\sum_{s_d} Y_i/\pi_i\right) \Big/ \left(\sum_{s_d} 1/\pi_i\right).$$

We write s_d for the part of the sample s that coincides with U_d, and n_d for the size of s_d, $d = 1, \ldots, D$, writing D for the total number of domains such that U_d's are disjoint, coincident with U when amalgamated over all the U_d's $d = 1, \ldots, D$. We suppose D is very large and so even for large $n = \sum_{d=1}^{D} n_d$, the values of n_d for numerous values of d turn out to be quite small, and even nil for many of them. Thus the sample base of t_d or t'_d happens in practice to be so small that they may not serve any useful purpose, having inordinately large magnitudes and unstable estimators for their variances, leading to inconsequential confidence intervals, which in most cases fail to cover the true domain totals. Similar and more acute happens to be the problem of estimating the domain means $\bar{T}_d = T_d/N_d$, writing domain size as N_d, which often is unknown. Hence the problem of **small domain statistics**, and a special method of estimation is needed for the parameters relating to small domains, which are often geographical areas and hence are called small areas or local areas. In this section, we will briefly discuss a few issues involved in small area or local area estimation.

Often a population containing numerous domains of interest is also divisible into a small number of disjoint groups $U_{.1}, \ldots, U_{.G}$, say G in practice not exceeding 20 so that U may be supposed to be cross-classified into DG cells U_{dg}, $d = 1, \ldots, D$ and $g = 1, \ldots, G$, of sizes N_{dg} such that $\sum_g N_{dg} = N_d$, $\sum_d N_{dg} = N_{.g}$ and $\sum_g \sum_d N_{gd} = \sum_d \sum_g N_{dg} = \sum_d N_d = \sum_g N_{.g} = N$. Of course the union of U_{dg} over d is $U_{.g}$ and that over g is U_d. If the sample is chosen from U disregarding U_g's the latter are just the post-strata in case $N_{.g}$'s are known, as will be supposed to be the case; often N_{dg}'s themselves are reliably known from a recent past census or from administration or registration data sources in problems of local area estimation. These post-strata may stand for age, sex, or racial classifications in usual practices. If the population is divided again into strata for sampling purposes, then we have classifications leading to the entities for which we have the following obvious notations. The hth stratum is $U_{..h}$, of size $N_{..h}$, the size of cell U_{dgh} is N_{dgh}, $N = \sum_d \sum_g \sum_h N_{dgh} = \sum_g \sum_h N_{.gh} = \sum_d \sum_h N_{d.h} = \sum_d \sum_g N_{dg}$, etc. Correspondingly, $N, n_{dgh}, n_{.gh}, n_{d.h}, n_{dg}$ will denote sizes of the samples $s, s_{dgh}, s_{.gh}, s_{d.h}, s_{dg}$, etc. Further, we shall write H_d to denote the set of design strata having a non-empty intersection with the domain U_d. The problem is now to estimate the domain total

$$T_d = \Sigma_{H_d} \Sigma_{U_{d.h}} Y_k$$

and the expansion or direct estimators for it are

$$t_d = \Sigma_{H_d} \Sigma_{sd.h} Y_k/\pi_k$$

or

$$t'_d = N_d \left(\Sigma_{H_d} \Sigma_{sd.h} Y_k/\pi_k\right)/\left(\Sigma_{H_d} \Sigma_{sd.h} 1/\pi_k\right)$$

based on a stratified sample. These estimators make a minimal use of data that may be available and for most domains, being based on too-scanty survey data, are too inefficient to be useful. So ways and means are to be explored to effect improvements upon them by broadening their databases and borrowing strengths from data available on similar domains and secondary external sources.

One procedure is to use poststratified estimators if auxiliary data, for example, values X_i on a correlated variable, are available for every unit for each cell U_{dgh}. Then the following estimators of T_d may be employed based on poststratification:

$$t_{pdx} = \sum_g \left[\left(\Sigma_{U_{dg}} X_k\right) \left(\Sigma_{sdg} Y_k/\pi_k\right) \left(\Sigma_{sdg} X_k/\pi_k\right)\right]$$

$$t_{pdxsc} = \sum_g \left[\left(\Sigma_{H_d} \sum_{U_{dgh}} X_k \left(\Sigma_{H_d} \Sigma_{sdgh} Y_k/\pi_k\right) / \left(\Sigma_{H_d} \Sigma_{sdgh} X_k/\pi_k\right)\right)\right]$$

$$t_{pdxss} = \sum_g \Sigma_{H_d} \left(\Sigma_{sdgh} Y_k/\pi_k\right) \left[\frac{\Sigma_{U_{dgh}} X_k}{\Sigma_{sdgh} X_k/\pi_k}\right].$$

These are ratio-type poststratified estimators, the latter two being, respectively, combined-ratio and separate-ratio types based on stratified sampling. In case X_k's are not available but the sizes N_{dg} and, in case of stratified sampling, the sizes N_{dgh}, are known, then we have the simpler count-type poststratified estimators based on SRSWORs from U or $U_{..h}$'s:

$$t_{pdc} = \sum_g N_{dg} \overline{y}_{dg},$$

$$t_{pdcsc} = \sum_g \Sigma_{H_d} N_{dgh} \left(\Sigma_{H_d} N_{..h} \frac{n_{dgh}}{n_{..h}} \overline{y}_{dgh}\right) \Big/ \left(\Sigma_{H_d} N_{..h} \frac{n_{dgh}}{n_{..h}}\right)$$

$$t_{pdcss} = \sum_g \Sigma_{H_d} N_{dgh} \overline{y}_{dgh}.$$

10.4.2 Synthetic Estimators

Since n_{dg} and n_{dgh}'s are very small, if we may believe that the g groups have been so effectively formed that in respect of the characteristics of interest y there is homogeneity within each separate group across the domains, then the following broad-based estimators for T_d may be useful

$$t_{csd} = \sum_g N_{dg} \left(\Sigma_{s.g} Y_k/\pi_k\right)/\left(\Sigma_{s.g} 1/\pi_k\right)$$

$$t_{cscd} = \sum_g \left(\Sigma_{H_d} N_{dgh}\right)\left(\Sigma_{H_d} \Sigma_{s.gh} Y_k/\pi_k\right)/\left(\Sigma_{H_d} \Sigma_{s.gh} 1/\pi_k\right)$$

$$t'_{cssd} = \sum_g \Sigma_{H_d} N_{dgh} \left(\Sigma_{s.gh} Y_k/\pi_k\right)/\left(\Sigma_{s.gh} 1/\pi_k\right)$$

called the **count-synthetic estimators** for unstratified, **stratified-combined**, and **stratified-separate** sampling, respectively. The corresponding **ratio-synthetic** estimators for unstratified and stratified sampling are:

$$t_{Rsd} = \sum_g X_{dg} \left(\Sigma_{s.g} Y_k/\pi_k\right)/\left(\Sigma_{s.g} X_k/\pi_k\right)$$

$$t_{Rscd} = \sum_g \left(\Sigma_{H_d} X_{dgh}\right) \frac{\left(\Sigma_{H_d} \Sigma_{s.gh} Y_k/\pi_k\right)}{\left(\Sigma_{H_d} \Sigma_{s.gh} X_k/\pi_k\right)}$$

$$t_{R_{ssd}} = \sum_g \Sigma_{H_d} X_{dgh} \left(\Sigma_{s.gh} Y_k/\pi_k\right)/\left(\Sigma_{s.gh} X_k/\pi_k\right).$$

For SRSWOR from U and independent SRSWORs from $U_{..h}$, we have the six simpler synthetic estimators

$$t_1 = \sum_g N_{dg} \overline{y}_{.g}$$

$$t_2 = \sum_g X_{dg} \frac{\overline{y}_{.g}}{\overline{x}_{.g}},$$

$$t_3 = \sum_g \Sigma_{H_d} N_{dgh} \left(\Sigma_{H_d} \frac{N_{..h}}{n_{..h}} n_{.gh} \overline{y}_{.gh}\right) \Big/ \left(\Sigma_{H_d} \frac{N_{..h}}{n_{..h}} n_{.gh}\right)$$

$$t_4 = \sum_g \Sigma_{H_d} N_{dgh} \overline{y}_{.gh}$$

$$t_5 = \sum_g \Sigma_{H_d} X_{dgh} \frac{\left(\Sigma_{H_d} \frac{N_{..h}}{n_{..h}} n_{.gh} \overline{y}_{.gh}\right)}{\left(\Sigma_{H_d} \frac{N_{..h}}{n_{..h}} n_{.gh} \overline{x}_{.gh}\right)}$$

$$t_6 = \sum_g \Sigma_{H_d} X_{dgh} \frac{\overline{y}_{.gh}}{\overline{x}_{.gh}}.$$

Since the sample sizes $n_{\cdot gh}$ compared to n_{dgh} and $n_{\cdot g}$ compared to n_{dg} are large, the synthetic estimators are based on much broader sample survey databases than the poststratified estimators, and hence have much smaller variances. But if the construction of the post-strata is not effective so that the characteristics across the domains within respective post-strata are not homogeneous, the synthetic estimators are likely to involve considerable biases. As a result, reduction of variances need not in practice be enough to offset the magnitudes of squared biases to yield values of mean square errors within reasonable limits. Also estimating their biases and MSEs is not an easy task. Incidentally, a simple count-synthetic estimator based on SRSWOR, for $\overline{T}_d = \frac{T_d}{N_d}$ is

$$\bar{t}_{csd} = \sum_g \frac{N_{dg}}{N_d} \bar{y}_{\cdot g} = \sum_g P_{dg} \bar{y}_{\cdot g},$$

such that $0 < P_{dg} < 1, \sum_g P_{dg} = 1$. An alternative count-synthetic estimator for $\overline{T}_d$, namely,

$$\bar{t}_{csd} = \sum_g \frac{N_{dg}}{N_{\cdot g}} \bar{y}_{\cdot g} = \sum_g W_{dg} \bar{y}_{\cdot g}$$

with $0 < W_{dg} < 1, \sum_d W_{dg} = 1$ has also been studied in the literature and shows different properties.

10.4.3 Model-Based Estimation

An alternative procedure of small area estimation involving a technique of borrowing strength is the following. Suppose $T_d, d = 1, \ldots, D$ are the true values for large number, D, of domains of interest and, employing suitable sampling schemes, estimates t_d for $d \in s_0$ are obtained, where s_0 is a subset of m domains. Now, suppose auxiliary characters $x_j, j = 1, \ldots, K$ are available with known values $X_{jd}, d = 1, \ldots, D$. Then, postulating a linear multiple regression

$$T_d = \beta_0 + \beta_1 X_{1d} + \ldots + \beta_K X_{Kd} + \epsilon_d; \; d = 1, \ldots, m$$

one may write for $d \in s_0$

$$t_d = \beta_0 + \beta_1 X_{1d} + \ldots + \beta_K X_{Kd} + e_d + \epsilon_d$$

writing $e_d = t_d - T_d$, the error in estimating T_d by t_d. Now applying the principle of least squares utilizing the sampled values, one may get estimates $\widehat{\beta}_j$ for $j = 0, 1, \ldots, K$ based on (t_d, X_{jd}) for $d \epsilon s_0$ and $j = 1, \ldots, K$, assuming $m > K + 1$. Then, we may take $\sum_0^K \widehat{\beta}_j X_{jd} = \widehat{T}_d$ as estimates for T_d not only for $d \epsilon s_0$ but also for the remaining domains $d \notin s_0$.

This method has been found by ERICKSEN (1974) to work well in many situations of estimating current population figures in large numbers of U.S. counties and in correcting census undercounts. An obvious step forward is to combine the estimators t_d with $\widehat{T}_d$ for $d = 1, \ldots, m$ to derive estimators that should outperform both t_d and $\widehat{T}_d, d = 1, \ldots, m$. Postulating that e_d's and ϵ_d's are mutually independent and separately iid random variates respectively distributed as $N(0, \sigma^2)$ and $N(0, \tau^2)$, following GHOSH and MEEDEN (1986) one may derive weighted estimators

$$t_d^* = \frac{\tau^2}{\sigma^2 + \tau^2} t_d + \frac{\sigma^2}{\sigma^2 + \tau^2} \widehat{T}_d, d = 1, \ldots, m$$

provided σ and τ are known. If they are unknown, they are to be replaced by suitable estimators. Thus, here we may use JAMES–STEIN or empirical Bayes estimators of the form

$$\widehat{t}_d = \widehat{W} t_d + (1 - \widehat{W}) \widehat{T}_d$$

with $0 < \widehat{W} < 1$, such that according as $t_d (\widehat{T}_d)$ is more accurate for T_d, the weight $\widehat{W}$ goes closer to 1(0). These procedures we have explained and illustrated in section 4.2. PRASAD (1988) is an important reference.

A compelling text on small area estimation is J. N. K. RAO (2002); MUKHOPADHYAY (1998) is an immediately earlier text. In the context of small area estimation some of the concepts need to be mentioned as below. A direct estimator for a domain parameter is one that uses the values of the variable of interest relating only to the units in the sample that belong to this particular domain. An indirect estimator for a domain parameter of interest is one that uses values of the variables of interest in the sample of units even outside this specific domain. As illustrations, let us consider the generalized regression (GREG) estimator for a dth domain total

Y_d of a variable of interest ($d = 1, \ldots, D$), viz.

$$t_{gd} = \sum_{i \in s} \frac{y_i}{\pi_i} I_{d_i} + \left(X_d - \sum_{i \in s} \frac{x_i}{\pi_i} I_{di} \right) b_{Qd}$$

writing

$$\begin{aligned} I_{di} &= 1 \quad \text{if} \quad i \in U_d \\ &= 0, \quad \text{else,} \end{aligned}$$

$X_d = \sum_1^N x_i I_{di}$, x a variable well associated with y, $Q_i (> 0)$ a preassigned real number and

$$b_{Qd} = \frac{\sum_{i \in s} y_i x_i Q_i I_{di}}{\sum_{i \in s} x_i^2 Q_i I_{di}}$$

This t_{gd} may be treated as a model-motivated, rather than model-assisted, as per SÄRNDAL, SWENSSON and WRETMAN's (SSW, 1992) terminology, estimator or predictor for Y_d suggested by the underlying model for which we may write

$$M_1 : y_i = \beta_d x_i + \epsilon_i, \; i \in U_d, \; d = 1, \ldots, D.$$

The regression coefficient β_d in this model is estimated by b_{Qd} and used in t_{gd}. The t_{gd} is a direct estimator and it does not borrow any strength from outside the domain. If M_1 is replaced by the model:

$$M_2 : \; y_i = \beta x_i + \epsilon_i, \; i \in U,$$

then t_{gd} may more reasonably be replaced by

$$t_{sgd} = \sum_{i \in s} \frac{y_i}{\pi_i} I_{di} + \left(X_d - \sum_{i \in s} \frac{x_i}{\pi_i} I_{di} \right) b_Q$$

taking

$$b_Q = \frac{\sum_{i \in s} y_i x_i Q_i}{\sum_{i \in s} x_i^2 Q_i}.$$

This t_{sgd} borrows strength from outside the domain U_d because in b_Q values of y_i are used for i in s that are outside $s_d = s \cap U_d$ and hence it is an indirect estimator. So, we call it a synthetic GREG estimator in contrast to the nonsynthetic GREG estimator t_{gd}, which is a direct estimator.

Let us write

$$t_{gd} = \sum_{i \in s} \frac{y_i}{\pi_i} g_{sdi},$$

$$g_{sdi} = \left[1 + \left(X_d - \sum_{i \in s} \frac{x_i}{\pi_i} I_{di} \right) \frac{x_i Q_i \pi_i}{\sum_{i \in s} x_i^2 Q_i I_{sdi}} \right] I_{di},$$

$$t_{sgd} = \sum_{i \in s} \frac{y_i}{\pi_i} G_{sdi},$$

$$G_{sdi} = \left[I_{di} + \left(X_d - \sum_{i \in s} \frac{x_i}{\pi_i} I_{di} \right) \frac{x_i Q_i \pi_i}{\sum_{i \in s} x_i^2 Q_i} \right]$$

$$e_{di} = (y_i - b_{Qd} x_i), e_{sdi} = (y_i - b_Q x_i)$$

Then, following SÄRNDAL (1982), two estimators for each of the mean square errors (MSE) of t_{gd} and of t_{sgd} about Y_d are available as

$$m_{kd} = \sum\sum_{i<j \in s} \left(\frac{\pi_i \pi_j - \pi_{ij}}{\pi_{ij}} \right) \left(\frac{a_{ki} e_{di}}{\pi_i} - \frac{a_{kj} e_{dj}}{\pi_j} \right)^2,$$

$$k = 1, 2; \; a_{1i} = I_{di}, a_{2i} = g_{sdi}$$

$$m_{skd} = \sum\sum_{i<j \in s} \left(\frac{\pi_i \pi_j - \pi_{ij}}{\pi_{ij}} \right) \left(\frac{b_{ki} e_{sdi}}{\pi_i} - \frac{b_{kj} e_{sdj}}{\pi_j} \right)^2,$$

$$k = 1, 2; \; b_{1i} = I_{di}, b_{2i} = G_{sdi}, i \in s$$

In order to borrow further strength in estimation, let us illustrate a way by a straightforward utilization of the above models M_1 and M_2 further limited respectively as follows:

M_1' : Model M_1 with $\in_i \overset{ind}{\sim} N(0, A)$

M_2' : Model M_2 with $\in_i \overset{ind}{\sim} N(0, A)$

with A as an unspecified non-negative real constant. Let us further postulate:

I. $t_{gd} / Y_d \overset{ind}{\sim} N(\beta_d X_d, v_d)$

$Y_d \overset{ind}{\sim} N(\beta_d X_d, A)$

and

$$\text{II.} \quad t_{sgd}/Y_d \sim N(\beta X_d, v_d), Y_d \overset{ind}{\sim} N(\beta X_d, A)$$

with v_d as either m_{kd} in case I and as m_{skd} in case II.

Considering case II it follows that

$$\begin{pmatrix} t_{sgd} \\ Y_d \end{pmatrix} \sim N_2\left(\begin{pmatrix} \beta X_d \\ \beta X_d \end{pmatrix}, \begin{pmatrix} A+v_d & A \\ A & A \end{pmatrix}\right);$$

Consequently,

$$Y_d | t_{sgd} \sim N\left(\beta X_d + \frac{A}{A+v_d}(t_{sgd} - \beta X_d), \frac{Av_d}{A+v_d}\right)$$

So,

$$\hat{Y}_{Bd} = \left(\frac{A}{A+v_d}\right) t_{sgd} + \left(\frac{v_d}{A+v_d}\right) \beta X_d$$

is the Bayes estimator (BE) for Y_d. This is true for any t_d if the model is valid for t_d and not just for t_{sgd}. But as A and B are unknown, $\hat{Y}_{Bd}$ is not usable.

Let

$$\widetilde{\beta} = \frac{\sum_{d=1}^{D} t_{sgd} X_d/(A+v_d)}{\sum_{d=1}^{D} X_d^2/(A+v_d)} \qquad (10.1)$$

and

$$\sum_{d=1}^{D} (t_{sgd} - \widetilde{\beta} X_d)^2/(A+v_d) \text{ be equated to } (D-1). \qquad (10.2)$$

Solving Eq. (10.1) and Eq. (10.2) by iteration starting with $A = 0$ in Eq. (10.1), let us find $\hat{A}$ as an estimator for A and

$$\hat{\beta} = \frac{\sum_{d=1}^{D} t_{sgd} X_d/(\hat{A}+v_d)}{\sum_{d=1}^{D} X_d^2/(\hat{A}+v_d)}.$$

Taking $\hat{\beta}, \hat{A}$ as estimators of β, A by the method of moments it is usual to take

$$\hat{Y}_{EBd} = \left(\frac{\hat{A}}{\hat{A}+v_d}\right) t_{sgd} + \left(\frac{v_d}{\hat{A}+v_d}\right) \hat{\beta} X_d$$

as the empirical Bayes estimator for Y_d. FAY and HERRIOT (1979) is the relevant reference. PRASAD and RAO (1990) have given the following estimator for $\hat{Y}_{EBd}$ as

$$m_d = m_{1d} + m_{2d} + 2m_{3d},$$

where

$$m_{1d} = \frac{\hat{A}v_d}{\hat{A}+v_d} = r_d v_d, \text{ say}, r_d = \frac{\hat{A}}{\hat{A}+v_d}$$

$$m_{2d} = \frac{(1-r_d)^2 X_d^2}{\sum_{d=1}^{D}\left(\frac{X_d^2}{\hat{A}+v_d}\right)},$$

$$m_{3d} = \frac{v_d^2}{(\hat{A}+v_d)^3}\left[\frac{2}{D}\sum_{d=1}^{D}(\hat{A}+v_d)^2\right]$$

GHOSH (1986) and GHOSH and LAHIRI (1987) have discussed asymptotical optimality properties of empirical Bayes estimators (EBE) valid when D is large.

In an unrealistic special case when $v_d = v$ for every $d = 1, 2, \ldots, D$, we have

$$\hat{Y}'_{Bd} = \left(\frac{A}{A+v}\right) t_{sgd} + \left(\frac{v}{A+v}\right)\beta X_d$$

$$\widetilde{\beta}' = \left(\sum_{d=1}^{D} t_{sgd} X_d\right) \bigg/ \sum_{d=1}^{D} X_d^2.$$

Also

$$E\left[\sum_{d=1}^{D}(t_{sgd} - \widetilde{\beta}' X_d)^2/(A+v)\right] = \frac{1}{D-1}$$

Writing

$$S = \sum_{d=1}^{D}(t_{sgd} - \widetilde{\beta}' X_d)^2$$

we have

$$\frac{1}{A+v} = E\left(\frac{1}{S}\right)/(D-3).$$

So, $\frac{D-3}{S}$ is an unbiased estimator for $\frac{1}{A+v}$. Consequently, one may employ for Y_d the JAMES–STEIN (1961) estimator

$$\hat{Y}_{Jsd} = \left(1 - \frac{(D-3)v}{S}\right) t_{sgd} + \left(\frac{(D-3)v}{S}\right)\widetilde{\beta}' X_d.$$

This has the property that

$$E\left[\sum_{d=1}^{D}(\hat{Y}_{Jsd} - Y_d)^2\right] \leq E\left[\sum_{d=1}^{D}(t_{sgd} - Y_d)^2\right].$$

Obviously $\hat{Y}_{EBd}$ is more realistic than $\hat{Y}_{JSd}$, and hence the latter is discarded in practice. We have illustrated how small domain statistics are derived by way of borrowing strength from the geographically neighboring domains. An approach of borrowing from past data on the same domain for which a parameter needs to be estimated and also on the neighboring domains is possible. An effective way to do this is by Kalman filter technique as succinctly described by MEINHOLD and SINGPURWALLA (1983) and CHAUDHURI and MAITI (1994, 1997), two relevant references.

10.5 CONDITIONAL INFERENCE

In the design-based approach, usually the inferential basis for survey data analysis is provided by conceptually repeated selection of samples. Performance characteristics of sampling strategies are assessed on averaging out certain functions of samples and parameters over all possible samples bearing positive selection probabilities. In the predictive approach and Bayesian inference, the assessment is conditional on the realized sample without speculation of any kind as to what would have happened if, instead of the sample at hand, some other samples might have been drawn, distorting the current sample configurations. But recently some information is available in survey sampling literature on possible conditional inference even within the ambit of classical design-based repeated sampling approach. We intend to refer to some of them here in brief as the issue is relevant in the contexts of poststratified sampling and small area estimation.

Suppose for a sample s of size n taken at random from a population $U = (1, \ldots, i, \ldots, N)$ of N units with H post-strata of known sizes N_h an observed sample configuration is $\underline{n} = (n1, \ldots, n_h, \ldots, n_H)$, n_h $(\geq 0, \sum_1^H n_h = n)$ denoting the numbers of units of s coming from the hth post-stratum, $h = 1, \ldots, H$. Then, in evaluating the performances of

$$t_1 = \sum_h W_h \overline{y}_h$$

where $\overline{y}_h$ is the mean of the n_h sample observations if $n_h \geq 1$, and 0 otherwise,

$$t_2 = \sum_h W_h \overline{y}_h I_h / E(I_h)$$

or of

$$t_3 = \sum W_h \overline{y}_h I_h / E(I_h) / \sum W_h I_h / E(I_h)$$

in estimating $\widehat{Y}$, where $W_h = \frac{N_h}{N}$, $\overline{y}_h$ as before if $n_h \geq 1$ and otherwise $\overline{y}_h = \widehat{Y}_h$, the hth post-stratum mean, and $I_h = 1(0)$ if $n_h \geq 1$ $(= 0)$ and

$$E(I_h) = \mathit{Prob}(I_h = 1) = 1 - \binom{N - N_h}{n} \bigg/ \binom{N}{h},$$

the questions are the following. Is it right to evaluate $t_j, j = 1, 2, 3$ in terms of overall expectations $E = E(t_j)$ and MSEs $M = E(t_j - \widehat{Y})^2$ or the conditional expectations $E_c(t_j | \underline{n}) = E_c$ and conditional MSEs

$$M_c(t_j | \underline{n}) = E_c \left[(t_j - \overline{Y})^2 | \underline{n} \right] = M_c,$$

given the realized $\underline{n}$ for the sample s at hand? A consensus is not easy to reach, but it seems that currently the balance has tilted in favor of the opinions that (a) for future planning of similar surveys, for example, in allocating a sample size consistently with a given constrained budget, the parameters E and M are more relevant than E_c and M_c while (b) in analyzing the current data through point estimation along with a measure of its error and in interval estimation, the relevant parameters are E_c and M_c. Admitting (b), one should construct conditional rather than unconditional confidence intervals utilizing sample-based estimators $\widehat{M}_c$ for M_c rather than $\widehat{M}$ for M.

For example, noting that

$$M = \sum W_h^2 S_h^2 \left[E\left(\frac{1}{n_h}\right) - \frac{1}{N_h} \right],$$

$$S_h^2 = \frac{1}{N_h^{-1}} \sum_1^{N_h} (Y_{hk} - \overline{Y}_h)^2, \; W_h = \frac{N_h}{N}$$

and

$$M_c = \sum W_h^2 S_h^2 \left(\frac{1}{n_h} - \frac{1}{N_h} \right),$$

writing

$$s_h^2 = \frac{1}{n_h - 1} \sum_1^{n_h} (Y_{nk} - \overline{y}_h)^2$$

if $n_h > 1$ and 0, otherwise, it seems more plausible to construct a confidence interval $t_1 \pm \tau_{\alpha/2}\sqrt{\widehat{M}_c}$ where

$$\widehat{M}_c = \sum W_h^2 s_h^2 \left(\frac{1}{n_h} - \frac{1}{N_h} \right)$$

rather than $t_1 \pm \tau_{\alpha/2}\sqrt{M_c}$ where

$$\widehat{M} = \sum W_h^2 s_h^2 \left[E\left(\frac{1}{n_h}\right) - \frac{1}{N_h} \right].$$

Similarly, in comparing the performances as point estimators of t_1 with a comparable overall sample mean $\overline{y}_s = \frac{\sum n_h \overline{y}_h}{n}$, it is more meaningful to compare M_c instead of M with $M_c' = E_c[(\overline{y}_s - \overline{Y})^2 | \underline{n}]$ instead of with $M' = E[(\overline{y}_s - \overline{Y})^2]$. In small area estimation throughout conditional MSEs, domain estimators are usually considered relevant and confidence statements are to be based on suitable estimators of these conditional MSEs. In each case the crux of the matter is that one must find a suitable ancillary statistic $a = a(d)$ given the survey data $d = (i, Y_i | i \epsilon s)$, such that the probability distribution of $a(d)$ is independent of $\underline{Y}$ and then one should condition on $a(d)$ for given d in proceeding with a conditional inferential approach in survey sampling. For further illuminations one should consult HOLT and SMITH (1979) and J. N. K. RAO's (1985) works on this topic.

Chapter 11

Analytic Studies of Survey Data

Suppose $y, x_1, \ldots, x_k$ are real variables with values $Y_i, X_{ji}, j = 1, \ldots, k; i = 1, \ldots, N$, assumed on the units of $U = (1, \ldots, i, \ldots, N)$, labeled $i = 1, \ldots, N$. If the survey data $d = (s, Y_i, X_{ji} | i \epsilon s)$, provided by a design p, are employed in inference about certain known functions of Y_i, X_{ji}, for $i = 1, \ldots, k; i = 1, \ldots, N$ then we have what is called a **descriptive study**. For example, we may intend to estimate the totals $Y = \sum_i^N Y_i, X_j = \sum_1^N X_{ji}, j = 1, \ldots, k$ or corresponding means or ratios along with their variance or mean square error estimators and set up confidence intervals concerning these estimand parameters. Or we may be interested to examine the values of correlation coefficients between pairs of variables or multiple correlation coefficients of one variable on a set of variables, or may like to estimate the regression coefficient of y on $x_1, \ldots, x_k$, and so on. Then the parameters involved are also defined on the values Y_i, X_{ji} for $i = 1, \ldots, N$, and our analysis is descriptive.

Often, however, the parameters of concern relate to aggregates beyond those defined exclusively on the population $U = (1, \ldots, N)$ at hand with values Y_i, X_{ji} currently assumed by y, x_j's on the members of U. More specifically, consider a superpopulation setup so that $(Y_i, Y_{1i}, \ldots, X_{ki})$ is regarded as a particular realization of a random vector with $k + 1$ real-valued coordinates. Then the survey data may be employed to infer about the parameters of the superpopulation model, in which case we say that we have **analytic studies**.

In this chapter we briefly discuss theoretical developments available from the literature about how to utilize survey data in examining correlation and regression coefficients of random variables under postulated models. It is important to decide whether a purely design-based (p-based) or a purely model-based (m-based) approach or a combination of both (pm-based) is appropriate to be able to end up with the right formulation of inference problems, choose correct criteria for choice of strategies, appropriate point and interval estimators, along with suitable measures of error and coverage probabilities. These issues are briefly narrated in section 11.2.

In section 11.1 we take up another, more elementary, problem of handling surveys. Suppose, in terms of certain characteristics, the individuals in $U = (1, \ldots, i, \ldots, N)$ are assignable to a number of disjoint categories, and on the basis of ascertainments from a sample s of individuals chosen with probability $p(s)$ we obtain a sample frequency distribution of individuals falling into these categories. Then we may be interested to use this observed sample frequency distribution to test hypotheses concerning the corresponding superpopulation probabilities. Our hypotheses to be tested may concern agreement with a postulated set of category probabilities or independence among two-way cross-classified distributions. For these problems of tests for goodness of fit, homogeneity, and independence, classical theories of statistics are well-known. These classical theories are developed under the assumption that the observations are independent and identically distributed (iid, in brief). But when samples are chosen from finite populations, they are selected in various alternative ways like SRSWOR, with non-negligible sampling fractions, stratified sampling with equal or unequal probabilities of selection, cluster sampling, multistage sampling, and various varying probability sampling schemes. Any sampling different from SRSWR from an unstratified population will be referred to as **complex sampling**. So, it is important to examine whether the classical analytical procedures available for iid observations continue to remain valid under violation of this basic assumption and, if not, to study the nature of the effect of complex sampling and, in case the effects are drastic, what kind of modifications may be needed to restore their validity.

11.1 DESIGN EFFECTS ON CATEGORICAL DATA ANLYSIS

11.1.1 Goodness of Fit, Conservative Design-Based Tests

Suppose a character may reveal itself in $k+1$ distinct forms $1,\ldots,i,\ldots,k+1$ with respective probabilities $p_1,\ldots,p_i,\ldots,$ p_k, p_{k+1}, $(0 \le p_i \le 1, \sum_1^{k+1} p_i = 1)$, which are unknown. Let a sample s of size n be drawn with probability $p(s)$ from $U = (1,\ldots,N)$ such that each population member bears one of these disjoint forms of this character. Let $\widehat{p}_i$ with $0 \le \widehat{p}_i \le 1$ denote suitable consistent estimators for $p_i, i = 1,\ldots,k+1$ based on such a sample s. Suppose $p_{i0}, i = 1,\ldots,k+1$ are certain preassigned values of p_i, $i = 1,\ldots,k+1$. We may be interested to test the goodness of fit null hypothesis

$$H_0 : p_i = p_{i0},\ i = 1,\ldots,k+1$$

against the alternative $H : p_i \ne p_{i0}$ for at least one $i = 1,\ldots,$ $k+1$. Let us write

$$\underline{p} = (p_1,\ldots,p_k)',$$
$$\underline{\widehat{p}} = (\widehat{p}_1,\ldots,\widehat{p}_k)',$$
$$\underline{p}_0 = (p_{10},\ldots,p_{k0})'.$$

We shall assume that n is large and, under H_0, the vector $\sqrt{n}(\underline{\widehat{p}} - \underline{p}_0)$ has an asymptotically normal distribution with a k-dimensional null mean vector $\underline{o} = \underline{o}_k$ and an unknown variance–covariance matrix $V = V_{k\times k}$, that is, symbolically,

$$\sqrt{n}(\underline{\widehat{p}} - \underline{p}_0) \sim N_k(\underline{o}, V).$$

Writing $V = (V_{ij})$, let $\widehat{V}_{ij}$, based on s, be consistent for V_{ij} and assume that $\widehat{V} = (\widehat{V}_{ij}) = \widehat{V}_{k\times k}$ is nonsingular. Then, the well-known Wald statistic,

$$X_W = n(\underline{\widehat{p}} - \underline{p}_0)'\widehat{V}^{-1}(\underline{\widehat{p}} - p_0)$$

is useful to test the above-mentioned $H_0 : \underline{p} = \underline{p}_0$. Under the assumptions stated, this X_W is distributed asymptotically as a chi-square variable χ_k^2 with k degrees of freedom (df) if H_0 is true.

Let $Z_i, i = 1, \ldots, k$ be k independent variables distributed as $N(0, 1)$. Then $Z_i^2, i = 1, \ldots, k$ are independent chi-square variables with 1 df each so that $\sum_1^k Z_i^2$ is a variable distributed as a chi-square with k df. Hence, for large n, we write,

$$X_W \sim \sum_1^k Z_i^2.$$

In using X_W we need to have $\widehat{V}$ and $\widehat{V}^{-1}$. But in large-scale surveys, at most, $\widehat{V}_{ii}$'s are published, and even if $\widehat{V}_{ij}$'s for $i \neq j$ are available, $\widehat{V}^{-1}$ is often found to have considerable instability when the number of categories is large, the number of clusters is small, and the sample size per category is small. So, alternatives to X_W are desirable to test for goodness of fit.

A well-known alternative statistic to test H_0 is the Pearsonian chi-square statistic

$$X = X_p = n\sum_1^{k+1}(\widehat{p}_i - p_{io})^2/p_{io}$$

or a modified version of it, namely,

$$X_M = n\sum_1^{k+1}(\widehat{p}_i - p_{io})^2/\widehat{p}_i$$

which, for large n, is asymptotically equivalent to X_p. Let us write

$$P = \text{Diag}(\underline{p}) - \underline{p}\,\underline{p}' \quad \text{and} \quad P_0 = \text{Diag}(\underline{p}_0) - \underline{p}_0, \underline{p}_0'.$$

Then it follows that

$$X = n(\underline{\widehat{p}} - \underline{p}_0)'P_0^{-1}(\underline{\widehat{p}} - \underline{p}_0).$$

Of course, $P = P_0$ if H_0 is true.

If one takes an SRSWR in n drawn and denotes by n_i the sample frequencies of individuals bearing the form i, then the vector $\underline{n} = (n_1, \ldots, n_k)'$ has a multinomial distribution with expectation $\underline{p}$ and dispersion matrix P; therefore, in this context SRSWR is referred to as multinomial sampling. If H_0 is true, then X has asymptotically the distribution χ_k^2. Thus, under H_0, for a general scheme of sampling, we may write $X_W \sim \chi_k^2 = \sum_1^k Z_i^2$ and for multinomial sampling

$$X = X_p \sim X_M \sim \chi_k^2 = \sum_1^k Z_i^2.$$

But, for sampling schemes other than the multinomial, one cannot take X under H_0 as a χ_k^2 variable. These cases require a separate treatment as briefly discussed below.

Let $D = P_0^{-1}V$ and $\lambda_1 \geq \lambda_2 \ldots \geq \lambda_k$ be the eigenvalues of D. Each of the λ_i's may be seen to be non-negative. RAO and SCOTT (1981) have shown that under H_0, the Pearsonian statistic X is distributed asymptotically as $\Sigma \lambda_i Z_i^2$ and we write

$$X \sim \sum_1^k \lambda_i Z_i^2.$$

In case of multinomial sampling it may be checked that $D = I = I_k$ the identity matrix of order k and $\lambda_i = 1$ for each $i = 1, \ldots, k$.

The ratio of the variance of an estimator based on a given complex sampling design to the variance of a comparable estimator based on SRSWR, with the same sample sizes for both, has been denoted by KISH (1965) as the **design effect** (deff) of the complex sampling design. Now, RAO and SCOTT (1981) noted that

$$\lambda_1 = \sup_{\underline{c}} \frac{\underline{c}'V\underline{c}}{\underline{c}'P\underline{c}}, \quad \lambda_k = \inf_{\underline{c}} \frac{\underline{c}'V\underline{c}}{\underline{c}'P\underline{c}},$$

for an arbitrary k vector $\underline{c} = (c_1, \ldots, c_k)'$ of real coordinates so that

$$\underline{c}'V\underline{c} = Var\left(\sum_1^k c_i \widehat{p}_i\right)$$

for a complex sampling design p and

$$\underline{c}'P\underline{c} = Var\left(\sum_1^k c_i\widehat{p}_i\right)$$

for SRSWR. So, following KISH's definition, RAO and SCOTT (1981) give the name **generalized design effects** (generalized deff) to the λ_i's above such that $\lambda_1(\lambda_k)$ is the maximal (minimal) generalized deff.

If one may correctly guess the value of λ_1, then X/λ_1 provides a conservative test for H_0 treating χ_k^2 under H_0, that is, the procedure of rejecting H_0 when X/λ_1 exceeds $\chi^2_{k,\alpha}$, achieves a significance level (SL) less than the nominal level of α. Thus the price paid in replacing the available level $-\alpha$ test based on X_W by one based on the simpler statistic X is that we achieve a lower SL. By contrast, if we reject H_0 on observing $X \geq \chi^2_{k,\alpha}$ then in many cases the achieved SL will far exceed α.

If SRSWOR in n draws is used, then

$$V = \left(1 - \frac{n}{N}\right)P_0,\ D = \left(1 - \frac{n}{N}\right)I_k.$$

Thus, here $\lambda_1 = (1 - \frac{n}{N})$ for every $i = 1, \ldots, k$. In this case RAO and SCOTT's (1981) modification of X_P is $X_{RS} = X/(1 - \frac{n}{N})$, which under H_0 has the asymptotic distribution of χ_k^2. The test of H_0 consists of rejecting it if $X_{RS} > \chi^2_{k,\alpha}$ achieves asymptotically the SL α as desired and RAO and SCOTT (1981) have shown that in case $(1 - \frac{n}{N})$ is not negligible relative to unity, this test acquires substantially higher power than the Pearson test procedure, keeping the SL for both fixed at a desired level α.

If the complex design corresponds to the stratified random sampling with proportional allocations, then it is not difficult to check that $\lambda_1 \leq 1$, implying that $X \leq \sum_1^k Z_i^2$. So, the Pearson test with no modifications remains a conservative test in this situation. FELLEGI's (1978) observation that the limiting value of $E(X)$ is less than k in this case was a pointer to this test being a conservative one as demonstrated by RAO and SCOTT (1981).

If the number of strata is only two, then the asymptotic distribution of X is that of $\chi^2_{k-1} + (1-a)\chi_1^2$, where χ^2_{k-1} and χ_1^2

are independent and

$$a = W_1(1 - W_1)\sum_{i=1}^{k+1}(p_{i1} - p_{i2})^2/p_{i0} \leq 1$$

is the trace of the matrix

$$W_1(1 - W_1)P^{-1}(\underline{p}_1 - \underline{p}_2)(\underline{p}_1 - \underline{p}_2)'.$$

Here W_1 is the first stratum proportion, p_{ih} is the probability of category i for stratum h, and $\underline{p}_h = (p_{1h}, \ldots, p_{kh})'$, $h = 1, 2$. If k is large, there is little error in approximating X by χ_k^2 because $\chi_{k-1}^2 + \chi_1^2 = \chi_k^2$.

Let a two-stage sampling scheme be adopted, choosing primary sampling units (psu) out of R available psus with replacement with selection probabilities proportional to the numbers $M_1, M_2, \ldots, M_R$ of secondary sampling units (ssu) contained in them. Assume r draws are made, and every time a psu is chosen an SRSWR of ssus is taken from it in m draws, giving a total sample size $n = mr$. Let $p_{it}(i = 1, \ldots, k + 1; t = 1, \ldots, R)$ be the probabilities of category i in psu t and define

$$W_t = M_t/\sum_1^R M_t$$

$$p_i = \sum_1^R W_t p_{it}, \; i = 1, \ldots, k + 1,$$

$$\underline{p} = (p_1, \ldots, p_k)', \; \underline{p}_t = (p_{1t}, \ldots, p_{kt})'$$

Then, one may check that

$$V = P_0 + (m - 1)\sum W_t(\underline{p}_t - \underline{p}_0)(\underline{p}_t - \underline{p}_0)' = P_0 + (m - 1)A,$$

Let $B = P_0^{-1}A$ and $\rho_i(i = 1, \ldots, k)$ be the eigenvalues of B. Then the eigenvalues λ_i of V satisfy $\lambda_i = 1 + (m - 1)\rho_i$. These ρ_i's are interpreted as generalized measures of homogeneity. Supposing $\rho_1 \geq \ldots \geq \rho_k$, if a value of ρ_1 can be guessed a conservative test for $H_0 : \underline{p} = \underline{p}_0$ may be based on the statistic $X/[1 + (m - 1)\rho_i]$ because this, under H_0, is asymptotically less than $\sum_1^k Z_i^2$. Since $\rho_1 \leq 1$, a test based on X/m is always conservative.

11.1.2 Goodness of Fit, Approximative Design-Based Tests

Whatever the eigenvalue λ_i of $D = P_0^{-1}V$, let

$$\overline{\lambda} = \sum_1^k \lambda_i/k,\ a^2 = \frac{1}{(\overline{\lambda})^2}\sum_1^k(\lambda_i - \overline{\lambda})^2/k,\ b = \frac{k}{1+a^2}.$$

It follows that under H_0 and under large sample approximation,

$$E(X/\overline{\lambda}) = k = E\sum_1^k Z_i^2$$

$$V(X/\overline{\lambda}) = 2k(1+a^2) > 2k = V\left(\sum_1^k Z_i^2\right).$$

Also,

$$\overline{\lambda} = \frac{tr\,(P_0^{-1}V)}{k} = \frac{tr\,(D)}{k} = \sum_1^{k+1} V_{ii}/p_i,$$

where V_{ii} are the diagonal elements of $V = (V_{ij})$.

Let

$$d_i = \frac{V_{ii}}{p_i(1-p_i)} = \frac{V_{ii}/n}{p_i(1-p_i)/n} = \frac{V_p(\widehat{p}_i)}{V_{srs}(\widehat{p}_i)}$$

be the deff for $\widehat{p}_i$, writing V_p, V_{srs} as variances for a given design p and for SRSWR, respectively. Then,

$$\overline{\lambda} = \frac{1}{k}\sum_1^{k+1} d_i(1-p_i).$$

Now, if suitably consistent estimators $\widehat{V}_{ii}$ of V_{ii} and $\widehat{d}_i$ of d_i are available, then one may get an estimate $\widehat{\lambda}$ of $\overline{\lambda}$ and $X_F = X/\widehat{\lambda}$ is a suitable modification of Pearson's statistic X. If one rejects H_0 on finding $X/\widehat{\overline{\lambda}} > \chi^2_{k,\alpha}$, then one's achieved SL value for large samples should be close to the nominal level α, provided the λ_i's do not have wide variations among themselves. X_F is known as RAO and SCOTT's **first-order correction** of X.

Using the estimators $\widehat{\lambda}_i$ for λ_i and $\widehat{\overline{\lambda}} = \frac{1}{k}\sum_1^k \widehat{\lambda}$ for $\overline{\lambda}$ one may get estimators

$$\widehat{a}^2 = \frac{1}{(\widehat{\overline{\lambda}})^2}\sum_1^k (\widehat{\lambda}_i - \widehat{\overline{\lambda}})^2/k \quad \text{for } a^2$$

$$\widehat{b} = \frac{k}{(1+\widehat{a}^2)} \quad \text{for } b$$

and then use the **second-order correction**

$$X_S = X_F/(1+\widehat{a}^2)$$

and reject H_0 at level of significance α if $X_S \geq \chi^2_{\widehat{b},\alpha}$, where $\chi^2_{\widehat{b},\alpha}$ is such that for a chi-square variable $\chi^2_{\widehat{b}}$ with $\widehat{b}$ df

$$\text{Prob}\left[\chi^2_{\widehat{b}} \geq \chi^2_{\widehat{b},\alpha}\right] = \alpha.$$

This approximation given by RAO and SCOTT (1981) is based on the result of SATTERTHWAITE (1946) that the distribution of X/λ may be approximated by that of $(1+a^2)\chi^2_b$. But one may check that $\sum_1^k \lambda_i^2 = \sum_i^{k+1}\sum_j^{k+1} V_{ij}/p_i p_j$ and so one needs $\widehat{V}_{ij}$ to calculate $\widehat{a}$.

Even if $\widehat{V}_{ij}$ are available such that the procedure is applicable, it may not be stable enough. The effect of instability is failure to achieve the desired value of SL. FAY (1985) and THOMAS and RAO (1987) have reported that if $\widehat{V}_{ij}$'s are not stable, then, in spite of its asymptotic validity, a test based on X_W also often fails to achieve the intended SL values. But the test based on X_F is often found good unless $\widehat{\lambda}_i$'s vary considerably, as RAO and SCOTT (1981) have illustrated that SLs achieved by X_F remain within the range 0.05–0.056, whereas those based on uncorrected X vary over 0.14–0.77, while the desired level is 0.05.

FELLEGI (1980) recommended another correction for X given by $X/\overline{d}$, where

$$\widehat{\overline{d}} = \frac{1}{k+1}\sum_1^{k+1} \widehat{d}_i$$

Some further corrections of the above test procedures proposed in the literature enjoin consulting Fisher's F table rather than chi-square tables. THOMAS and RAO (1987) and RAO and

THOMAS (1988) are good references for these studies. The tests of goodness of fit may also be based on the well-known likelihood ratio statistic

$$G = 2n \sum_{1}^{k+1} \widehat{p}_i \log(\widehat{p}_i / p_{io}).$$

In addition, FAY (1985) has given test procedures based on jackknifed chi-square statistics, which fare better than X_F in case of wide fluctuations among $\widehat{\lambda}_i$'s.

11.1.3 Goodness-of-Fit Tests, Based on Superpopulation Models

ALTHAM (1976) made a model-based approach in this two-stage setup. An extended version of that due to RAO and SCOTT (1981) consists of defining indicator variables Z_{tji} that equal 1(0) if jth ssu of ith psu bears category i (else) and choosing r psus out of R psus of sizes M_t and m_t ssus out of M_t ssus in tth psu is sampled. Let $\underline{n} = (n_1, \ldots, n_{k+1})$ where

$$n_i = \sum_{t=1}^{r} \sum_{j=1}^{m_t} Z_{tji}, \; i = 1, \ldots, k+1.$$

Let

$$E_m(Z_{tji}) = p_i, \text{cov}_m(Z_{tji}, Z_{tj'i}) = q_{ij} \text{ say, for every } j' \neq j.$$

These conditions lead to

$$E_m(n_i) = np_i, \; V_m(n_i) = np_i(1-p_i) + \left(\sum m_t^2 - n\right) q_{ii},$$

$$\text{cov}_m(n_i, n_j) = -np_i p_j + \left(\sum m_t^2 - n\right) q_{ij}, \; i \neq j.$$

Let $Q = (q_{ij})$, $G = P^{-1}Q$, $\rho_1 \geq \rho_2 \geq \ldots \geq \rho_K$ the eigenvalues of G, $m_0 = \Sigma m_t^2/n$, $\lambda_i = 1 + (m_0 - 1)\rho_i$. Then $\rho_1 < 1$ and $X/\lambda_1 = X/m_0$ provides a basis for a conservative test. If $\rho_i = \rho$ for every $i = 1, \ldots, k$, then in case ρ may be correctly guessed, a test for the goodness of fit is based on $X/[1 + (m_0 - 1)\rho]$. If $M_t = M$ and $m_t = m$ for every t then X/m provides a conservative test.

BRIER (1980) postulates a slightly altered model for the above two-stage setting. Suppose m_{ti} is the number of sampled ssus bearing the form i of the character $\underline{m_t} = (m_{t1}, \ldots, m_{t,k+1})'$

and let $\underline{p}_t = (p_{t1}, \ldots, p_{t,k+1})$, $\sum_1^{k+1} p_{ti} = 1, 0 < P_{ti} < 1, i = 1, \ldots, k+1$. Let $\underline{p}_t$ have the Dirichlet's distribution with a density

$$f\,(p_{t1}, \ldots, p_{t,k+1}) = \frac{\Gamma(\nu)}{\overset{k+1}{\underset{1}{\pi}}\Gamma(\nu p_i)} \overset{k+1}{\underset{1}{\pi}} p_i^{\nu p_i - 1},$$

where $\nu > 0, 0 < p_i < 1, \sum_1^{k+1} p_i = 1$ and $\Gamma(x) = \int_0^\infty \bar{e}^u u^{x-1} du$. Also, given a realization $\underline{p}_t$ from the density, it is postulated that $\underline{m}_t$ has a multinomial distribution.

In the special case for which $m_t = m$ for every t, the resulting compound Dirichlet multinomial distribution of $\underline{m}_t$ yields a test based on the modification $\overline{X} = \frac{X(1+\nu)}{(m+\nu)}$ of X as an asymptotically good test for the goodness of fit. It is based on a constant deff model and it achieves the nominal SL for large samples. Another alternative to it, namely $X^* = \frac{1+\nu}{m_0+\nu} X$ where $m_0 = \sum m_t^2/n$, when m_t's may be unequal, is also asymptotically valid. To apply these tests one needs to estimate ν, and procedures are given by RAO and SCOTT (1981).

From the above discussion, it is apparent that it is not easy in practice to find λ_i's in order to be able to work out a test that rejects H_0 if $X > \chi^2_{k,\alpha}$ for a preassigned α. Using methods given by SOLOMON and STEPHENS (1977) it is possible to work these out for trial values of λ_i's just to see how the attained values of SL compare with a nominal value of α fixed at 0.05. RAO and SCOTT (1979, 1981), HOLT, SCOTT and EWINGS (1980), HIDIROGLOU and RAO (1987), RAO (1987), and others have shown that, for stratified or clustered sampling schemes, the Pearson chi-square statistic X_P frequently leads to SLs in the range of 20–40%, and not infrequently about 70%, as opposed to the nominal level of 5%. Hence, the effect of designs on blindly applied classical test procedures may be disastrous.

11.1.4 Tests of Independence

In the context of categorical data analysis, one problem is of testing for independence in two-way contingency tables with cell probabilities P_{ij}, $i = 1, \ldots, r+1$; $j = 1, \ldots, c+1$ with

$\widehat{p}_{ij}$'s as their consistent estimators based on a suitably taken sample of size n chosen according to a certain design p. Let

$$
\begin{aligned}
P_{io} &= \sum_{j=1}^{c+1} p_{ij}, \\
P_{0j} &= \sum_{i=1}^{r+1} p_{ij}, \\
h_{ij} &= p_{ij} - p_{io}p_{ij}, \\
\underline{p} &= (p_{11}, p_{12}, \ldots, p_{1c+1}, p_{21}, \ldots, p_{2c+1}, \ldots, p_{r+1c})' \\
\underline{h} &= (h_{11}, h_{12}, \ldots, h_{1c}, h_{21}, \ldots, h_{2c}, \ldots, h_{rc})' \\
\underline{\widehat{p}}_r &= (\widehat{p}_{10}, \ldots, \widehat{p}_{ro})',\ \underline{P}_r = Diag(\underline{p}_r) - \underline{p}_r\underline{p}_r' \\
\underline{\widehat{p}}_c &= (\widehat{p}_{01}, \ldots, \widehat{p}_{0c})', \underline{P}_c = Diag(\underline{p}_c) - \underline{p}_c\underline{p}_c'
\end{aligned}
$$

and define analogously

$$\widehat{p}_{10},\ \widehat{p}_{0j},\ \underline{\widehat{p}}_c,\ \underline{\widehat{p}}_r,\ \underline{\widehat{p}},\ \widehat{\underline{P}_c},\ \widehat{P_r}\underline{\widehat{h}}$$

Note that $\underline{p}$ and $\underline{\widehat{p}}$ have $(r+1)(c+1) - 1$ components, while $\underline{h}$ and $\underline{\widehat{h}}$ have rc components.

Writing $V/n(\widehat{V}/n)$ for the covariance (estimated) matrix of $\underline{\widehat{p}}$, the covariance (estimated) matrix of $\underline{\widehat{h}}$ will be $\frac{1}{n}H'VH$ (resp. $\frac{1}{n}\widehat{H}\text{–}\widehat{V}\widehat{H}$) where

$$H = \partial\underline{h}/\partial\underline{p}$$

is the matrix of partial derivatives of $\underline{h}$ wrt $\underline{p}$ and $\widehat{H}$ is defined by replacing p_{ij} in H by $\widehat{p}_{ij}$.

To test for independence of the two characters in terms of which the individuals have been classified into $(r+1)(c+1)$ categories is to test the null hypothesis

$$H_0 : p_{ij} = p_{i0}p_{0j} \text{ for every } i = 1, \ldots, r \text{ and } j = 1, \ldots, c$$

against an alternative that $h_{ij} = p_{ij} - p_{io}p_{oj}$ is non-zero for at least one pair (i, j).

The Wald statistic for this null hypothesis of independence is

$$X_W = n\underline{\widehat{h}}'(\widehat{H}'\widehat{V}\widehat{H})^{-1}\underline{\widehat{h}}$$

and the Pearson statistic is

$$X_I = n\underline{\widehat{h}}'\left(\widehat{\underline{P}_r}^{-1} \otimes \widehat{\underline{P}_c}^{-1}\right)\underline{\widehat{h}}.$$

Here, $\otimes$ denotes the Kronecker product of two matrices. Under H_0, X_W is asymptotically χ^2_{rc} distributed, while X_1 is asymptotically distributed as the variable $\sum_1^T \delta_i Z_i^2$ where $T = rc$, the δ_i's are the eigenvalues of $(\underline{P}_r^{-1} \otimes \underline{P}_c^{-1})(H'V H)$ such that $\delta_1 \geq \ldots \geq \delta_T$ and the Z_i^2's are independent χ_1^2 variables.

Here the δ_i's may be interpreted as the deffs corresponding to estimators of p_{ij}'s as functions of h_{ij}'s. As in the case of goodness of fit problems, X_1/δ_1 provides a conservative test for independence if δ_1 can be guessed or reliably estimated. If a complex design corresponds to stratified random sampling with proportional allocations, then $\delta_1 \leq 1$ and X_1 provides a conservative test. Unfortunately, simple alternative useful tests modifying X_I in this case are not yet available, as in the case of goodness of fit problems. But, as a saving grace, the deviations of SL values achieved by the Pearsonian statistic X_I from the nominal value $\alpha = 0.05$, while rejecting H_0 in case $X_I \geq \chi^2_{T,\alpha}$, are not so alarming as in the case of goodness of fit problems.

11.1.5 Tests of Homogeneity

Next we consider the problem of testing homogeneity of two populations both classified according to the same criterion into $k + 1$ disjoint categories on surveying both the populations on obtaining two independent samples of sizes n_1 and n_2 from the two populations following any complex designs.

Let $p_{ji}, i = 1, \ldots, k + 1; j = 1, 2\,(0 < p_{ij} < 1, \sum_1^{k+1} p_{ji} = 1, j = 1, 2)$ be the unknown proportions of individuals of the jth $(j = 1, 2)$ population bearing the form i $(i = 1, \ldots, k+1)$ of the classificatory character. Let $\underline{p}_j = (p_{j1}, \ldots, p_{j,k})', j = 1, 2$. Let $\widehat{p}_{ji}$ be suitably consistent estimators of p_{ji} based on the respective samples from the two populations. Let V_j/n_j, $(j = 1, 2)$ denote the variance–covariance matrices (of order $k \times k$) corresponding to $\widehat{p}_{ji}$'s admitting consistent estimators $\widehat{V}_j/n_j, (j = 1, 2)$. We will write

$$\widehat{n}_j = (\widehat{p}_{j1}, \ldots, \widehat{p}_{jk})', \; j = 1, 2.$$

The problem is to test the null hypothesis

$$H_0 : \underline{p}_1 = \underline{p}_2 = \underline{p}, \text{ say,}$$

writing $\underline{p} = (p_1, \ldots, p_k)'$ corresponding to the supposition that, under H_0, the common values of p_{ji} for $j = 1, 2$ are $p_i, i = 1, \ldots, k+1$. Let

$$P = Diag(\underline{p}) - \underline{p}\,\underline{p}', D_j = P^{-1}V_j,$$

$$\widehat{D} = (D_1/n_1 + D_2/n_2)/(1/n_1 + 1/n_2), \; \overline{n} = \frac{1}{1/n_1 + 1/n_2},$$

$$\widehat{p}_{oi} = (n_1\widehat{p}_{1i} + n_2\widehat{p}_{2i})/(n_1 + n_2),$$

$$\underline{\widehat{p}}_0 = (\widehat{p}_{01}, \ldots, \widehat{p}_{0,k})', \; \widehat{P_0} = Diag(\widehat{p}_0) - \widehat{p}_0\widehat{p}_0'.$$

Then the Wald statistic for the test of the above H_0 concerning homogeneity of two populations is

$$X_W = (\underline{\widehat{p}}_1 - \underline{\widehat{p}}_2)' \left(\frac{\widehat{V_1}}{n_1} + \frac{\widehat{V_2}}{n_2} \right)^{-1} (\underline{\widehat{p}}_1 - \underline{\widehat{p}}_2).$$

Under H_0, X_W has an asymptotic χ_k^2 distribution. The Pearson statistic for the test of this H_0 on homogeneity of two populations is

$$X_H = \overline{n}(\underline{\widehat{p}}_1 - \underline{\widehat{p}}_2)'\widehat{P_0}^{-1}(\underline{\widehat{p}}_1 - \underline{\widehat{p}}_2).$$

Writing λ_i as the eigenvalues of $\widehat{D}$, the generalized deff matrix, SCOTT and RAO (1981) and RAO and SCOTT (1981) note that under H_0, for large $n_j (j = 1, 2), X_H$ is asymptotically, distributed as $\sum_1^k \lambda_i Z_i^2$. They have noted that, for clustered designs, the SLs achieved on rejecting H_0 in case $X_H > \chi_{k,\alpha}^2$ deviate drastically from the nominal value α. For example, against a desired $\alpha = 0.05$, SL values for several clustered two-stage sampling designs actually achieved vary over the range 0.17 to 0.51, as may be checked with SCOTT and RAO (1981).

Extensions to the case of $j > 2$, that is, more than two populations, have also been covered by RAO and SCOTT (1981). In dealing with multi-way classifications, RAO and SCOTT (1984) have studied the goodness of fit problem postulating log-linear models. In this context, also, they have observed that a relevant Pearson statistic motivated by multinomial sampling is inappropriate when the sample is actually based on a complex design. They demonstrated that the large sample distribution of Pearson's statistic in this case, under the null hypothesis of a log-linear model, is that of a linear combination of independent χ_1^2 variables, with the compounding coefficients amenable to interpretations in terms of deffs. They have also demonstrated that conclusions derived from the wrong supposition that the Pearsonian statistic has a chi-square distribution yield SL values widely discrepant from the desired nominal ones. In this case, they also further presented simple corrective measures presuming the availability of suitable estimates of deffs of individual cell estimates or of certain marginal totals.

In fitting logistic and logit models while analyzing variation in estimated proportions associated with a binary response, variable similar problems are also encountered when one takes recourse to complex designs involving cluster sampling in particular, and devices available with a similar approach are reported in the literature. The details are available from RAO and SCOTT (1987), RAO and THOMAS (1988), ROBERTS, RAO and KUMAR (1987), and the references cited therein. We also omit developments originated from likelihood ratio statistics and FAY's (1985) works on jackknifed versions of Pearsonian chi-squared tests, which are generally improvements over RAO and SCOTT's (1981) first-order corrections in case estimated eigenvalues of deff matrices fluctuate too much.

11.2 REGRESSION ANALYSIS FROM COMPLEX SURVEY DATA

On regression analysis of data available through complex designing, the first problem is to fix the target parameters to infer about, the second to settle for an inferential approach. Further, there are problems of choosing the correct regressor variables and deciding on the question of whether to include design variables among the regressors or to keep them separate. We briefly report on these issues in what follows, of course, as usual drawing upon a vast literature already grown around them.

11.2.1 Design-Based Regression Analysis

Suppose $\underline{Y} = (Y_1, \ldots, Y_N)'$ is the $N \times 1$ vector of values for the N units of a finite population $U = (1, \ldots, N)$ on a dependent variable y and $\underline{X}_N$ an $N \times r$ matrix of values for these N units on r regressor variables $x_1, \ldots, x_r$. With a strictly finite population setup one may take

$$B = (\underline{X}'_N \underline{X}_N)^{-1} \underline{X}'_N \underline{Y}$$

as the parameter of interest. Let s be a sample of size n drawn from U following any scheme of sampling corresponding to a design p admitting inclusion probabilities

$$\pi_i = \sum_{s \ni i} p(s) > 0$$

$$\pi_{ij} = \sum_{s \ni i,j} p(s) > 0.$$

Let $\underline{X}_s$ be an $n \times r$ submatrix of $\underline{X}_N$ containing the values of $x_j (j = 1, \ldots, r)$ on only the n sampled units of U occurring in s and $\underline{Y}_s$ the $n \times 1$ subvector of $\underline{Y}_N$ including the y values for the units only in s. Let $\underline{W}_N$ be an $N \times N$ diagonal matrix with diagonal entries as W_i's and $\underline{W}_s$ an $n \times n$ submatrix of it involving W_i's for $i \epsilon s$ as its diagonal entries. Similarly, let $\underline{\pi}_N$, $\underline{\pi}_s$ stand for them, respectively, when W_i equals π_i, for $i = 1, \ldots, N$. Then, replacing every term of the form $\sum_{i \epsilon s} u_i W_i$ or, in particular, by $\sum_1^N u_i$ occurring in the $r \times 1$ vector B of unknown regression parameters of y on $x_1, \ldots, x_r$ by a term of the form $\sum_{i \epsilon s} \frac{u_i}{\pi_i}$, one approach is to estimate B by

$$\underline{\hat{B}}_W = (\underline{X}'_s \underline{W}_s \underline{X}_s)^{-1} (\underline{X}'_s \underline{W}_s \underline{Y}_s)$$

or, in particular, by the Horvitz–Thompson type estimator

$$\underline{\hat{B}}_{\pi^{-1}} = \left(\underline{X}'_s \underline{\pi}_s^{-1} \underline{X}_s\right)^{-1} \left(\underline{X}'_s \underline{\pi}_s^{-1} \underline{Y}_s\right).$$

We will assume the existence of the inverse matrices whenever employed. In the above, the rationale behind the use of B is that this choice minimizes the quantity

$$\underline{e}'_N \underline{e}_N$$

where $\underline{e}_N$ is defined by

$$\underline{Y}_N = \underline{X}_N B^* + \underline{e}_N$$

Thus B above provides the least squares solution for B^*. If, however, the dispersion of $\underline{e}_N$ is of an enormous magnitude, then B, in spite of providing a least squares fit, may not be very useful in explaining the relationship of y on $x_1, \ldots, x_r$. A practice of treating B as the target parameters is adopted by KISH and FRANKEL (1974), JÖNRUP and RENNERMALM (1976), SHAH, HOLT and FOLSOM (1977), and others. Admitting this B as a parameter of interest, estimators of variances of $\underline{\widehat{B}}_W$ and $\underline{\widehat{B}}_{\pi^{-1}}$ may be worked out, applying the techniques of (a) linearization based on Taylor expansion of nonlinear functions, (b) balanced repeated replication (BRR), (c) jackknifing, and (d) bootstrap. Details are available from KISH and FRANKEL (1974). In case the population is clustered, with high positive intracluster correlations and cluster sample designs employed, then they have shown that the variances of $\underline{\widehat{B}}_{\pi^{-1}}$ or $\underline{\widehat{B}}_W$ are inflated compared to what might have happened if they were based on SRSWR. Consequently, confidence intervals based on such strategies have poor coverage probabilities.

11.2.2 Model- and Design-Based Regression Analysis

Let us consider the usual model-based superpopulation approach. Then $\underline{X}_N$ is an $N \times r$ matrix of fixed real values assumed on the variables $x_1, \ldots, x_r$. But $\underline{Y}_N$ is regarded as a realization of an $N \times 1$ random vector of variables also denoted by $Y_1, \ldots, Y_N$, which have a joint probability distribution. E_m and V_m are used as operators for model-based expectation and variance–covariance:

$$E_m(\underline{Y}_N \mid \underline{X}_N) = \underline{X}_N \beta$$
$$V_m(\underline{Y}_N \mid \underline{X}_N) = \sigma^2 \underline{V}_N,$$

where β is an $r \times 1$ vector of unknown parameters and $\sigma (> 0)$ is an unknown constant. In particular $\underline{V}_N$ may equal I_N, the $N \times N$ identity matrix. Let

$$\underline{Y}_N = \underline{X}_N \beta + \underline{\epsilon}_N$$

with $\underline{\epsilon}_N$ as the $N \times 1$ vector of errors, for which

$$E_m(\underline{\epsilon}_N \mid \underline{X}_N) = 0.$$
$$V_M(\underline{\epsilon}_N \mid \underline{X}_N) = \sigma^2 \underline{V}_N.$$

In order to apply the principle of least squares to estimate β from a sample chosen from U, it is necessary that, for the subvectors and submatrices $\underline{Y}_s, \underline{X}_s, \underline{\epsilon}_s$ corresponding to $\underline{Y}_s, \underline{X}_s, \underline{\epsilon}_N$, respectively, we must have $E_m(\underline{\epsilon}_s \mid \underline{X}_s) = 0$. One way to ensure this for every s with $p(s) > 0$ is to suppose that all the variables in terms of which selection probabilities $p(s)$ are determined are covered within $x_1, \ldots, x_r$ and $p(s)$ is not influenced by the values of the dependent variable y. Later on, we will consider certain exceptional situations.

Under the above formulation, if all the values of $\underline{Y}_N, \underline{X}_N$ are available and $\underline{V}_N$ is completely known, then

$$\widehat{\beta}_G = \left(\underline{X}'_N \underline{V}_N^{-1} \underline{X}_N\right)^{-1} \left(\underline{X}'_N \underline{V}_N^{-1} \underline{Y}_N\right)$$

is the generalized least squares (GLS) estimator (GLSE) for the target parameter β. In case $\underline{V}_N = I_N$, $\widehat{\beta}_G$ is identical with the ordinary least squares estimator (OLSE)

$$\widehat{\beta}_0 = (\underline{X}'_N \underline{X}_N)^{-1} (\underline{X}'_N \underline{Y}_N).$$

But these estimators are available only if a census, rather than a sample survey, is undertaken in order to fit a regression line as modeled above. So, the problem is to use the sample survey data $\underline{Y}_s, \underline{X}_s$ to obtain a suitable estimator for $\widehat{\beta}_G$ or $\widehat{\beta}_0$, whichever is appropriate. For simplicity, let us assume that $\underline{V}_N$ is known and write $\underline{V}_s$ for the submatrix of $\underline{V}_N$ consisting of the elements corresponding to units in s.

Let us consider the estimators

$$\widehat{\beta}_1 = (\underline{X}'_s \underline{X}_s)^{-1} (\underline{X}'_s \underline{Y}_s),$$
$$\widehat{\beta}_2 = (\underline{X}'_s W_s \underline{X}_s)^{-1} (\underline{X}'_s W_s \underline{Y}_s)$$
$$\widehat{\beta}_3 = \left(\underline{X}'_s \underline{\pi}_s^{-1} \underline{X}_s\right)^{-1} \left(\underline{X}'_s \underline{\pi}_s^{-1} \underline{Y}_s\right)$$
$$\widehat{\beta}_4 = \left(\underline{X}'_s \underline{V}_s^{-1} \underline{X}_s\right)^{-1} \left(\underline{X}'_s \underline{V}_s^{-1} \underline{Y}_s\right).$$

First we note that

$$E_m(\widehat{\beta}_G) = E_m(\widehat{\beta}_0) = \beta$$
$$E_m(\widehat{\beta}_1) = E_m(\widehat{\beta}_2) = E_m(\widehat{\beta}_3) = E_m(\widehat{\beta}_4) = \beta$$

that is, each of the estimators $\widehat{\beta}_i$; $i = 1, 2, 3, 4$ is model-unbiased for β.

Further,

$$V_m(\widehat{\beta}_i) \leq V_m(\widehat{\beta}_1) \quad \text{for} \quad i = 1, 2, 3$$

The estimator $\widehat{\beta}_3$ is asymptotically unbiased and consistent. If $\underline{V}$ is diagonal and $\pi_i \propto V_{ii}$, then $\widehat{\beta}_3 = \widehat{\beta}_4$.

Among model-unbiased estimators $\widehat{\beta}_s$ of β or equivalently among model-unbiased predictors $\widehat{\beta}_s$ of $\widehat{\beta}_0$ or $\widehat{\beta}_G$ according as $\underline{V}_N = I_N (\neq I_N)$, consider those that are asymptotically design-unbiased or design-consistent for $\widehat{\beta}_0$ (or $\widehat{\beta}_G$) such that the magnitudes of $E_m E_p(\widehat{\beta}_s - \widehat{\beta}_0)^2$ or $E_m E_p(\widehat{\beta}_s - \widehat{\beta}_G)^2$ are suitably controlled. Since the population sizes in case of large-scale surveys are usually very large, the quantities $E_m(\widehat{\beta}_0 - \beta)^2$ and $E_m(\widehat{\beta} - \widehat{\beta})^2$ may disregard the differences between the target parameters β and $\widehat{\beta}_0$ (or β and $\widehat{\beta}_G$), and a predictor $\widehat{\beta}_s$ with small $E_m E_p(\widehat{\beta}_s - \widehat{\beta}_0)^2$ or $E_m E_p(\widehat{\beta}_s - \widehat{\beta}_G)^2$ may be supposed to achieve a small $E_m E_p(\widehat{\beta}_s - \beta)^2$. After such a predictor $\widehat{\beta}_s$ is found, it is an important issue as to whether to use suitable estimators for $E_m(\widehat{\beta}_s - \widehat{\beta}_0)^2$ and $E_m(\widehat{\beta}_s - \widehat{\beta}_G)^2$ for deriving what HARTLEY and SIELKEN (1975) call **tolerance intervals** of $\widehat{\beta}_0$ and $\widehat{\beta}_G$. While setting up confidence intervals for β, the question is whether to use an estimator of $E_m(\widehat{\beta}_s - \beta)^2$ or of $E_m E_p(\widehat{\beta}_s - \beta)^2$. Clear-cut solutions are not available. But let us discuss some of the developments reported in the literature.

We shall write

$$\widehat{\sigma}^2 = \frac{1}{(n-r)}(\underline{Y}_s - \underline{X}_s\widehat{\beta}_s)'(\underline{Y}_s - \underline{X}_s\widehat{\beta}_s)$$

where $\widehat{\beta}_s$ stands for the least squares estimator for β under an appropriate model, that is, $\widehat{\beta}_s$ is either $\widehat{\beta}_1$ or $\widehat{\beta}_4$. Then, an estimator for $E_m(\widehat{\beta}_4 - \beta)^2$ is $\widehat{\sigma}^2(\underline{X}_s'\underline{V}_s^{-1}\underline{X}_s)^{-1}$ and that for $E_m(\widehat{\beta}_1 - \beta)^2$ is $\widehat{\sigma}^2(\underline{X}_s'\underline{X}_s)^{-1}$.

Note that $E_m(\widehat{\beta}_2 - \beta)^2$ equals

$$\sigma^2(\underline{X}_s' W_s \underline{X}_s)^{-1}(\underline{X}_s' W_s V_s W_s \underline{X}_s)(\underline{X}_s' W_s \underline{X}_s)^{-1} = \sigma^2 \underline{Z}_s,$$

and hence an estimator for it should be taken as $\widehat{\sigma}^2\underline{Z}_s$. But since standard computer packages like SPSS, BMDP, etc., report values of $(\underline{X}_s'\underline{V}_s^{-1}\underline{X}_s)^{-1}$ as an estimate for $E_m(\widehat{\beta}_4 - \beta)^2$, often $\widehat{\sigma}^2(\underline{X}_s'W_s\underline{X}_s)^{-1}$ is derived as an estimate for $E_m(\widehat{\beta}_2 - \beta)^2$, substituting W_s for $\underline{V}_s^{-1}$ in the former. But this practice is unwarranted by theory. In the absence of the correction, the confidence interval based on such an erroneous variance estimator often turns out to yield poor coverage probabilities.

HARTLEY and SIELKEN (1975) observe that $E_m(\widehat{\beta}_1 - \widehat{\beta}_0) = 0$, $V_m(\widehat{\beta}_1 - \beta_0) = \sigma^2[(\underline{X}_s'\underline{X}_s)^{-1} - (\underline{X}_N'\underline{X}_N)^{-1}]$ in case $\underline{V}_N = I_N$ and, assuming normality, treat

$$\underline{c}'(\widehat{\beta}_1 - \widehat{\beta}_0)/\widehat{\sigma}\left\{\underline{c}'(\underline{X}_s'\underline{X}_s)^{-1} - (\underline{X}_N'\underline{X}_N)^{-1}\underline{c}\right\}^{1/2}$$

as a STUDENT's t variable with $(n-1)$ degrees of freedom, leading to confidence intervals for $\underline{c}'\widehat{\beta}_0$, which they call tolerance intervals because $\underline{c}'\widehat{\beta}_0$ is a random variable for a chosen $r \times 1$ vector $\underline{c}$.

The literature mainly gives accounts of asymptotic design-based properties of consistency and extents of biases of the four estimators $\widehat{\beta}_j$, $j = 1, \ldots, 4$ and coverage properties of confidence intervals based on estimated design mean square errors or model mean square errors of these estimators taken either as estimators of β or as predictors of $\widehat{\beta}_0$ or $\widehat{\beta}_G$. For details, one may consult FULLER (1975), SMITH (1981), PFEFFERMANN and SMITH (1985), NATHAN (1988), and references cited therein. BREWER and MELLOR (1973), HOLT and SCOTT (1981), and HOLT and SMITH (1976) are interesting further references in this context.

11.2.3 Model-Based Regression Analysis

In the above, we really considered a two-step randomization: the finite population is supposedly a realization from an infinite hypothetical superpopulation with reference to which a regression relationship is postulated connecting a dependent variable and a set of independent regressor variables. Then, from the given or realized finite population a sample is randomly drawn because the population is too large to be completely investigated. The sample is then utilized to make inference with reference to the two-step randomization. But now let us consider a purely model-based approach that takes account of the structure of the finite population at hand by postulating an appropriate model.

Suppose for a sample of c clusters from a given finite population, observations are taken on a dependent variable y and a set of independent regressor variables $x_1, \ldots, x_r$ for independently drawn samples of second stage units (SSUs) of sizes m_i from the respective sampled clusters labeled $i = 1, \ldots, c$ so that $\sum_1^c m_i = n$, the total sample size. Let $\underline{Y}_n$ be an $n \times 1$ vector of observations on y, successive rows in it giving values on the m_i observations in the order $i = 1, \ldots, c$ and the observations X_j's, $j = 1, \ldots, r$ be also similarly arranged in succession. Now it is only to be surmised that the observations within the same cluster should be substantially well and positively correlated compared to those across the clusters. So, after postulating a regression relation of $\underline{Y}_n$ on $\underline{X}_n$, which is an $n \times r$ matrix, the successive rows in it arranging the values for the clusters taken in order $i = 1, \ldots, c$, which states that

$$E_m(\underline{Y}_n) = \underline{X}_n \beta$$

where β is an $r \times 1$ vector of unknown regression parameters, one should carefully postulate about the distribution of the error vector

$$\underline{\epsilon}_n = \underline{Y}_n - \underline{X}_n \beta.$$

One obvious postulation is that $E_m(\underline{\epsilon}_n \,|\, \underline{X}_n) = 0$ and the variance–covariance matrix of $\underline{\epsilon}_n$ is such that $V_m(\underline{Y}_n) = \sigma^2 V$, where V is a block diagonal matrix with the ith block $V_i = I_{mi} + \rho J_{mi}$, where I_{mi} is the $m_i \times m_i$ identity matrix, J_{mi} the $m_i \times m_i$ matrix with each entry as unity and ρ the intraclass correlation for each cluster.

If ρ is known and we may identify the cluster from which each observation comes, then the best linear unbiased estimator (BLUE) for β is the GLSE, which is

$$\widehat{\beta}_{opt} = \left(\underline{X}_n' V^{-1} \underline{X}_n\right)^{-1} \left(\underline{X}_n' V^{-1} \underline{Y}_n\right).$$

But in practice it is simpler to employ the ordinary least square estimator (OLSE), namely

$$\widehat{\beta}_{ols} = (\underline{X}'_n\underline{X}_n)^{-1}(\underline{X}'_n\underline{Y}_n).$$

Both are model-unbiased estimators for β but

$$E_m(\widehat{\beta}_{opt} - \beta)^2 < E_m(\widehat{\beta}_{ols} - \beta)^2.$$

The least squares unbiased estimator for σ^2 is

$$\widehat{\sigma}^2 = \frac{1}{(n-r)}\underline{Y}'_n(I_n - P_0)\underline{Y}_n$$

where $P_0 = \underline{X}_n(\underline{X}'_n\underline{X}_n)^{-1}\underline{X}'_n$ and the appropriate least squares estimator for $E_m(\beta_{ols} - \beta)^2$ is

$$\widehat{\sigma}^2(\underline{X}'_n - \underline{X}_n)^{-1}(\underline{X}'_n V\,\underline{X}_n)(\underline{X}'_n\underline{X}_n)^{-1} = \widehat{\sigma}^2(\underline{X}'_n\underline{X}_n)^{-1}C,$$

In evaluating an estimator for $E_m(\widehat{\beta}_{ols} - \beta)^2$ while using the standard computer program packages like SAS, SPSS, and BMDP, one often disregards the correction term C, which reflects the effect of clustering and plays the role analogously to that of KISH's deffs in case of the design-based regression studies. SCOTT and HOLT (1982) first pointed out the importance of the role of this correction term C, which should not be disregarded.

11.2.4 Design Variables

Next we consider an important situation where, besides the regressor variables, there exist another set of variables that are utilized in determining the selection probabilities, called the **design variables**. For example, one may plan to examine how expenses on certain items of consumption, the dependent variable y, vary with the annual income, the single regressor variable x. Then, if accounts of the taxes paid by the relevant individuals in the last financial year, values of a variable z, are available, this information can be utilized in stratifying the population accordingly. Then z is a design variable obviously well-correlated with x and y.

Following the works of NATHAN and HOLT (1980), HOLT, SMITH and WINTER (1980), and PFEFFERMANN and HOLMES

(1985) let us consider the simple case of a single dependent (endogeneous) variable y, a single regressor (exogeneous, independent) variable x, and a single design variable z. Assume the regression model

$$y = \alpha + \beta x + \epsilon$$

with $E_m(\epsilon \mid x) = 0, V_m(\epsilon \mid x) = \sigma^2 (\sigma > 0)$. Suppose a random sample s of size n is taken following a design p using the values $Z_1, Z_2, \ldots, Z_N$ of z and define

$$\nu_z = \frac{1}{N} \sum_1^N Z_i, \ \sigma_z^2 = \frac{1}{N-1} \sum_1^N (Z_i - \nu_z)^2.$$

Also, let $\overline{y}, \overline{x}, \overline{z}$ denote sample means of $y, x, z, s_y^2, s_x^2, s_z^2$ the sample variances and s_{yx}, s_{yz}, s_{xz} the sample covariances. The problem is to infer about β, the regression coefficient of y on x under the model-based approach.

Consider the ordinary least squares estimator (OLSE),

$$b = s_{yx}/s_x^2.$$

Its performance depends essentially on the relation between the design variable z and the variables x, y in the regression model. In the simplest case x, y, z might follow a trivariate normal distribution. DEMETS and HALPERIN (1977) have shown that, under this assumption, b is biased. Following ANDERSON's (1957) missing value approach, they derive an alternative estimator, which is the maximum likelihood estimator (MLE) for β, namely,

$$\widehat{\beta} = \left[s_{yx} + \frac{s_{yz} s_{xz}}{s_z^2} \left(\frac{\sigma_z^2}{s_z^2} - 1 \right) \right] \Big/ \left[s_x^2 + \frac{s_{xz}^2}{s_z^2} \left(\frac{\sigma_z^2}{s_z^2} - 1 \right) \right].$$

NATHAN and HOLT (1980) have relaxed the normality assumption and postulated only a suitable linear regression connecting y, x, z. They have found that, even then, $\widehat{\beta}$ is asymptotically unbiased in the sense that for large n we have approximately

$$E_p E_m \widehat{\beta} = \beta.$$

But $E_m\widehat{\beta} = \beta$ holds asymptotically only if s_z^2 equals σ_z^2. Writing

$$\overline{y}^* = \frac{1}{N}\sum_s \frac{Y_i}{\pi_i}, \overline{x}^* = \frac{1}{N}\sum_s \frac{X_i}{\pi_i}, \overline{z}^* = \frac{1}{N}\sum_s \frac{Z_i}{\pi_i},$$

$$s_{yx}^* = \frac{1}{N}\sum_s \frac{Y_iX_i}{\pi_i} - \frac{\overline{y}^*\overline{x}^*}{\sum_s \frac{1}{N\pi_i}}, s_{xz}^*, s_{yz}^* \text{ likewise,}$$

$$s_y^{*2} = \frac{1}{N}\sum_s \frac{X_i^2}{\pi_i} - \frac{(\overline{y}^*)^2}{\sum_s \frac{1}{N\pi_i}}, s_x^{*2}, s_z^{*2} \text{ likewise,}$$

an alternative design-weighted estimator is also proposed for β, namely,

$$\widehat{\beta}^* = \left[s_{yx}^* + \frac{s_{yz}^*s_{xz}^*}{s_z^{*2}}\left(\frac{\sigma_z^2}{s_z^{*2}} - 1\right)\right] \Big/ \left[s_x^{*2} + \frac{s_{xz}^{*2}}{s_z^{*2}}\left(\frac{\sigma_z^2}{s_z^{*2}} - 1\right)\right]$$

and it may be seen that

$$E_mE_p(\widehat{\beta}^*)$$

is asymptotically equal to β, that is, $\widehat{\beta}^*$ is asymptotically unbiased.

For any estimator e for β, considering the criterion

$$\begin{aligned} E_mE_p(e-\beta)^2 &= E_mE_p\left[(e - E_p(e) + (E_p(e) - \beta)\right]^2 \\ &= E_mV_p(e) + E_m(E_p(e) - \beta)^2 \end{aligned}$$

and supposing that for large samples $E_p(e)$ should be close to β for many appropriate choices of e, one may neglect the second term here. Then, if an estimator for $V_p(e)$, namely $v_p(e)$ with $E_p(v_p(e))$, close to $V_p(e)$ at least for large samples be available, it may be a good idea to employ $v_p(e)$ as an estimator for the overall MSE $E_mE_p(e-\beta)^2$ and use $v_p(e)$ in constructing confidence intervals. In terms of this approach, a comparison among b, $\widehat{\beta}$ and $\widehat{\beta}^*$ is available in the literature, showing that $\widehat{\beta}$ is the most promising, followed by $\widehat{\beta}^*$. It must be noted, however, that $\widehat{\beta}(\widehat{\beta}^*)$ coincides with (or approximates) b if $s_z^2(s_z^{*2})$ matches (or approximately matches) σ_z^2. Thus, the design variable is important in yielding alternative estimators even with a model-based approach, and the values of the design variable may be suitably used in achieving required properties for the simple statistic, namely b, for example, by bringing s_z^2 or s_z^{*2}

close to σ_z^2, the latter being known. Then it is not necessarily the design but the values of the design variable that may affect the performance of model-based regression analysis.

11.2.5 Varying Regression Coefficients for Clusters

So far we have considered fitting a single regression equation applicable to the entire aggregate, whether it is a finite population or a hypothetical modeled population that is infinite. Now we consider a population divisible into strata or clusters for which we postulate a regression relationship to connect a dependent variable y and a regressor variable x such that regression curves may be supposed to vary over the clusters or the strata.

First we consider the case where there are N clusters with ith cluster ($i = 1, \ldots, N$) having M_i units so that $\sum_1^N M_i = M$ is the total number of individuals in a finite population for which Y_{ij} is the value of a dependent variable y on the jth member of ith cluster ($j = 1, \ldots, M_i, i = 1, \ldots, N$). Following PFEFFERMANN and NATHAN (1981), we adopt a model-based approach postulating the model

$$Y_{ij} = \beta_i X_{ij} + \epsilon_{ij},$$

with $E_m(\epsilon_{ij} \mid x_{ij}) = 0$ and $E_m(\epsilon_{ij}^2 \mid x_{ij}) = \sigma_i^2$ and $E_m(\epsilon_{ij}\epsilon_{kl} \mid x_{ij}, x_{kl}) = 0$ if either $i \neq j$ or $k \neq l$ or both. Let a sample consist of n clusters out of N clusters and from ith cluster, if selected, m_i units be taken. KONJIN (1962) and PORTER (1973) considered estimating, respectively, $\frac{1}{M}\sum_1^N M_i\beta_i$ and $\frac{1}{N}\sum_1^N \beta_i$ for which solutions are rather easy utilizing the approach as in multistage sampling, especially if one employs design-based estimators, which approach these authors followed. But following SCOTT and SMITH (1969), the under-noted model-based approach is worth consideration that treats the following **random effects model**. Following them, PFEFFERMANN and NATHAN (1981) postulate the following model for the β_i's

$$\beta_i = \beta + v_i, \ i = 1, \ldots, N$$
$$E_m(v_i) = 0, \ V_m(v_i) = \delta^2 \quad \text{and} \quad C_m(v_i, v_j) = 0, \ i \neq j.$$

Writing s for a sample of n clusters and s_i for a sample of m_i units from ith cluster for i in s, and first supposing that σ_i and δ are known, PFEFFERMANN and NATHAN (1981) give the following estimator β_i^* for $\beta_i, i = 1, \ldots, N$, namely

$$\beta_i^* = \lambda_i \widehat{\beta}_i + (1 - \lambda_i)\widehat{\beta}, \; i = 1, \ldots, N$$

where

$$\begin{aligned}
\lambda_i &= \delta^2 \Big/ \left[\delta^2 + \sigma_i^2 \Big/ \sum_{j \in s_i} x_{ij}^2 \right] \quad \text{for } i \in s; \\
&= 0 \quad \text{for } i \notin s, \\
\widehat{\beta}_i &= \sum_{j \in s_i} y_{ij} x_{ij} \Big/ \sum_{j \in s_i} x_{ij}^2 \quad \text{for} \quad i \in s \\
&= 0 \quad \text{for } i \notin s \\
\widehat{\beta} &= \sum_{i \in s} \lambda_i \widehat{\beta}_i \Big/ \sum_{i \in s} \lambda_i .
\end{aligned}$$

Then

$$\widehat{\sigma}_i^2 = \frac{1}{(m_i - 1)} \sum_{s_i} (y_{ij} - \widehat{\beta}_i x_{ij})^2$$

is taken as an estimator for $\sigma_i^2, i \in s$. Let

$$\widetilde{\lambda}_i = \frac{\delta^2}{\delta^2 + \left(\widehat{\sigma}_i^2 / \sum_{j \in s_i} x_{ij}^2 \right)},$$

then δ^2 is estimated by $\widehat{\delta}^2$ which is the largest solution of

$$\frac{1}{(n-1)} \sum_{i \in s} \left(\widehat{\beta}_i - \sum_{i \in s} \widetilde{\lambda}_i \widehat{\beta}_i \right) \Big/ \sum_{i \in s} \widetilde{\lambda}_i)^2 = \delta^2 .$$

Then, writing

$$\begin{aligned}
\widehat{\lambda}_i &= \frac{\widehat{\delta}^2}{(\widehat{\delta}^2 + \widehat{\sigma}_i^2 / \sum_{s_i} x_{ij}^2)}, \\
\widetilde{\beta} &= \sum_{s} \widehat{\lambda}_i \widehat{\beta}_i \Big/ \sum_{s} \widehat{\lambda}_i
\end{aligned}$$

the final estimator for β_i is

$$\widehat{\beta}_i = \widehat{\lambda}_i \widehat{\beta}_i + (1 - \widehat{\lambda}_i)\widehat{\beta}, \; i = 1, \ldots, N.$$

Chapter 12

Randomized Response

Suppose a survey is required to deal with sensitive issues like the extent to which habits of drunken driving, tax evasion, gambling, etc., are prevalent in a certain community in a given time period. The entire survey need not be exclusively concerned with such stigmatizing items of query, but some of the structured questions in an elaborate survey questionnaire may cover a few specimens like these. It is likely that an investigator will hesitate to raise such delicate questions, and people when so addressed may refuse to reply or supply evasive or false answers. As a possible way out one may try to replace a direct response (DR) query by a randomized response (RR) survey. We discuss briefly how it can be planned and implemented and indicate some possible consequences.

12.1 SRSWR FOR QUALITATIVE AND QUANTITATIVE DATA

12.1.1 Warner Model

First let us consider the pioneering work in this area by WARNER (1965), who dealt with a qualitative character like alcoholism, which appears only in two mutually exclusive forms.

Suppose A denotes a stigmatizing character and $\overline{A}$ its complement. Let in a given community of people the unknown proportion of persons bearing the form A of the character be π_A and $1 - \pi_A$ be the proportion of persons bearing $\overline{A}$. Our problem is to estimate π_A and obtain an estimate of the variance of the estimate on taking a simple random sample (SRS) with replacement (WR) in n draws. If a DR survey is undertaken and every sampled person responds and each response is assumed to be truthful, then the proportion of Yes response to the question

Do you bear $\overline{A}$?

$p_Y = n_Y/n$, where n_Y = (Yes) responses in the sample would give an unbiased estimator of π_A with a variance

$$V(p_Y) = \frac{\pi_A(1 - \pi_A)}{n} = V_D$$

admitting an unbiased variance estimator

$$v_D = \frac{p_Y(1 - p_Y)}{n - 1}.$$

But if we believe that there may be a substantial nonresponse as well as incorrect response, then this estimate cannot do, as it is grossly biased and unreliable.

Instead, let us ask a sampled person

Do you bear A?

with a probability P and the negation of it, that is,

Do you bear A?

with the complementary probability $Q = 1 - P$, choosing a suitable positive proper fraction P. The answer Yes or No is then requested of the respondent in a truthful manner, assuring him or her that the interrogator does not know to which of the two complementary questions the given answer relates.

A possible device is to offer to the respondent a pack of identical-looking cards, a proportion P of which is marked as A and the rest as $\overline{A}$ with the instruction that the respondent, after thoroughly shuffing the pack, would choose one, unnoticed by the investigator, and record in the questionnaire the truthful Yes or No response that corresponds to the type of card. Thus a Yes response may refer to his/her bearing A or $\overline{A}$ with the variation of the type of card he/she happens to choose.

If this RR procedure is adopted, on the basis of the SRSWR of size n, the proportion of Yes response will unbiasedly estimate $\pi_y \equiv$ the probability of Yes response, which equals

$$\pi_y = P\pi_A + (1-P)(1-\pi_A) = (1-P) + (2P-1)\pi_A.$$

So, using the sample proportion p_{yr} of Yes responses, we get an unbiased estimator $\widehat{\pi}_A$ of π_A as

$$\widehat{\pi}_A = \frac{p_{yr}(1-P)}{(2P-1)}, \text{ provided } P \neq \frac{1}{2}.$$

Then,

$$V(\widehat{\pi}_A) = \frac{1}{(2P-1)^2} V(p_{yr}) = \frac{\pi_y(1-\pi_y)}{n(2P-1)^2} = V_R, \text{ say,}$$

which simplifies to

$$\begin{aligned} V_R &= \frac{\pi_A(1-\pi_A)}{n} + \frac{P(1-P)}{n(2P-1)^2} \\ &= \frac{\pi_A(1-\pi_A)}{n} + \frac{1}{n}\left[\frac{1}{16(P-1/2)^2} - \frac{1}{4}\right]. \end{aligned}$$

Clearly, comparing V_R with V_D, one notes the loss in efficiency in resorting to RR and how the loss in efficiency decreases as P approaches either 0 or 1. But the values of P close to 0 or 1 should not be acceptable to an intelligent respondent who, for the sake of protected privacy, would prefer a value of P close to $1/2$, which leads to increasing loss in efficiency. An unbiased estimator for V_R is obviously

$$\begin{aligned} v_R &= \frac{p_{yr}(1-p_{yr})}{(n-1)(2P-1)^2} \\ &= \frac{1}{(n-1)}\left[\widehat{\pi}_A(1-\widehat{\pi}_A) + \left\{\frac{1}{16(P-1/2)^2} - \frac{1}{4}\right\}\right]. \end{aligned}$$

12.1.2 Unrelated Question Model

The attributes A and $\overline{A}$ may both be sensitive, for example, affiliation to two rival political blocks. An alternative RR device for estimating π_A in this dichotomous case is described below.

Suppose B is another innocuous character unrelated to the sensitive attribute A, for example, B may mean preference for fish over chicken and $\overline{B}$ its complement. Assume further that the proportion of persons bearing B is a known number π_B. Then, for an SRSWR in n draws a sampled respondent is requested to report Yes or No truthfully about bearing A with a probability P and about bearing B with the complementary probability $Q = 1 - P$. The sample proportion p_{yr} of Yes responses is an unbiased estimator for

$$\pi_y = P\pi_A + (1 - P)\pi_B .$$

Since π_B is supposed known and P is preassigned, an unbiased estimator for π_A is

$$\widehat{\pi}_A = [p_{yr} - (1 - P)\pi_B / P,$$

provided $P \neq 0$.

One way to have π_B known is to adopt the following modified device where a respondent is asked to (1) report Yes or No truthfully about bearing A with probability P_1, (2) report Yes with a probability P_2 and (3) report No with a probability P_3, choosing numbers P_1, P_2, P_3 such that $0 < P_1,\ P_2,\ P_3 < 1$ and $P_1 + P_2 + P_3 = 1$, using a pack of cards of three types mixed in proportions $P_1 : P_2 : P_3$. Then,

$$\pi_y = P_1\pi_A + P_2 = P_1\pi_A + \left(\frac{P_2}{P_2 + P_3}\right)(1 - P_1)$$

and the known quantity $\frac{P_2}{P_2+P_3}$ may be supposed to play the role of π_B.

However, a better way to deal with the case when π_B is unknown is to draw two independent SRSWRs of sizes n_1 and n_2 and for the two samples use separate probabilities P_1, P_2 with which a response is to relate to A. Then, the sample proportions p_{yr} for the two samples, p_1, p_2 of Yes responses are respectively unbiased estimators (independent) of

$$\pi_{y1} = P_1\pi_A + (1 - P_1)\pi_B \quad \text{and} \quad \pi_{y2} = P_2\pi_A + (1 - P_2)\pi_B .$$

Then

$$\widehat{\pi}_A = [(1 - P_2)p_1 - (1 - P_1)p_2]/(P_1 - P_2)$$

is an unbiased estimator of π_A provided $P_1 \neq P_2$. Then,

$$V(\hat{\pi}_A) = \big[(1-P_2)^2 \pi_{y1}(1-\pi_{y1})/n_1 \\ + (1-P_1)^2 \pi_{y2}(1-\pi_{y2}/n_2\big]/(P_1-P_2)^2$$

and an unbiased estimator for it is

$$v(\hat{\pi}_A) = \left[(1-P_2)^2 \frac{p_1(1-p_1)}{n_1-1} + (1-P_1)^2 \frac{p_2(1-p_2)}{n_2-1}\right] \Big/ (P_1-P_2)^2.$$

With this scheme, problems are to choose $P_1 \neq P_2$ to achieve high efficiency but both close to 1/2 to induce a sense of protected privacy in a respondent and thus enhance prospects for trustworthy cooperation. Also, the ratio n_1/n_2 must be rightly chosen subject to a preassigned value for $n_1 + n_2 = n$ consistently with a given budget. The literature contains results with varied and detailed discussions, and one may refer to CHAUDHURI and MUKERJEE (1988) and the appropriate references cited therein.

Another slight variation of the above procedure introduces a third innocuous character C unrelated to the sensitive attribute A, and two independent SRSWRs of sizes n_1, n_2 are taken as above. But in the first sample, RR queries are made about A and B as above, but also a DR query is made about bearing C. The second sample is used to make an RR query concerning A and C but a DR query about B. Writing π_C as the unknown proportion bearing C and probability (sample proportion) for the two samples for Yes responses based on RR, DR as

$$\pi_{Ryi}(p_{Ryi}),\ \pi_{Dyi}(p_{Dyi}),\ i = 1, 2,$$

we have the probabilities and unbiased estimators as follows

$$\pi_{Ry1} = P_1\pi_A + (1-P_1)\pi_B,\ \pi_{Dy1} = \pi_C$$
$$\pi_{Ry2} = P_2\pi_A + (1-P_2)\pi_C,\ \pi_{Dy2} = \pi_B$$
$$\hat{\pi}_C = p_{Dy1},\ \hat{\pi}_B = p_{Dy2},$$
$$\hat{\pi}_{A1} = \frac{P_{Ry1} - (1-P_1)\hat{\pi}_B}{P_1},\ \hat{\pi}_{A2} = \frac{p_{Ry2} - (1-P_2)\hat{\pi}_C}{P_2}.$$

A combined weighted estimator $\pi_A^* = W\hat{\pi}_{A1} + (1-W)\hat{\pi}_{A2}$ may then be determined with W chosen to minimize $V(\pi_A^*)$ and then replacing the unknown parameters in the optimal W by their sample-based estimates.

12.1.3 Polychotomous Populations

Many alternative devices are available for the purpose we are discussing. We will mention selectively a few more. Suppose a population may be classified into several mutually exclusive and exhaustive categories according to a sensitive characteristic. For example, women may be classified according to the number of self-induced abortions so far implemented. Suppose, in general $\pi_i, i = 1, \ldots, k, \sum_k^k \pi_i = 1$, denote the unknown proportions of individuals belonging to k disjoint and exhaustive categories according to a stigmatizing character. In order to estimate π_i on taking an SRSWR of a given size n, let us apply the following device. Suppose small marbles or beads of k distinct colors numbering $m_i, i = 1, 2, \ldots, k, \sum_k^k m_i = m$ are put into a flask with a long neck marked $1, \ldots, m$ spaced apart to accommodate one bead each when turned upside down with the mouth tightly closed. Each color represents a category and a sampled person is requested to shake the flask thoroughly, unobserved by the investigator, and to record on the questionnaire the number on the flask-neck accomodating the bottommost bead of the color of his/her category when turned upside down. Writing λ_j as the probability of reporting the value j, P_{ij} as the probability of reporting j when the true category is i, and p_j as the sample proportion of RR as j, we have p_j as an estimator for λ_j given by

$$\lambda_j = \sum_{i=1}^{k} P_{ij}\pi_i, \; j = 1, \ldots, J, \text{ where } J = m - \min_{1 \le i \le k} m_i + 1.$$

Here P_{ij} is easy to calculate for the given m_i's, $i = 1, \ldots, k$. For example,

$$P_{11} = \frac{m_1}{m},$$

$$P_{21} = \frac{m_2}{m},$$

$$P_{12} = \frac{m - m_1}{m} \cdot \frac{m_1}{m - 1},$$

$$P_{23} = \frac{m - m_2}{m} \cdot \frac{m - m_2 - 1}{m - 1} \cdot \frac{m_2}{m - 2}$$

and so on. The values of m_i should be kept small and distinct for simplicity. Yet $J > k$. One good choice is $m_i = i; i = 1, \ldots, k$, in which case $J = m = k(k+1)/2$. So, π_i is to be estimated as $\widehat{\pi}_i$ on solving

$$p_j = \sum_{1}^{p} P_{ij}\widehat{\pi}_i$$

but a unique solution is not possible. One procedure recommended in the literature is to apply the theory of linear models. The solution requires evaluation of generalized inverses and is complicated and unlikely in practice to yield $\widehat{\pi}_i$ within the permitted range [0, 1].

12.1.4 Quantitative Characters

If x denotes the amount spent last month on alcohol, amount earned in clandestine manners, etc., so that we may anticipate its range and form equidistant intervals, then, applying the above technique, it is easy to estimate the relative frequencies π_j together with the moments of the corresponding distribution. A simpler alternative is described below.

Consider the mean $\mu = \sum_1^k j\pi_j$ of a variable x with values $j = 1, \ldots, k$ and let a disc be divided into k equal cross-sections marked $1, 2, \ldots, k$ in the clockwise direction. Also suppose there is a pointer revolving along the clockwise direction indicating one of the cross-sections where it stops after a few revolutions. Then for an SRSWR in n draws we may request a sample person to revolve the pointer, unobserved by the investigator, and report Yes (No) if the pointer, after revolution, stops in a section marked i such that $i \leq j$, where j is his true value.

Then, writing P_y as the probability of a Yes response and p_y as the sample proportion of Yes responses, we have

$$P_y = \frac{1}{k}\sum_{1}^{k} j\pi_j = \frac{\mu}{k}$$

and so kp_y provides an estimator for μ. The variance of this estimator $\widehat{\mu} = kp_y$ is then $V(\widehat{\mu}) = k^2V(p_y) = \frac{k^2}{n}(\frac{\mu}{k})(1-\frac{\mu}{k})$ and an unbiased estimator for this variance is

$$v = \frac{k^2}{n-1}\left(\frac{\widehat{\mu}}{k}\right)\left(1-\frac{\widehat{\mu}}{k}\right) = \frac{k^2}{n-1}p_y(1-p_y).$$

A more straightforward RR method of estimating the mean μ_x of a sensitive variable x is obtained by an extension of a method we discussed in what precedes in estimating an attribute parameter. Let y be an innocuous variable unrelated to x with an unknown expected value μ_y. Then, we may take two independent SRSWRs of sizes $n_i, i = 1, 2, n_1 + n_2 = n$ and request every sampled person j for the ith ($i = 1, 2$) sample to report a value of x, say X_j with a probability P_i and his/her true value of y, Y_j with the complementary probability $Q_i = 1 - P_i$ without divulging to the interviewer the variable on which he/she is reporting. Writing the value reported, that is, the RR as Z_{ji} on z_i, a random variable thus generated for the ith sample, we may use the sample mean $\overline{z}_i$ of the RRs to estimate the mean μ_{zi} of z_i which is given by

$$\mu_{zi} = P_i\mu_x + (1-P_i)\mu_y,\ i = 1, 2,\ P_1 \neq P_2.$$

Then,

$$\mu_x = [(1-P_2)\mu_{z1} - (1-P_1)\mu_{z2}]/(P_1 - P_2)$$

and hence

$$\widehat{\mu}_x = \left[(1-P_2)\,\overline{z}_1 - (1-P_2)\,\overline{z} - 2\right]/(P_1 - P_2)$$

is an unbiased estimator for μ_x. Writing

$$s_{zi}^2 = \frac{1}{(n_i - 1)}\sum_{j=1}^{n_i}(z_{ji} - \overline{z}_i)^2$$

an unbiased estimator for $V(\widehat{\mu}_x)$ is given by

$$v = \left[(1-P_2)^2 s_{z1}^2/n_1 + (1-P_1)^2 s_{z2}^2/n_2\right]/(P_1 - P_2)^2.$$

In the next section, we consider a strictly finite population setup allowing sample selection with unequal probabilities.

12.2 A GENERAL APPROACH

12.2.1 Linear Unbiased Estimators

Let a sensitive variable y be defined on a finite population $U = (1, \ldots, N)$ with values $Y_i, i = 1, \ldots, N$, which are supposed to be unavailable through a DR survey. Suppose a sample s of size n is chosen according to a design p with a selection probability $p(s)$. In order to estimate $Y = \sum_1^N Y_i$, let an RR as a value Z_i be available on request from each sampled person labeled i included in a sample. Before describing how a Z_i may be generated, let us note the properties required of it. We will denote by $E_R(V_R, C_R)$ the operator for expectation (variance, covariance) with respect to the randomized procedure of generating RR. The basic RRs Z_i should allow derivation by a simple transformation reduced RRs as R_i's satisfying the conditions

(a) $E_R(R_i) = Y_i$
(b) $V_R(R_i) = \alpha_i Y_i^2 + \beta_i Y_i + \theta_i$ with $\alpha_i (> 0), \beta_i, \theta_i$'s as known constants
(c) $C_R(R_i, R_j) = 0$ for $i \neq j$
(d) estimators $v_i = a_i R_i^2 + b_i R_i + C_i$ exist, a_i, b_i, c_i known constants, such that $E_R(v_i) = V_R(R_i) = V_i$, say, for all i.

We will illustrate only two possible ways of obtaining Z_i's from a sampled individual i on request. First, let two vectors $\underline{A} = (A_1, \ldots, A_T)'$ and $\underline{B} = (B_1, \ldots, B_L)'$ of suitable real numbers be chosen with means $\overline{A} \neq 0, \overline{B}$ and variances σ_A^2, σ_B^2. A sample person i is requested to independently choose at random a_i out of $\underline{A}$ and b_i out of $\underline{B}$, and report the value $Z_i = a_i Y_i + b_i$. Then, it follows that $E_R(Z_i) = \overline{A} Y_i + \overline{B}$, giving

$$R_i = (Z_i - \overline{B})/\overline{A}$$

such that

$$E_R(R_i) = Y_i,$$
$$V_R(R_i) = \left(Y_i^2 \sigma_A^2 + \sigma_B^2\right) \Big/ \left(\overline{A}\right)^2 = V_i,$$
$$C_R(R_i, R_J) = 0, \quad i \neq j$$

and

$$v_i = \left(\sigma_A^2 R_i^2 + \sigma_B^2\right) / \left(\sigma_A^2 + \overline{A}^2\right)$$

has

$$E_R(v_i) = V_i.$$

As a second example, let a large number of real numbers X_j, $j = 1, \ldots, m$, not necessarily distinct, be chosen and a sample person i be requested to report the value Z_i where Z_i equals Y_i with a preassigned probability C, and equals X_j with a probability q_j, which is also preassigned, $j = 1, \ldots, m$ such that

$$C + \sum_{j=1}^{m} q_j = 1.$$

Then,

$$E_R(Z_i) = CY_i + \sum_{1}^{m} q_j X_j = CY_i + (1 - C)\mu, \text{ say,}$$

writing $\mu = \sum_1^m q_j X_j / \sum_1^m q_j$. Then, $R_i = [Z_i - (1 - C)\mu]/C$ has $E_R(R_i) = Y_i$. Also,

$$\begin{aligned} V_R(R_i) &= V_R(Z_i)/C^2 = V_i \\ &= \left[C(1 - C)Y_i^2 - 2C(1 - C)\mu Y_i + \left(\sum q_j X_j^2\right)\right. \\ &\quad \left. - (1 - C)^2 \mu^2\right]/C^2 \end{aligned}$$

which admits an obvious unbiased estimator v_i.

Thus we may assume the existence of a vector $\underline{R} = (R_1, \ldots, R_N)'$ derivable from RRs Z_i corresponding to the vector $\underline{Y} = (Y_1, \ldots, Y_N)'$. Let $t = t(s, \underline{Y}) = \sum b_{si} I_{si} Y_i$ be a p-based estimator for Y, assuming that Y_i for $i \in s$ is ascertainable admitting the MSE

$$M_p = M_p(t) = E_p(t - Y)^2 = \sum_i \sum_j d_{ij} Y_i Y_j$$

where

$$d_{ij} = E_p(b_{si} I_{si} - 1)(b_{sj} I_{sj} - 1).$$

Assume further that there exist non-zero constants W_i's such that $Y_i / W_i = C$ for every $i = 1, \ldots, N$ and $C \neq 0$ implies

$M_p = 0$. Then M_p reduces to

$$M_p = -\sum\sum_{i<j} d_{ij} W_i W_j \left(\frac{Y_i}{W_i} - \frac{Y_j}{W_j}\right)^2$$

as was discussed in chapter 2. Now, since Y_i's are supposedly not realizable, we cannot use t in estimating Y, nor can we use

$$m_p = -\sum\sum_{i<j} d_{sij} I_{sij} W_i W_j \left(\frac{Y_i}{W_i} - \frac{Y_j}{W_j}\right)^2$$

to unbiasedly estimate M_p. So, let us replace Y_i in t by R_i to get

$$e = e(s, \underline{R}) = t(s, \underline{Y})|_{\underline{Y}=\underline{R}} = \sum b_{si} I_{si} R_i.$$

Then, $E_R(e) = t$ and hence, in case t is p unbiased for Y, that is, $E_P(t) = \sum_s p(s) t(s, \underline{Y}) = Y$, then

$$E(e) = E_p E_R(e) = E_p(t) = Y,$$

writing $E_p(V_p)$ from now on again as operator for design expectation (variance) and

$$E = E_{pR} = E_p E_R$$

as an overall operator for expectation with respect to randomized response and design. Similarly, we will write

$$V = V_{pR} = E_p[V_R] + V_p[E_R]$$

as the operator for overall variance, first over RR followed by design. In case $E_p E_R(e) = Y$, we call e an unbiased estimator for Y. With the assumptions made above, now we may work out the overall MSE of e about Y, namely,

$$\begin{aligned}
M &= E(e - Y)^2 = E_p E_R\left[(e - t) + (t - Y)\right]^2 \\
&= M_p(t) + E_p E_R \left[\sum b_{si} I_{si}(R_i - Y_i)\right]^2 \\
&= -\sum\sum_{i<j} d_{ij} W_i W_j \left(\frac{Y_i}{W_i} - \frac{Y_j}{W_j}\right)^2 + E_p \sum b_{si}^2 I_{si} V_i \\
&= -\sum\sum_{i<j} d_{ij} W_i W_j \left(\frac{Y_i}{W_i} - \frac{Y_j}{W_j}\right)^2 + \sum_1^N V_i E_p\left(b_{si}^2 I_{si}\right).
\end{aligned}$$

It then follows that

$$m = -\sum_{i<j}\sum d_{sij} I_{sij} W_i W_j \left[\left(\frac{R_i}{W_i} - \frac{R_j}{W_j} \right)^2 - \left(\frac{v_i}{W_i^2} + \frac{v_j}{W_j^2} \right)^2 \right] + \sum_i v_i b_{si}^2 I_{si}$$

may be taken as an unbiased estimator for M because it is not difficult to check that

$$E(m) = E_p E_R(m) = M \quad \text{if} \quad E_p(d_{sij} I_{sij}) = d_{ij}.$$

12.2.2 A Few Specific Strategies

Let us illustrate a few familiar specific cases. Corresponding to the HTE $\bar{t} = \bar{t}(s, \underline{Y}) = \sum_i \frac{Y_i}{\pi_i} I_{si}$, we have the derived estimator $e = (s, \underline{R}) = \sum_i \frac{R_i}{\pi_i} I_{si}$ for which

$$M = -\sum_{i<j}\sum (\pi_i \pi_j - \pi_{ij})(Y_i/\pi_i - Y_j/\pi_j)^2 + \sum_i \frac{V_i}{\pi_i}$$

and

$$m = \sum_{i<j}\sum \left(\frac{\pi_i \pi_j - \pi_{ij}}{\pi_{ij}} \right) \left(\frac{R_i}{\pi_i} - \frac{R_j}{\pi_j} \right)^2 + \sum \frac{v_i}{\pi_i} I_{si}.$$

To LAHIRI's (1951) ratio estimator $t_L = Y_i / \sum_s P_i$ based on LAHIRI-MIDZUNO-SEN (LMS, 1951, 1952, 1953) scheme corresponds the estimator

$$e_L = \sum_s R_i / \sum_s P_i$$

($0 < P_i < 1$, $\Sigma_1^N P_i = 1$) for which

$$M = \sum_{i<j}\sum a_{ij} \left(1 - \frac{1}{C_1} \sum_s \frac{I_{sij}}{P_s} \right) + \sum V_i E_p(I_{si}/P_s^2),$$

where

$$C_r = \binom{N-r}{n-r}, r = 0, 1, 2, \ldots, P_s = \sum_s P_i, a_{ij} = P_i P_j (Y_i/P_i - Y_j/P_j)^2$$

$$m = \sum\sum P_i P_j I_{si} I_{sij} \left(\frac{N-1}{n-1} - \frac{1}{P_s}\right) \Big/$$

$$P_s \left[\left(\frac{R_i}{P_i} - \frac{R_j}{P_j}\right)^2 - \left(\frac{v_i}{p_i^2} + \frac{v_j}{p_j^2}\right)\right] + \sum v_i I_{si}/P_s^2$$

is unbiased for M. If t_L and e_L above are based on SRSWOR in n draws, then, M equals

$$M' = -\frac{1}{C_0}\left[\sum\sum_{i<j} a_{ij} \sum_s \left(\frac{I_{sij}}{p_s^2} - \frac{I_{sj}}{P_s} - \frac{I_{sj}}{P_s} + 1\right)\right.$$

$$\left. - \sum_i V_i \left(\sum_s I_{si}/P_s^2\right)\right]$$

and

$$m' = -\frac{N(N-1)}{n(n-1)C_0}\sum\sum_{i<j} \widehat{a}_{ij} I_{sij} \sum_s \left(\frac{I_{sij}}{p_s^2} - \frac{I_{si}}{P_s} - \frac{I_{sj}}{P_s} + 1\right)$$

$$+ \frac{1}{C_0}\frac{N}{n}\sum v_i I_{si}\left(\sum_s I_{si}/P_s^2\right)$$

writing

$$\widehat{a}_{ij} = \left\{\left(\frac{R_i}{P_i} - \frac{R_j}{P_j}\right)^2 - \frac{v_j}{P_i^2} + \frac{v_j}{P_j^2}\right)\right\} P_i P_j.$$

But the coefficients of a_{ij} in M' and of $\widehat{a}_{ij}$ in m' are so complicated that m' is hardly usable. Instead, we shall approximate

$$M' = E_p E_R \left(\sum_s R_i / \sum_s P_i - Y\right)^2$$

$$= E_p E_R \left[\sum_s (R_i - Y_i)/\sum_s P_i + \left(\sum_s Y_i \Big/ \sum_s P_i - Y\right)\right]^2$$

by

$$M' = \frac{N}{f}(1-f)\sum_1^N (Y_i - Y P_i)^2/(N-1)$$

$$\left. + E_p \left(\sum_s V_i \Big/ \left(\sum_s P_i\right)^2\right]\right.$$

writing $f = \frac{n}{N}$ as usual. An approximately unbiased estimator of M' is

$$m' = \frac{N}{f}(1-f)u(s) - \frac{N}{f}\frac{1-f}{(N-1)}\left[\sum_s v_i \frac{1}{f} + \frac{(\sum_s v_i)\left(\sum_s P_i^2\right)}{(\sum_s P_i)^2}\right.$$
$$\left. -2\sum_s P_i v_i / \sum_s P_i\right] + \left(\sum_s v_i\right) \bigg/ \left(\sum_s P_i\right)^2,$$

where

$$u(s) = \frac{1}{(n-1)}\sum_s \left(R_i - \frac{\overline{r}_s}{\overline{p}_s}Pi\right)^2$$

with

$$\overline{r}_s = \frac{1}{n}\sum_s R_i, \overline{p}_s = \frac{1}{n}\sum_s P_i.$$

Assume a PPSWR sample is drawn using normed size measures $P_i, (0 < P_i < 1, \Sigma P_i = 1)$, and each time a person appears in the sample, an independent RR r_k is obtained. Write y_k, r_k, and p_k for the corresponding Y_i, R_i, and P_i value for the individual i if chosen on the kth draw, then, corresponding to $t_{HH} = \frac{1}{n}\sum_{r=1}^{n}\frac{y_k}{p_k}$, the HANSEN–HURWITZ (1953) estimator for Y, the derived estimator is $e_{HH} = \frac{1}{n}\sum_{k=1}^{n}\frac{r_k}{p_k}$ having the variance

$$M = \frac{1}{n}\left(\sum_1^N \frac{Y_i^2}{P_i} - Y^2\right) + \frac{1}{n}\sum \frac{V_i}{P_i}$$

and an unbiased variance estimator is

$$m = \frac{1}{n(n-1)}\sum_1^n \left(\frac{r_k}{p_k} - \frac{1}{n}\sum_1^n \frac{r_k}{p_k}\right)^2.$$

Presuming that a person, on every reappearance in the sample, may understandably refuse to reapply the RR device and may be requested only to report one RR, then a less efficient estimator is

$$e'_{HH} = \frac{1}{n}\sum_s \frac{R_i}{P_i} f_{si},$$

f_{si} = frequency of i in s, with a variance

$$M' = \frac{1}{n}\left(\sum \frac{Y_i^2}{P_i} - Y^2\right) + \frac{1}{n}\sum \frac{V_i}{P_i} + \frac{n-1}{n}\sum V_i$$

and an unbiased estimator for it is

$$m' = \frac{1}{n(n-1)}\sum_1^N \left(\frac{R_i}{P_i} - e'_{HH}\right)^2 f_{si} + \frac{1}{n}\sum_1^N \frac{v_i}{P_i} f_{si}.$$

Corresponding to other standard sampling strategies due to DES RAJ (1956), RAO-HARTLEY-COCHRAN (1962), MURTHY (1957), and others, also similar RR-based estimators along with formulae for variance and estimators of variance are rather easy to derive.

12.2.3 Use of Superpopulations

In the case of DR surveys, models for $\underline{Y}$ are usually postulated to derive optimal strategies (p, t) with $t = t(s, \underline{Y})$ to control the magnitudes of $E_m E_p (t - Y)^2$ writing $E_m(V_m, C_m)$ for expectation (variance, covariance) operators with respect to the model. In the RR context, it is also possible to derive, under the same models, optimal sampling strategies (p, e), with $e = e(s, \underline{R})$ to control the magnitude of

$$E_m E(e - Y)^2 = E_m E_p E_R (e - Y)^2.$$

Here it is necessary to assume that (1) E_m, E_p and E_R commute and (2) that $E_p(e) = \Sigma p(s) e(s, \underline{R}) = \sum_1^N R_i = R$. Since

$$e(s, \underline{R}) = t(s, \underline{Y})|_{\underline{Y}=\underline{R}} = R,$$

the assumption (2) is rather trivial because in DR optimal p-based model optimal estimators t are subject to $E_p(t) = Y$.

We follow GODAMBE and JOSHI (1965), GODAMBE and THOMPSON (1977), and HO (1980) and postulate the model for which

$$E_m(Y_i) = \mu_i, V_m(Y_i) = \sigma_i^2$$

and the Y_i's are independent. Write

$$\bar{e} = \sum \frac{R_i}{\pi_i} I_{si},$$
$$e = e(s, \underline{R}) = \bar{e} + h,$$

with $h = h(s, \underline{R})$ subject to $E_p h = 0$. Define, in addition

$$e_0 = e_0(s, \underline{R}) = \sum_i \left(\frac{R_i - \mu_i}{\pi_i}\right) I_{si} + \mu,$$

$$h_0 = e_0 - \bar{e} = -\sum \frac{\mu_i}{\pi_i} I_{si} + \mu,$$

where $\mu = \sum_1^N \mu_i$, and check that

$$\begin{aligned} M = E_m E(e - Y)^2 &= E_m E_p V_R(\bar{e}) + E_m E_p V_R(h) \\ &+ E_p V_m(E_R \bar{e}) E_p V_m(E_R h) \\ &+ E_p(E_m E_R e - \mu)^2 - V_m(Y) \end{aligned}$$

$$\begin{aligned} \widehat{M} = E_m E(\bar{e} - Y)^2 &= E_m E_p V_R(\bar{e}) + E_p V_m(E_R \bar{e}) \\ &+ E_p(E_m E_R \bar{e} - \mu)^2 V_m(Y) \end{aligned}$$

and

$$M_0 = E_m E(e_0 - Y)^2 = E_m \left(\sum \frac{V_i}{\pi_i}\right) + \sum \sigma_i^2 \left(\frac{1}{\pi_i} - 1\right)$$

on observing, in particular, that

$$V_R(h_0) = 0, \;\; V_m(E_R h_0) = 0, \;\; E_m E_R(e_0) = \mu.$$

So, as an analogous result of Ho (1980) for the DR case, we derive that an optimal strategy involves e_0 based on any design p. But since, in practice, μ_i may not be fully known, this optimal strategy is not practicable in general. Assuming that $\mu_i = \beta X_i$ with $X_i(> 0)$ known but $\beta(> 0)$ unknown, restricting within fixed (a) sample size designs p_n and in particular adopting a design p_{nx} for which $\pi_i = nX_i/X, X = \sum_1^N X_i$, one gets $e_0 = \bar{e}$ and

$$E_m E_{p_{nx}} E_R(e - Y)^2 \geq E_m E_{p_{nx}} E_R(\bar{e} - Y)^2$$

that is, the class $(p_{nx}, \bar{e})$ is optimal among (p_{nx}, e). If in addition $\sigma_i = \sigma X_i (\sigma > 0)$, then, writing $p_{nx\sigma}$ as a p_n design with $\pi_i = \frac{nX_i}{X} = \frac{n\sigma_i}{\sum \sigma_i}$, we have

$$\begin{aligned} E_m E_{p_n} E_R(e - Y)^2 &\geq E_m \sum \frac{V_i}{\pi_i} + \frac{(\sum \sigma_i)^2}{n} - \sum \sigma_i^2 \\ &= E_m E_{p_{nx\sigma}} E_R(\bar{e} - Y)^2. \end{aligned}$$

Thus, $(p_{nx\sigma}, \bar{e})$ is optimal among (p_n, e).

We may observe at the end that in the developments of RR strategies, we have followed closely the procedure of multistage sampling. An important distinction is that, in multistage sampling estimating the variance of an estimator $\widehat{Y}_i$ for fsu total Y_i is an important problem, while in the RR context the problem of estimating unbiasedly the variance of R_i as an estimator of Y_i does not exist, at least if one employs the techniques we have illustrated.

12.2.4 Application of Warner's (1965) and Other Classical Techniques When a Sample Is Chosen with Unequal Probabilities with or without Replacement

Let, for a person labeled i in $U = (1, \ldots, N)$, $y_i = 1$ if i bears a sensitive characteristic A, $= 0$ if i bears the complementary characteristic A^c. Then, $Y = \Sigma y_i$ denotes, for a given community, the total number of people bearing A needed to be estimated.

Let every person sampled participate in WARNER's RR programme in an independent way. Let

$$I_i = 1 \text{ if } i \text{ answers Yes on applying Warner's device}$$
$$= 0 \text{ if } i \text{ answers No}$$

Then,

$$Prob[I_i = 1] = E_R(I_i) = py_i + (1-p)(1-y_i)$$

yielding

$$r_i = \frac{I_i - (1-p)}{2p-1},$$

provided $p \neq \frac{1}{2}$, as an unbiased estimator for y_i because $E_R(r_i) = y_i$ for every i in U. Also,

$$V_R(r_i) = \frac{1}{(2p-1)^2} V_R(I_i) = V_i = \frac{p(1-p)}{(2p-1)^2}$$

since

$$V_R(I_i) = E_R(I_i)(1 - E_R(I_i)) = p(1-p)$$

on noting that $y_i^2 = y_i$. So, if

$$t = t(s, \underline{Y}) = \sum_{i \in s} y_i b_{si} = \sum_{i=1}^{N} y_i b_{si} I_{si}$$

subject to

$$E_p(b_{si} I_{si}) = 1 \forall i,$$

then,

$$e = e(s, \underline{R}) = \Sigma r_i b_{si} I_{si}$$

writing

$$\underline{Y} = (y_1, \ldots, y_i, \ldots, y_N), \ \underline{R} = (r_1, \ldots, r_i, \ldots, r_N),$$

satisfies

$$E(e) = E_p E_R(e) = E_p \Sigma y_i b_{si} I_{si} = Y$$

and also,

$$E(e) = E_R E_p(e) = E_R(\Sigma r_i) = Y$$

Again,

$$\begin{aligned} V(e) &= E_p V_R(e) + V_p E_R(e) \\ &= E_p(\Sigma V_i b_{si}^2 I_{si}) + V_p(t) \end{aligned} \tag{12.1}$$

and also,

$$\begin{aligned} V(e) &= E_R V_p(e) + V_R E_p(e) \\ &= E_R V_p(e) + V_R(\Sigma r_i) \\ &= E_R V_p(e) + \Sigma V_i, \end{aligned} \tag{12.2}$$

following CHAUDHURI, ADHIKARI and DIHIDAR (2000a). Consulting CHAUDHURI and PAL (2002), we may write

$$V_p(t) = -\sum_{i<j}\sum w_i w_j \left(\frac{y_i}{w_i} - \frac{y_j}{w_j} \right)^2 + \Sigma \frac{y_i^2}{w_i} \alpha_i$$

with $w_i (\neq 0)$ arbitrarily assignable,

$$d_{ij} = E_p(b_{si} I_{si} - 1)(b_{sj} I_{sj} - 1)$$

and

$$\alpha_i = \sum_{j=1}^{N} d_{ij},$$

and

$$V_p(e) = V_p(t)|_{\underline{Y}=\underline{R}} = -\sum\sum_{i<j} d_{ij} w_i w_j \left(\frac{r_i}{w_i} - \frac{r_j}{w_j}\right) + \Sigma \frac{r_i^2}{w_i} \alpha_i$$

Let it be possible to find d_{sij}'s free of $\underline{Y}$, $\underline{R}$, such that

$$E_p(d_{sij} I_{sij}) = d_{ij}, \ I_{sij} = I_{si} I_{sj}, \ I_{si} = 1 \quad \text{if} \quad i \in s, \ \pi_i > 0 \ \forall i.$$

Then,

$$v_p(t) = -\sum\sum_{i<j} d_{sij} I_{sij} w_i w_j \left(\frac{y_i}{w_i} - \frac{y_j}{w_j}\right)^2 + \Sigma \frac{y_i^2}{w_i} \alpha_i \frac{I_{si}}{\pi_i}$$

and

$$v_p(e) = v_p(t)|_{\underline{Y}=\underline{R}}$$

satisfy respectively

$$E_p v_p(t) = V_p(t)$$

and

$$E_p v_p(e) = V_p(e).$$

Then,

$$v_1 = v_p(e) + \Sigma V_i b_{si} I_{si}$$

satisfies $E(v_1) = V(e)$, vide Eq. (12.2). Since

$$E_R v_p(e) = v_p(t) - \sum\sum_{i<j} d_{sij} I_{sij} w_i w_j \left(\frac{V_i}{w_i^2} + \frac{V_j}{w_j^2}\right) + \Sigma \frac{V_i}{w_i} \alpha_i \frac{I_{si}}{\pi_i}$$

it follows from Eq. (12.1) above that

$$v_2 = v_p(e) + \sum\sum_{i<j} d_{sij} I_{sij} w_i w_j \left(\frac{V_i}{w_i^2} + \frac{V_j}{w_j^2}\right) + \Sigma \left(b_{si}^2 - \frac{\alpha_i}{w_i \pi_i}\right) I_{si}$$

is an unbiased estimator of $V(e)$ because

$$E(v_2) = E_p E_R(v_2) = V(e).$$

REMARK 12.1 *For* WARNER*'s RR scheme, V_i is known. But in other schemes, V_i may have to be estimated from the sample by some statistic $\hat{V}_i$, which has to be substituted for V_i in the above formulae for v_1 and v_2.*

If, as in RAJ (1968) and RAO (1975),

$$V_p(t) = \sum_i a_i y_i^2 + \sum\sum_{i \neq j} a_{ij} y_i y_j$$

and

$$v'_p(t) = \Sigma y_i^2 a_{si} I_{si} + \sum\sum_{i \neq j} y_i y_j a_{sij} I_{sij}$$

such that $E_p(a_{si} I_{si}) = a_i$ and

$$E_p(a_{sij} E_{sij}) = a_{ij},$$

then if $\hat{V}_i$ be an unbiased estimator for $V_i = V_R(r_i)$, then two alternative unbiased estimators for $V(e)$ turn out as

$$v'_1 = v'_p(e) + \Sigma \hat{V}_i b_{si} I_{si}$$

and

$$v'_2 = v'_p(e) + \Sigma \hat{V}_i (b_{si}^2 - a_{si}) I_{si}$$

writing

$$v'_p(e) = v'_p(t)|_{\underline{Y}=\underline{R}}$$

This is because it is easy to check that

$$Ev'_1 = V(e) \text{ of Eq. (12.2)}$$

and

$$Ev'_2 = V(e) \text{ of Eq. (12.1) above.}$$

For the well-known unrelated question RR model of HORVITZ et al. (1967), for any sampled person i, four independent RRs are needed according to the following devices.

Let I_i, I'_i be distributed independently and identically such as $I_i = 1$ if i draws at random a card from a box with a proportion p_1 of cards marked A and the remaining ones as marked B, and the card type drawn matches his/her actual trait A or B, $= 0$, else.

Similarly, let J_i and J_i' be independently and identically distributed random variables generated in the same manner as I_i, I_i', with the exception that p_1 is replaced by p_2 $(0 < p_1 < 1, 0 < p_2 < 1, p_1 \neq p_2)$.

Letting

$$
\begin{aligned}
y_i &= 1 \text{ if } i \text{ bears the sensitive trait } A \\
&= 0, \text{ else}
\end{aligned}
$$

and

$$
\begin{aligned}
x_i &= 1 \text{ if } i \text{ bears an unrrelated innocuous trait } B \\
&= 0, \text{ else}
\end{aligned}
$$

we may check that

$$
\begin{aligned}
E_R(I_i) &= p_1 y_i + (1 - p_1)x_i = E_R(I_i') \\
E_R(J_i) &= p_2 y_i + (1 - p_2)x_i = E_R(J_i')
\end{aligned}
$$

leading to

$$
r_i' = \frac{(1 - p_2)I_i - (1 - p_1)J_i}{(p_1 - p_2)} \cdot \ni \cdot E_R(r_i') = y_i
$$

and

$$
r_i'' = \frac{(1 - p_2)I_i' - (1 - p_1)J_i'}{(p_1 - p_2)} \cdot \ni \cdot E_R(r_i'') = y_i
$$

so that $r_i = \frac{1}{2}(r_i' + r_i'')$ satisfies $E_R(r_i) = y_i$ and $\hat{V}_i = \frac{1}{4}(r_i' - r_i'')^2$ satisfies $E_R(\hat{V}_i) = V_R(r_i) = V_i$. So, for $e = \Sigma r_i b_{si} I_{si}$ one may easily work out v_1, v_2, v_1', v_2'.

Chapter 13

Incomplete Data

13.1 NONSAMPLING ERRORS

The chapters that precede this develop theories and methods of survey sampling under the suppositions that we have a **target population** of individuals that can be identified and, using labels for identification of the units, we choose a sample of units of a desired size and derive from them values of one or more variables of interest. However, to execute a real-life sample survey, one usually faces additional problems. Corresponding to a target population one has to demarcate a **frame population**, or **frame** for short, which is a list of sampling units to choose from, or a map in case of geographical coverage problems. The target and the frame often do not exactly coincide. For example, the map or list may be outdated, may involve duplications, may overlap, and may together under or over cover the target. Corresponding to a frame population one has the concept of a **survey population**, which consists of the units that one could select in case of a 100 percent sampling. These two also need not coincide because during the field enquiry one may discover that some of the frame units may not qualify as the members of the target population and hence have to be discarded to keep close to the target. The field investigation values may be unascertainable for certain sections of the survey populations, or, even if ascertained, may have to be dropped because of inherent inconsistencies or palpable inaccuracies at the processing stage. Consequently, the sample data actually processed may logically yield conclusions concerning an **inference population**, which may differ from the survey population. MURTHY (1983) elegantly enlightens on these aspects.

The units from which one may gather variate values of interest, irrespective of accuracies, are called the **responding** units, the corresponding values being the **responses**; those that fail to yield responses constitute the **nonrespondents**. Some of the nonrespondents may, as a matter of fact, refuse to respond, giving rise to what are called **refusals**, while some, although identified and exactly located, may not be available for response during the field investigation, giving rise to the phenomenon of **not-at-homes**.

The discrepancies between the recorded responses and the corresponding true values are called **response errors**, or **measurement errors**. These errors are often correlated and arise because of faulty reporting by the respondents or because of mistaken recording by the agents of the investigator, namely the interviewers, coders, and processors. Interpenetrating network of subsampling is one of several procedures to provide estimators for correlated response variances arising because of interviewer (and/or coder-to-coder) variations. Further sophisticated model-based approaches making use of the techniques of variance components analysis and Minque (Minimum normed quadratic unbiased estimator) procedures are reported in the recent literature.

As a consequence of measurement inaccuracies, estimators based on processed survey data will deviate from the estimand parameters even if they are based on the whole population. The deviations due to sampling are called sampling errors, and the residual deviations are clubbed together under the title **nonsampling errors**.

If an estimator for a finite population mean (or total) is subject to an appreciable nonsampling error, then its mean square error about the true mean (or total) will involve not only a sampling error but also a component of nonsampling error. Consequently, estimators of sampling mean square errors discussed in the previous chapters will underestimate the overall mean square errors. Hence, the estimators in practice will not be as accurate as claimed or expected solely in terms of sampling error measures, and the confidence intervals based on them may often fail to cover the estimand parameters with the nominal confidence proclaimed. So, it is necessary to anticipate possible effects of nonsampling errors while undertaking a large-scale sample survey and consider taking precautionary measures to mitigate their adverse effects on the inferences drawn.

Another point to attend to in this context is that exclusively design-based inference is not possible in the presence of nonsampling errors. In the design-based approach, irrespective of the nature of variate values, inferences are drawn solely in terms of the selection probabilities, which are completely under the investigator's control. But nonresponse due to refusal unavailability, or ascertainment errors cannot be under the investigator's complete command. In order to draw inferences in spite of the presence of nonsampling errors, it is essential to speculate about their nature and magnitude and possible alternative and cumulative sources. Therefore, one needs to postulate models characterizing these errors and use the models to draw inferences.

In the next few sections we give a brief account of various aspects of nonsampling errors, especially of errors due to inadequate coverage of an intended sample due to nonresponse leading to the incidence of what we shall call incomplete data.

13.2 NONRESPONSE

To cite a simple example, suppose that unit i, provided it is included in a sample s, responds with probability q_i, q_i not depending on s or $\underline{Y} = (Y_i, \ldots, Y_N)$. Suppose n units are drawn by SRSWOR and define

$$M_i = \begin{cases} 1 & \text{if unit } i \text{ is sampled and responds} \\ 0 & \text{otherwise} \end{cases}$$

Consider the arithmetic mean

$$\overline{y} = \frac{\sum_1^N M_i Y_i}{\sum_1^N M_i}$$

of all observations as an estimator of $\overline{Y}$. Then

$$EM_i = \frac{n}{N} q_i$$

and $E\overline{y}$ is asymptotically equal to

$$\frac{\sum q_i Y_i}{\sum q_i}$$

The bias

$$\sum \left(\frac{q_i}{\sum q_i} - \frac{1}{N} \right) Y_i$$

is negligible only if approximately

$$q_i = \frac{1}{N} \sum q_i.$$

Even if the last equality holds for $i = 1, 2, \ldots, N$ the variance of $\overline{y}$ is inflated by the reduced size of the sample of respondents. So it behooves us to pay attention to the problem of nonresponse in sample surveys. The nonresponse rate depends on various factors, namely the nature of the enquiry, goodwill of the investigating organization, range of the items of enquiry, educational, socioeconomic, racial, and occupational characteristics of the respondents, their habitations and sexes, etc. In case of surveys demanding sophisticated physical and instrumental measurements, as in agricultural and forest surveys covering inaccessible areas, various other factors like, sincerity and diligence of the investigator's agents and their preparedness and competence in doing the job with due care and competence, are essential. With the progress of time, unfortunately, rates of nonresponse are advancing, and rates of refusals among the nonresponses are gradually increasing faster and faster in most of the countries where sample surveys and censuses are undertaken.

In order to cope with this problem in advanced countries enquiries are mostly being done through telephone calls rather than through mailing questionnaires or direct face-to-face interviews. One practice to realize a desired sample size is to resort to quota sampling after deep stratification of the population. In quota sampling from each stratum, a required sample size is realized by contacting the sampling units in each stratum in succession following a preassigned pattern, and sampling in each stratum is terminated as soon as the predetermined quota of sample size is fulfilled and nonresponses and refusals in course of filling up the quota are just ignored. This is a nonprobability sampling and hence is not favored by many survey sampling experts.

Randomized response technique is also a device purported to improve on the availability of trustworthy response relating to sensitive and ticklish issues on which data are difficult to come by, as we have described in detail in chapter 12.

Another measure to reduce nonresponse is to **callback** either all or a suitable subsample of nonrespondents at successive repeat calls. We postpone to section 13.3 more details about the technique.

Sometimes during the field investigation itself, each nonresponse or refusal case after a reasonable number of callbacks and persuasive efforts fails to elicit response is replaced by a sampling unit found cooperative but outside the selected sample of units, although of course within the frame. Such a replacement unit is called a **substitute**. Anticipating possibilities of nonresponse, in practice, a preplanned procedure of choosing the substitutes as standbys or backups is usually followed in practice. In substitution it is, of course, tacitly assumed that the values for the substituting units closely resemble those for the ones correspondingly substituted. Success of this procedure depends strongly on the validity of this supposition.

As is evident from the text thus far developed, an estimator for a finite population total or mean is a weighted sum of the sampled values, the weights being determined in terms of the features of the sampling design and/or characteristics of the models if postulated to facilitate inference making. In case there is nonresponse, and hence a reduced effective size of the data-yielding sample, an obvious step to compensate for missing data is to revise the original sample weights. The sample weights are devised to render an estimator reasonably close to the estimand parameter. Since some of the sample values are missing due to nonresponse, the weights to be attached to the available respondent sample units need to be stepped up to bring the estimator reasonably close to the parameter. So, **weighting adjustment** is a popular device to compensate for missing data in sample surveys. In effect, in employing this technique, the nonresponses are treated as alike as the responses such that this technique also is tacitly based upon the assumption that the respondents and nonrespondents have similar characteristics and the nonrespondents are missing just at random.

In large-scale surveys the assumption of missingness at random is untenable. To overcome this difficulty, utilizing available background information provided by data on auxiliary correlated variables with values available on both the respondents and the nonrespondents, the population is divided into strata or into post-strata, in this case called **adjustment classes** or **weighting classes**, so that within a class the respondents and the nonrespondents may be presumed to have similar values on the variables of interest. Thus, missingness at random assumption is not required to be valid for the entire population, but only separately within the weighting classes. The nonresponse rates will vary appreciably across these classes. Then, weighting adjustment technique to compensate for nonresponse is applied using differential weight adjustments across the classes, the weights within each class being stepped up in proportion to the inverse of the rate of response.

HARTLEY (1946), followed by POLITZ and SIMMONS (1949, 1950), proposed to gather from each available respondent the number out of the five previous consecutive days he/she was available for a response. If someone was available on $h(h = 0, 1, 2, 3, 4, 5)$ days $\frac{h+1}{6}$ was used as an estimated probability of his/her response and $\frac{6}{h+1}$ was used as a weight for every respondent of the type $h(h = 0, 1, \ldots, 5)$. Here 1 is added because on the day of his/her actual interview he/she is available to report. This device, however, only takes care of not-at-homes, not the refusals. Also, no information is gathered on the actual not-at-homes on the day of the enquiry.

Weighting adjustment techniques, described in sections 13.4 and 13.5, are usually applied to tackle the problem of unit nonresponse, that is, when no data are available worth utilization on an entire unit sampled. But if, for a sampled unit, data are available on many of the items of enquiry but are missing on other items, then an alternative technique called **imputation** is usually employed. Imputation means filling in a missing record by a plausible value, which takes the place of the one actually missed by virtue of presumed closeness between the two. Various imputation procedures are currently being employed in practice, to be discussed in brief in section 13.7.

Another device to improve upon the availability of required data or cutting down the possibility of incomplete data is the technique of network sampling. A group of units that are eligible to report the values of a specific unit is called a **network**. A group of units about which a specific unit is able to provide data is called a cluster. In traditional surveys, the network and cluster relative to a given unit are both identical with the given unit itself. But in network sampling various rules are prescribed following which various members of networks and clusters are utilized in gathering information on sampled units. More details are discussed in section 13.6.

13.3 CALLBACKS

HANSEN and HURWITZ (1946) gave an elegant procedure for callbacks to tackle nonresponse problems later modified by SRINATH (1971) and J. N. K. RAO (1973), briefly described below. The population is conceptually dichotomized with $W_1(W_2 = 1 - W_1)$ and $\widehat{Y}_1(\widehat{Y}_2 = [\widehat{Y} - W_1\widehat{Y}_1]/W_2)$ as the proportion of respondents (nonrespondents) and mean of respondents (nonrespondents) and an SRSWOR of size n yields proportions $w_1 = n_1/n$ and $w_2 = 1 - w_1 = 1 - n_1/n = n_2/n$ of respondents and nonrespondents, respectively. Choosing a suitable number

$K > 1$ an SRSWOR of size $m_2 = n_2/K$, assumed to be an integer, is then drawn from the initial n_2 sample nonrespondents. Supposing that more expensive and persuasive procedures are followed in this second phase so that each of the m_2 units called back now responds, let $\overline{y}_1$ and $\overline{y}_{22}$ denote the first-phase and second-phase sample means based respectively on n_1 and m_2 respondents. Then, $\overline{Y}$ may be estimated by $\overline{y}_d = w_1\overline{y}_1 + w_2\overline{y}_{22}$, and the variance

$$V(\overline{y}_d) = (1 - f)\frac{S^2}{n} + W_2\frac{(K-1)}{n}S_2^2$$

by

$$\begin{aligned} v_d &= (1-f)\left(\frac{n_1 - 1}{n-1}\right) w_1 \frac{s_1^2}{n_1} \\ &+ \frac{(N-1)(n_2-1) - (n-1)(m_2-1)}{N(n-1)} w_2 \frac{s_{22}^2}{m_2} \\ &+ \frac{N-n}{N(n-1)}\left[w_1(\overline{y}_1 - \overline{y}_d)^2 + w_2(\overline{y}_{22} - \overline{y}_d)^2\right]. \end{aligned}$$

Here $f = \frac{n}{N}$; S^2 is the variance of the population of N units using divisor $(N-1)$, S_2^2, the variance of the population of nonrespondents, using divisor $(N_2 - 1)$, writing $N_i = NW_i (i = 1, 2)$, s_1^2, s_{22}^2 the variances of the sampled respondents in the first and second phases, using divisors $(n_1 - 1)$ and $(m_2 - 1)$, respectively.

Choosing a cost function $C = C_0 n + C_1 n_1 + C_2 m_2$ where C_0, C_1, C_2 are per unit costs of drawing and processing the initial, first-phase, and second-phase samples respectively of sizes n, n_1, and m_2 optimal choices of K and n that minimize the expected costs

$$E(C) = C_0 n + C_1 n W_1 + C_2 n W_2/K$$

for a preassigned value V of $V(\overline{y}_d)$ are, respectively,

$$K_{opt} = \left[C_2(S^2 - W_2 S_2^2)/S_2^2(C_0 + C_1 W_1)\right]^{1/2}$$

and

$$n_{opt} = \frac{NS^2}{NV + S^2}\left[1 + (K_{opt} - 1)W_2 S_2^2/S^2\right].$$

The same K_{opt} but

$$n'_{opt} = CK_{opt} / \left[K_{opt}(C_0 + C_1 W_1) + C_2 W_2\right]$$

minimize $V(\bar{y}_d)$ for a preassigned value C of $E(C)$. These results are inapplicable without knowledge about the magnitudes of S^2, S_2^2, W_2.

BARTHOLOMEW (1961) suggested an alternative of calling back. EL-BADRY (1956), SRINATH (1971), and P. S. R. S. RAO (1983) consider further extensions of the HANSEN–HURWITZ (1946) procedure of repeating callbacks, supposing that successive callbacks capture improved fractions of responses, leaving hardcore nonrespondents in succession in spite of more and more stringent efforts.

Another callback procedure is to keep records on the numbers of callbacks required in eliciting responses from each sampled unit and study the behavior pattern of the estimator, for example, the sample mean based on the successive numbers of calls $i = 1, 2, 3, \ldots$, etc., on which they were respectively based. If the sample mean $\bar{y}_i$ based on responses procured up to the ith call for $i = 1, 2, 3, \ldots$ up to t shows a trend as i moves ahead, then, fitting a trend curve, one may read off from the curve the estimates that would result if further callbacks are needed to get 100 percent response, and, using the corresponding extrapolated estimates $\bar{y}_j$ for $j > t$, one may get an average of the $\bar{y}_i$'s for $i = 1, 2, \ldots, t, t+1, \ldots$ using weights as the actual and estimated response rates to get a final weighted average estimator for the population mean. This extrapolation procedure, however, is not very sound because not-at-home nonresponses and refusal nonresponses are mixed up in this procedure, although their characteristics may be quite dissimilar on an average.

13.4 WEIGHT ADJUSTMENTS

In POLITZ-SIMMONS divided into disjoint and exhaustive weighting classes, weights are taken as reciprocals of the estimated response probabilities. The response probabilities here are estimated from the data on frequency of at-homes determined from the respondents met on a single call. THOMSEN and

SIRING (1983) extend this, allowing repeated calls. Utilizing, background knowledge and data on auxiliary variables, the sample is poststratified into weighting or adjustment classes. On encountering nonrespondents, several callbacks are made.

They consider three alternative courses, namely (1) getting responses on the first call, (2) getting nonrespondents and a decision to revisit, and (3) getting nonrespondents and abandoning them. In case (2) in successive visits, also, one of these three alternative courses is feasible. For the sake of simplicity let us illustrate a simple situation where there are only two post-strata and up to three callbacks are permitted. Let for the hth post-stratum or weighting class ($h = 1, 2$) P_h, Q_h and A_h denote the probabilities of (a) getting a response on the first call, (b) getting a response from one who earlier nonresponded, and (c) of getting a nonresponse and not calling back, abandoning the nonrespondents. Here Q_h is permitted to exceed P_h because after the first failure, a special appointment may be made to enhance chances of success in repeated calls. Let A_h for simplicity be taken as a constant A over $h = 1, 2$. Then, letting n_h as the observed sample size from the hth post-stratum and f_{hj} as the frequency of observed responses from the hth post-stratum on the jth call ($j = 1, 2, 3$), postulating a trinomial distribution for f_{h1}, f_{h2}, f_{h3} for each $h = 1, 2$ one may apply the method of moments to estimate P_h, Q_h, A by solving the equations (for $h = 1, 2$)

$$f_{h1} = n_h P_h$$
$$f_{h2} = n_h(1 - P_h - A)Q_h$$
$$f_{h3} = n_h(1 - P_h - A)(1 - Q_h - A)Q_h.$$

Alternatively, one may also use the least squares method by postulating, for example,

$$f_{hj} = \alpha_h + \beta_h j + \epsilon_j$$

with α_h, β_h as unknown parameters, $h = 1, 2,\ j = 1, 2, 3$, $E(\epsilon_j) = 0$, $V(\epsilon_j) = \sigma^2(> 0)$, so that $E(f_{hj}) = \alpha_h + \beta_h, j = 1, 2, 3$. After obtaining estimates of probabilities of responses available on the first, second, and third calls from sampling units of respective post-strata, weight-adjusted estimates of population means and totals are obtained using weights as reciprocals of estimated response probabilities. Further generalizations necessitating quite complicated formulae are available in the literature. OH and SCHEUREN (1983) is an important reference.

We will now consider samples drawn with equal probabilities, that is, by epsem (equal probability selection methods). Suppose the population is divisible into H weighting classes, rather post-strata with known sizes N_h or weights $W_h = N_h/N$ for the respective post-strata with known sizes N_h or weights $W_h = N_h/N$ for the respective post-strata denoted by $h = 1, \ldots, H$. Let $N_h = R_h + M_h, R_h(M_h)$, denoting the unknown numbers of units who would always respond (nonrespond) to the data collection procedure employed. Let $\overline{Y}_{rh}, \overline{Y}_{mh}, \overline{R}_h, \overline{M}_h$ denote the means of the respondents, nonrespondents, and corresponding proportions of the hth class, $h = 1, \ldots, H$. Let $\overline{y}_r$ be the overall mean of the sampled respondents and $\overline{y}_{rh}$ the mean of the sampled respondents from the hth class ($h = 1, \ldots, H$). Then, the bias of $\overline{y}_r$ as an estimator for the population mean $\overline{Y}$ is

$$\begin{aligned} B(\overline{y}_r) &= \sum W_h \left(\overline{Y}_{rh} - \overline{Y}_r\right)\left(\overline{R}_h - \overline{R}\right)/\overline{R} \\ &\quad + \sum W_h \overline{M}_h \left(\overline{Y}_{rh} - \overline{Y}_{mh}\right) \\ &= A + B, \text{ say,} \end{aligned}$$

writing $\overline{Y}_r$ as the overall population mean of all the R respondents, $\overline{R} = \frac{R}{N}, R = \sum N_h R_h$. An alternative estimator for $\overline{Y}$ is $\overline{y}_p = \sum W_h \overline{y}_{rh}$, called the **population weighting adjusted estimator**, available in case W_h's are known. Its bias is

$$B(\overline{y}_p) = \sum W_h \overline{M}_h (\overline{Y}_{rh} - \overline{Y}_{mh}) = B.$$

A condition for unbiasedness of $\overline{y}_r$ is $\overline{Y}_r = \overline{Y}_m$, writing $\overline{Y}_m$ for the mean of overall nonrespondents in the population, while that for $\overline{y}_p$ is $\overline{Y}_{rh} = \overline{Y}_{mh}$ for each $h = 1, \ldots, L$. THOMSEN (1973, 1978) and KALTON (1983b) examined in detail relative merits and demerits of these two in terms of their biases, variances, mean square errors, and availability of variance estimators. Preference of one over the other here is not conclusive.

In case W_h's are unknown, using their estimators, namely $w_h = n_h/n$, the proportion of the sample falling in the respective weighting classes, an alternative sample weighted estimator for $\overline{Y}$ is $\overline{y}_s = \sum_h w_h \overline{y}_{rh}$. Its bias is $B(\overline{y}_s) = B = B(\overline{y}_p)$. One may consult KALTON (1983b) and KISH (1965) for further details about the formulae for variances of $\overline{y}_s$ and comparison of $\overline{y}_r$, $\overline{y}_p$ and $\overline{y}_s$ with respect to their biases and mean square errors and variance estimators.

Raking ratio estimation, or **raking**, is another useful weighting adjustment procedure to compensate for nonresponse when a population is cross-classified according to two or more characteristics. For simplicity, we shall illustrate a cross-tabulation with respect to only two characteristics, which respectively appear in H and L distinct forms. Suppose W_{hl} is the proportion of the population of size N falling in the (h, l)th cell, which corresponds to the hth form of the first character I, and the lth form of the second character, say, $\pi, h = 1, \ldots, H$ and $1 = 1, \ldots, L$. Let $W_h = \sum_{l=1}^{L} W_{hl}$ and $W_{.l} = \sum_h W_{hl}$ denoting, respectively, the two marginal distributions, be known, $h = 1, \ldots, H$ and $l = 1, \ldots, L$. Let, for a sample of size n from the population, the sample proportion in the (h, l)th cell be $P_{hl} = n_{hl}/n, n_{hl}$, denoting the number of sample observations falling in the (h, l)th cell. We shall assume an epsem sample. The sample marginal distributions are then specified by $p_{h\cdot} = \sum_l p_{hl}$ and $p_{.l} = \sum_h p_{hl}$ for $h = 1, \ldots, H$ and $l = 1, \ldots, L$, respectively. In the above, the population joint distribution (W_{hl}) is supposed to be unknown. The problem of raking is one of finding right weights so that when the sample cell relative frequencies are weighted up, then the two resulting marginal distributions of the weighted sample cell proportions respectively agree simultaneously with the known population marginal distributions. In order to choose such appropriate weighting factors one needs to employ an algorithm involving iteration, called the method of **iterated proportional fitting** (IPF). To illustrate this algorithm, suppose the initial choice of weights is $W_h/p_{h\cdot}$. Then, the weighted sample proportions, namely $t_{hl} = \frac{W_{h\cdot}}{p_{h\cdot}} p_{hl}$, lead to a marginal distribution

$$\left\{\sum_l \frac{W_{h\cdot}}{p_{h\cdot}} p_{hl} = W_{h\cdot}\right\}$$

which agrees with one of the population marginal distributions, namely, with $\{W_{h\cdot}\}$ but not with the other, namely $\{W_{.l}\}$. So, at the second iteration, if we use the new set of weights $W_{.l}/t_{.l}$ where $t_{.l} = \sum_h t_{hl}$, then the new set of weighted sample cell proportions, namely, $e_{hl} = \frac{W_{.l}}{t_{.l}} t_{hl}$, will yield a marginal distribution $\{\sum_h e_{hl}\} = \{W_{.l}\}$, which coincides with the other population marginal distribution but differs from the first marginal distribution. So, further iteration should be continued in turn to achieve conformity with the two marginal distributions with a high degree of accuracy. If the convergence is rapid the method is successful; if not, usually as specified, 4 or 6 iteration cycles are employed and the process is stopped. Suppose the terminating weighted sample proportions for the cells conforming closely with respect to their marginal distributions to the given population marginal distributions are given by $\{\overline{W}_{hl}\}$. Then $t_r = \sum_h \sum_l \overline{W}_{hl} \overline{y}_{rhl}$ with $\overline{y}_{rhl}$ as the sample mean based on the respondents out of the sampled units falling in the (h, l)th cell, is taken as the estimator for $\overline{Y}$. For further discussion on raking ratio method of estimation, one may consult KALTON (1983b) and BRACKSTONE and RAO (1979).

13.5 USE OF SUPERPOPULATION MODELS

Suppose $x_1, x_2, \ldots, x_k$ are k auxiliary variables correlated with the variable of interest with values $X_{ji}, i = 1, \ldots, 1, \ldots, N$, $j = 1, \ldots, k$. Let $\underline{X}$ be the $N \times k$ matrix with ith row $\underline{x}_i' = (x_{1i}, \ldots, x_{ki})$, $i = 1, \ldots, N$, $\underline{X}_s$ an $n \times k$ submatrix of $\underline{X}$ consisting of n rows with entries for i in a sample s chosen with probability $p(s)$ with inclusion probabilities $\pi_i > 0$, and $\underline{X}_r$ an $n_1 \times k$ submatrix of $\underline{X}_s$ consisting of $n_1(< n)$ rows corresponding to n_1 units of s which respond. Let $\underline{\beta} = (\beta_1, \ldots, \beta_k)'$ be a $k \times 1$ vector of unknown parameters and let

$$E_m(\underline{Y}) = \underline{X}\underline{\beta}, \ V_m(\underline{Y}) = \sigma^2 \underline{V}$$

where $\sigma(> 0)$ is unknown but $\underline{V}$ is a known $N \times N$ diagonal matrix and $\underline{Y} = (Y_1, \ldots, Y_n)'$ (cf. section 4.1.1). Then, an estimator based on s assuming full response is

$$t_s = \sum_{i=1}^{N} \widehat{\mu}_i$$

where

$$\widehat{\mu 1} = \underline{x_i}'\underline{\widehat{\beta_s}}$$

$$\widehat{\underline{\beta}_s} = \left(\underline{X}_s'\pi_s^{-1}\underline{V}_s^{-1}\underline{X}_s\right)^{-1}\left(\underline{X}_s'\pi_s^{-1}\underline{V}_s^{-1}\underline{Y}_s\right)$$

$\underline{\pi}_s$ = diagonal matrix with π_i for i in s in the diagonals

$\underline{V}_s$ = diagonal submatrix of $\underline{V}$ with entries for $i \in s$

$\underline{Y}_s = n \times 1$ subvector of $\underline{Y}$ containing entries for $i \in s$

and all the inverses are assumed to exist throughout.

This t_s may be expressed in the form

$$t_s = \underline{U}_s'\underline{Y}_s = \sum_{i \in s} U_{si} Y_i,$$

with U_{si} as the ith element of the $1 \times n$ vector

$$\underline{U}_s' = \underline{1}_N'\underline{X}(\underline{X}_s'\pi_s^{-1}\underline{V}_s'\underline{X}_s)^{-1}\underline{X}_s'\pi_s^{-1}\underline{V}_s^{-1}.$$

In case response is available on only a subsample s_1 of size $n_1(< n)$ out of s, then we employ the estimator

$$\tilde{t}_s = \sum_{i \in s_1} U_{si} Y_i + \sum_{i \in s - s_1} U_{si} \overline{Y}_i$$

where, with

$$\underline{X}_{s_1}', \underline{\pi}_{s_1}^{-1} \underline{V}_{s_1}^{-1}, \underline{Y}_{s_1}$$

as submatrices and subvectors corresponding to $\underline{X}_s'$, $\underline{\pi}_s^{-1}$, $\underline{V}_s^{-1}$, $\underline{Y}_s$, omitting from the latter the entries corresponding to the units in $s - s_1$,

$$\widehat{\beta}_{s_1} = \left(\underline{X}_{s_1}'\underline{\pi}_{s_1}^{-1}\underline{V}_{s_1}^{-1}\underline{X}_{s_1}\right)^{-1}\left(\underline{X}_{s_1}'\underline{\pi}_{s_1}^{-1}\underline{V}_{s_1}^{-1}\underline{Y}_{s_1}^{-1}\right),$$

$$\widehat{Y}_i = \underline{x}_i'\widehat{\beta}_{s_1}.$$

And it may be shown that

$$\tilde{t}_s = \sum_{i \in s_1} U_{s_1 i} Y_i = t_{s_1}, \text{ say,}$$

with

$$\underline{U}'_{s_1} = 1'_N \underline{X} \left(\underline{X}'_{s_1} \underline{\pi}_{s_1}^{-1} \underline{V}_{s_1}^{-1} \underline{X}_{s_1} \right)^{-1} \underline{X}'_{s_1} \underline{\pi}_{s_1}^{-1} \underline{V}_{s_1}^{-1}$$

and $U_{s_1 i}$ the ith element of the $1 \times n_1$ vector $\underline{U}'_{s_1}$. This seems intuitively sensible, and its properties of asymptotic design-unbiasedness in spite of model failure and under assumption of random missingness of records have been investigated by CASSEL, SÄRNDAL and WRETMAN (1983). An alternative procedure in this context of using generalized regression estimator (GREG estimator) in the presence of nonresponse is considered as follows by SÄRNDAL and HUI (1981) in case every unit is assumed to have a positive but unknown response probability.

Let $q_i = q_i(\underline{X}, \theta)(> 0)$ denote an unknown response probability of ith unit ($i = 1, \ldots, N$), which is permitted to depend on the known matrix $\underline{X}$ and on some unknown parameter $\underline{\theta} = (\theta_1, \ldots, \theta_\alpha)$. SÄRANDAL and HUI (1981) suggest estimating $\underline{\theta}$ in q_i using the likelihood

$$\prod_{i \in s_1} q_i \prod_{i \in s - s_1} (1 - q_i)$$

assuming a simple form of $q_i = q_i(\underline{X}, \underline{\theta}) = q_i(\underline{\theta})$. Suppose that maximum likelihood or other suitable estimators $\widehat{q}_i$ for q_i are available and denote by $\underline{Q}_N$ the diagonal matrix of order $N \times N$ with $\widehat{q}_i$'s, $i = 1, \ldots, N$ in the diagonal and by $\underline{Q}_s$, $\underline{Q}_{s_1}$ the diagonal submatrix of $\underline{Q}_N$ accommodating only the entries corresponding to i in s and i in s_1, respectively. SÄRNDAL and HUI (1981) suggest estimating $\underline{\beta}$ by

$$\widehat{\underline{\beta}}_q = (\underline{X}'_{s_1} \underline{\pi}_{s_1}^{-1} \underline{V}_{s_1}^{-1} \underline{Q}_{s_1}^{-1} \underline{X}_{s_1})^{-1} (\underline{X}'_{s_1} \underline{\pi}_{s_1}^{-1} \underline{V}_{s_1}^{-1} \underline{Q}_{s_1}^{-1} \underline{Y}_{s_1}),$$

and

$$Y = \sum_1^N Y_i \text{ by } t_{qg} = \sum_1^N \widehat{\mu}_{qi} + \sum_{s_1} \frac{\widehat{e}_{qi}}{\pi_i}$$

where

$$\widehat{\mu}_{qi} = \underline{x}'_i \widehat{\beta}_q, \widehat{e}_{qi} = Y_i - \widehat{\mu}_{qi}$$

and examine properties of this revised GREG estimator under several postulated models for q_i. One difficulty with this approach is that the same model connecting both the respondents and nonrespondents is required to be postulated to derive good properties of t_{qg}.

In section 3.3.2, we discussed GODAMBE and THOMPSON's (1986a) estimating equation

$$\sum_{i \in s} \frac{\phi_i(Y_i, \theta)}{\pi_i} = 0$$

in deriving optimal estimators based on survey data $d = (i, Y_i | i \in s)$. If the response probability $q_i (> 0)$ is known and s_r is the responding subset of s, then GODAMBE and THOMPSON (1986) recommend estimation on solving

$$\sum_{i \in s_r} \frac{\phi_i(Y_i, \theta)}{\pi_i q_i} = 0.$$

In case q_i's are unknown, they propose further modifications we omit.

13.6 ADAPTIVE SAMPLING AND NETWORK SAMPLING

Suppose we intend to estimate the unknown size μ of a domain in a given finite population of individuals, the domain being characterized by a specified trait that is rather infrequent. Let such a domain be denoted by

$$\Omega = (1, \ldots, \mu).$$

Suppose we have a frame of households

$$F = (H_1, \ldots, H_M)$$

and let I_{ij} denote the jth person of ith household H_i which consists of T_i household members, $j = 1, \ldots, T_i$, $i = 1, \ldots, M$, and let $T = \sum_1^M T_i$. We presume that, taking hold of individuals I_{ij} from the households H_i, we can construct networks to obtain information about the individual α ($\alpha = 1, \ldots, \mu$) in the domain Ω. In order to estimate μ let us, for example, choose a counting rule r, as follows, which will enable us to derive an estimator for μ on taking a sample of households from F and contacting members of selected households who may serve as informants about the members of the domain Ω.

Let

$$\delta_{\alpha ij}(r) = 1 \text{ is } I_{ij} \text{ if eligible by rule } r \text{ to report about } \alpha$$
$$= 0, \text{ else.}$$

Then

$$S_{\alpha i}(r) = \sum_{j=1}^{T_i} \delta_{\alpha ij}(r)$$

is the total number of members of H_i eligible by rule r to report about α and

$$S_{\alpha}(r) = \sum_{i=1}^{M} S_{\alpha i}(r)$$

the total number of members of all the households in the frame F eligible to report on α by rule r.

Let an SRSWR in m draws be taken out of F and define

$$a_i = 1 \text{ if } H_i \text{ is sampled, } i = 1, \ldots, M$$
$$= 0, \text{ else.}$$

Let some sampling weights $W_{\alpha ij}(\alpha = 1, \ldots, \mu, i = 1, \ldots, M, j = 1, \ldots, T_i)$ be chosen somehow and consider the weighted sum

$$\lambda_i(r) = \sum_{\alpha=1}^{\mu} \sum_{j=1}^{T_i} \delta_{\alpha ij}(r) W_{\alpha ij}$$

Then

$$\widehat{\mu}(r) = \frac{M}{m} \sum_{i=1}^{M} a_i \lambda_i(r)$$

is called the **multiplicity** estimator for μ. For the sake of unbiasedness we assume $\alpha = 1, 2, \ldots, \mu$

(a) $S_{\alpha}(r) > 0$

(b) $\sum_1^M \sum_{j=1}^{T_i} S_{\alpha ij}(r) W_{\alpha ij} = 1.$

One choice is $W_{\alpha ij} = 1/S_{\alpha(r)}$. Let $\frac{1}{M} \sum_{i=1}^{M} \lambda_i(r) = \overline{\lambda}(r)$. Then, the variance of $\widehat{\mu}(r)$ is

$$V(\widehat{\mu}(r)) = \frac{M^2}{M} V(\lambda(r)),$$

where

$$V(\lambda(r)) = \frac{1}{M}\sum_{1}^{M}(\lambda_i(r) - \overline{\lambda}(r))^2.$$

To see an advantage of network sampling instead of traditional sampling in this context, let us assume that

$$\sum_{\alpha=1}^{\mu}\sum_{j=1}^{T_i}\delta_{\alpha ij}(r) \leq 1 \text{ for every } i = 1, \dots, M,$$

that is, (1) no more than one individual of Ω will be enumerable at a household and (2) no individual will be enumerable more than once at a household. If $P = \mu/M$ is quite small, that is, the trait characterizing the domain Ω is relatively rare, then this assumption should be satisfied. Then, taking

$$W_{\alpha ij}(r) = \frac{1}{S_\alpha(r)},$$

it follows that

$$V(\lambda(r)) = P(K_{(r)} - P) = P(1-P) - P(1-K_{(r)})$$

where

$$K_{(r)} = \frac{1}{\mu}\sum_{\alpha=1}^{\mu} 1/S_\alpha(r).$$

Writing

$$\overline{S}(r) = \frac{1}{\mu}\sum_{1}^{\mu} S_\alpha(r)$$

it follows that

$$\frac{1}{\overline{S}(r)} \leq K(r) \leq 1$$

since $K(r)$ is the inverse of the harmonic mean of the $S_{\alpha(r)} \geq 1$.

For traditional surveys $K(r) = 1$ and $V(\lambda(r)) = P(1-P)$. Thus $P(1-K(r))$ represents the gain in efficiency induced by network sampling. Introducing appropriate cost consideration, SIRKEN (1983) has shown that in addition to efficiency, average cost of survey may also be brought down by network sampling in many practical situations.

S. K. THOMPSON (1990) introduced adaptive sampling, later further developed by THOMPSON (1992) and THOMPSON and SEBER (1996). CHAUDHURI (2000a) clarified that if a sample provides an unbiased estimator for a finite population total along with an unbiased estimator for the variance of this estimator, then this initial sample can be extended into an adaptive sample, capturing more sampling units with desirable features of interest, yet providing an unbiased estimator for the same population total along with an unbiased variance estimator for this estimator.

An important virtue of adaptive sampling compared to the initial one is its ability to add to the information content of the original sample, although not necessarily boosting an upward efficiency level unless one starts with a simple random sample.

Historically, adaptive sampling is profitably put to use in exploring mineral deposits, inhabitance of land and sea animals in unknown segments of vast geographical locations, and pollution contents in various environments in diverse localities. Recently, CHAUDHURI, BOSE and GHOSH (2004) have applied it in effective estimation of numbers of rural earners, principally through specific small-scale single industries in the unorganized sector abounding in unknown pockets.

Suppose $U = (1, \ldots, i, \ldots, N)$ is a finite population of a known number of units with unknown values y_i which are nonnegative but many are zero or low-valued, but some are large enough so that the population total $Y = \Sigma y_i$ is substantial and should be estimated through a judiciously surveyed sample. If a chosen sample contains mostly zero or low-valued units, then evidently it is unlikely to yield an accurate estimate. A way to get over this is the following approach.

Suppose every unit i in U has a well-defined neighborhood composed of itself and one or more other units. Any unit for which a certain prespecified condition c^*, concerning its y value is not satisfied is called an **edge unit**. Starting with any unit i for which c^* is satisfied, the same condition is to be tested for all the units in its neighborhood. This testing is to be continued for any unit in the neighborhood satisfying c^* and is to be terminated only on encountering those for which c^* is not satisfied. The set of all the distinct units thus tested constitutes a cluster $c(i)$ for i including i itself. Dropping the units of $c(i)$ with c^* unsatisfied the remainder of $c(i)$ is called **network** $A(i)$ of i. An edge unit is then called a **singleton network**. Treating the singleton network also, by courtesy, as networks, it follows that all the networks thus formed are nonoverlapping, and they together exhaust the entire population. Writing C_i the cardinality of $A(i)$ and writing

$$t_i = \frac{1}{C_i} \sum_{j \in A(i)} y_j$$

it follows that $T = \Sigma t_i$ equals $Y = \Sigma y_i$. Consequently, to estimate Y is same as to estimate T.

If $t = t(s, y_i | i \in s)$ is an unbiased estimate for Y, then $t(s, t_i | i \in s)$ is unbiased for T and hence for Y as well. Now, in order to ascertain $t(s, t_i | i \in s)$, it is necessary to survey all the units in $A(s) = \sum_{i \in s} A(i)$. This $A(s)$ as an extension of s is called an **adaptive sample**. This process of extending from s to $A(s)$ is called adaptive sampling. Obviously, this is an example of **informative sampling**, because to reach $A(s)$ from s one has to check the values of y_i for i in s and also in $c(i)$ for i in s.

Let us treat a particular and familiar case of t as

$$t_b = \sum y_i b_{si} I_{si} \quad \text{with} \quad E_p(b_{si}, I_{si}) = 1 \forall i \ldots \tag{13.1}$$

when s is chosen with probability $p(s)$ according to design p. Then,

$$V_p(t_b) = -\sum_{i<j}\sum d_{ij} w_i w_j \left(\frac{y_i}{w_i} - \frac{y_j}{w_j}\right)^2 + \sum_i \frac{y_i^2}{w_i} \alpha_i,$$

where $w_i (\neq 0)$ are constants, $\alpha_i = \sum_j d_{ij} w_j$ and

$$d_{ij} = E_p(b_{si} I_{si} - 1)(b_{sj} I_{sj} - 1).$$

An unbiased estimator for $V(t_b)$ is

$$v(t_b) = -\sum_{i<j}\sum d_{sij} I_{sij} w_i w_j \left(\frac{y_i}{w_i} - \frac{y_j}{w_j}\right)^2 + \sum_i \frac{y_i^2}{w_i} \alpha_i C_{si} I_{si}$$

on choosing constants C_{si}, d_{sij} free of $\underline{Y} = (y_1, \ldots, y_i, \ldots, y_N)$ such that $E_p(C_{si} I_{si}) = 1$ and $E_p(d_{sij} I_{sij}) = d_{ij}$, for example, $C_{si} = \frac{1}{\pi_i}, d_{sij} = \frac{d_{ij}}{\pi_{ij}}$ provided $\pi_{ij} = \sum_{s \ni ij} p(s) > 0 \forall i, j (i \neq j)$, in which case also $\pi_i > 0 \forall i$. Now for the adaptive sample $A(s)$ reached through s, one has only to replace y_i by t_i for $i \in s$ in t_b and $v(t_b)$ to get the appropriate revised estimators for adaptive sampling.

With a different kind of network formation we must consider network sampling, which is thoroughly distinct from adaptive sampling.

Suppose there are M identifiable units labeled $j = 1, \ldots, M$ called selection units (su). Also, suppose to each su is linked one or more observation units (ou), to each of which are linked one or more of the sus. Let N be the total number of such unknown ous with their respective values y_is with a total $Y = \sum_1^N y_i$, which is required to be estimated on drawing a sample s of sus and surveying and ascertaining the y_i values of all the ous linked to the sus thus sampled. This process of reaching all the ous linked to the initially sampled sus is called network sampling.

Here, a network means a set of ous and sus mutually interlinked. The link here is a reciprocal relationship. One ou linked to an su is linked to another ou, to which this su is linked and also several ous may be mutually linked directly as well. A hospital, for example, may be an su, and a heart patient treated in it may be an ou. Through a sample of hospitals exploiting the mutual and reciprocal links, we may capture a number of ous. Ascertaining their y values, for example, the number of days spent in hospitals for a heart patient, the expenses incurred for treatment there, etc., it may be possible to estimate the totals for all the patients who are the ous.

To see this, let us proceed as follows. Let A_j denote the set of ous linked to the jth su and m_i be the number of sus to which the ith ou is linked. Let

$$w_j = \sum_{i \in A_j} \frac{y_i}{m_i}.$$

Then,

$$W = \sum_{j=1}^{M} w_j = \sum_{j=1}^{M} \sum_{i \in A_j} \frac{y_i}{m_i} = \sum_{i=1}^{N} \frac{y_i}{m_i} \sum_{(j \mid A_j \ni i)} 1 = Y.$$

Thus, to estimate Y is to estimate W. So, using the data $(s, w_j \mid j \in s)$ one may employ an estimator $t = t(s, w_j | j \in s)$ for W and hence estimate Y, and also if a variance estimator for t is available in terms of w_j's, that automatically provides a variance estimator in terms of y_i's.

The main situation when network sampling is needed and appropriate is when the same observational unit is associated with more than one selection unit and vice versa, and it is not practicable to create a frame of the observation units to be able to choose samples out of them in any feasible manner.

An outstanding problem that needs to be addressed for adaptive as well as network sampling is that there is no built-in provision to keep a desirable check on the sample sizes in either of the two. SALEHI and SEBER (1997, 2002) have introduced some devices to keep in check the size of an adaptive sample. For network sampling, no such procedure seems to be available in the literature.

One easy solution for adaptive sampling is to take simple random samples without replacement (SRSWOR) $B(i)$ of suitable sizes $d_i(\leq C_i)$ independently for every i in s such that $\sum_{i \in s} d_i \leq L$, where L is a preassigned suitable number so that with the resources at hand, ascertainment may be accomplished for y_i within $B(s) = \cup_{i \in s} B(i)$. Then, instead of t_i one may calculate $e_i = \frac{1}{d_i} \sum_{j \in B(i)} y_j$ and employ an estimator for Y based on e_i for i in $B(s)$.

Similarly, in the case of network sampling one may confine surveying SRSWORs taken independently from A_j's, say, B_j's and ascertaining y_i's for $i \in B_j$ only with cardinality D_j of B_j's suitably chosen subject to an upper limit for $\sum_{j \in s} D_j$. Estimation in both adaptive and network sampling with sample sizes thus constrained may be comfortably accomplished. SIRKEN (1993) has certain results on efficiency of network sampling.

For adaptive sampling THOMPSON and SEBER (1996) have observed that, in case the original sample is an SRSWOR, increased efficiency is ensured for adaptive sampling, as is easy to see considering the analysis of variance, keeping in mind the between and within network sums of squares. But for general sampling schemes, no general claim is warranted about gain in efficiency through adaptive sampling.

The techniques of constraining the sizes of adaptive samples or network samples may essentially be interpreted as means of adjusting in estimation in the presence of partial nonresponse in surveys. This is because the nonresponding units in the samples from within each stratum may be assumed to have been actually drawn as simple random samples without replacement (SRSWOR) by design from the sample already drawn. Let us illustrate with an example.

Suppose an initial sample of size n has been drawn from a population by the RAO, HARTLEY, COCHRAN (RHC) scheme utilizing the normed size measures p_i $(0 < p_i < 1, \sum p_i = 1)$. From the n groups formed let us take an SRSWOR of m groups with m as an integer suitably chosen between 2 and $(n-1)$. Corresponding to the following entitites relevant to the full sample, namely,

$$t = \Sigma_n y_i \frac{Q_i}{p_i}, \; V(t) = A\left[\Sigma \frac{y_i^2}{p_i} - Y^2\right],$$

$$v(t) = B\left[\Sigma_n Q_i \frac{y_i^2}{p_i^2} - t^2\right], \quad A = \frac{\Sigma_n N_i^2 - N}{N(N-1)}, \quad B = \frac{\Sigma_n N_i^2 - N}{N^2 - \Sigma_n N_i^2}$$

we may work out the following based on the SRSWOR out of it

$$e = \frac{n}{m}\Sigma_m y_i \frac{Q_i}{p_i}, \; E_m(e) = t = \Sigma_n \xi_i, \; \xi_i = y_i \frac{Q_i}{p_i}, \; E_m, V_m$$

as expectation, variance operators with respect to SRSWOR in m draws from the RHC sample of size n, Σ_n sum over m groups,

$$V_m(e) = n^2\left(\frac{1}{m} - \frac{1}{n}\right)\frac{1}{(n-1)}\Sigma_n(\xi_i - t)^2,$$

$$v_m(e) = n^2\left(\frac{1}{m} - \frac{1}{n}\right)\frac{1}{(m-1)}\Sigma_n\left(\xi_i - \frac{\Sigma_m r_i}{m}\right)^2,$$

$$E_m v_m(e) = V_m(e)$$

Writing

$$w = B\left[\frac{n}{m}\Sigma_m Q_i \frac{y_i^2}{p_i^2} - (e^2 - v_m(e))\right],$$

an unbiased estimator for the variance of e turns out to be

$$v = v_m(e) + w = (1 + B)v_m(e) + B\left[\frac{n}{m}\Sigma_m Q_i \frac{y_i^2}{p_i^2} - e^2\right].$$

This approach may be pursued with other procedures of sample selection and also in more than one stage of sampling with equal and unequal selection probabilities at various stages.

13.7 IMPUTATION

If, on an item of enquiry in a sample survey, values are recorded in respect of a number r of sampled units, the so-called responses, while the values are missing in respect of the remaining $m = n - r$ sampled units, then for the sake of completeness of records to facilitate standard analysis of data, it is often considered useful not to leave the missing records blank but to ascribe somehow certain values to them deemed plausible on certain accountable grounds. This procedure of assigning values to missing records is called **imputation**. In computerized processing of huge survey data covering prodigious sizes of ultimate sampling units sampled related to numerous items of enquiry, it is found convenient to have a prescribed number of readings on each item rather than arbitrarily varying ones across the items induced by varying item-wise response rates. A simple procedure to facilitate this is imputation. The aim of imputation is, of course, to mitigate the effect of bias due to nonresponse. So, it is to be conceded that the acid test of its efficiency is the closeness of the values imputed to the true ones. Since the true values are unknown, one cannot prove the merits of this technique, if any. When implementing imputation, one should be careful to announce the extent of imputation executed in respect of each item subjected to this and explicitly indicate how it is done. Let us now mention a few well-known procedures of imputation. While applying an imputation process, the population is customarily considered divisible into a number of disjoint classes, called imputation classes. Several variables called **control** on matching on an item of interest available from the respondents' records are utilized in some form to be assigned to some of the nonresponding units on this item. The respondent for which a value is thus extracted to be utilized in assigning a value to a missing record for a nonrespondent is called a **donor** and the latter is called a **recipient**. Some of the imputation methods are:

(1) **Deductive imputation**

A missing record may sometimes be filled in correctly or with negligible error, utilizing available data on other related items, which, for the sake of consistency, itself may pinpoint a specific value for it as may be ascertained while applying edit checks at the start of processing of survey data. This is called **logical** or **consistency** or **deductive** imputation.

(2) **Cold deck imputation**

If records are available on the items of interest on the same sampled units from a recent past survey of the same population, then, based on the past survey, a cold deck of records is built up. Then, if for the current survey a record is missing for a sampled unit while one is available on it from the cold deck, then the latter is assigned to it. Cold deck imputation is considered unsuitable because it is not up-to-date and is superseded by the currently popular method of hot deck.

(3) **Mean value imputation**

Separately within each imputation class, the mean based on the respondents' value is assigned to each missing record for the nonrespondents inside the respective class. This mean value imputation has the adverse effect of distorting the distribution of the recorded values.

(4) **Hot deck**

First the imputation classes are prescribed. Using past or similar survey data a cold deck is initiated. For each class, for each item the current records are run through, a current survey value whenever available replacing a cold deck value while a cold deck value is retained for a unit which is missing for the current survey when the records are arranged in a certain order, fixing a single cold deck value for each class. For example, for an item suppose for the hth class x_h is a cold value obtained from past data. Suppose the sampled units are arranged in the sequence $i_1, i_2, i_3, i_4, i_5, i_6, i_7, i_8, i_9, i_{10}$ and the current values available are y_{i3}, y_{i6}, y_{i9} only and the remaining ones are unavailable. Then, the imputed values will be $z_{i1}, z_{i2}, z_{i3}, z_{i4}, z_{i5}, z_{i6}, z_{i7}, z_{i8}, z_{i9}, z_{i10}$ where $z_{i1} = z_{i2} = x_h, z_{i3} = y_{i3}, z_{i4} = y_{i3}, z_{i5} = y_{i3}, z_{i6} = y_{i6}, z_{i7} = y_{i6}, z_{i8} = y_{i6}, z_{i9} = y_{i9}$ and $z_{i10} = y_{i9}$. Two noteworthy limitations of the procedure are that (a) values of a single donor may be used with multiplicities and (b) the number of imputation classes should be small, for otherwise current survey donors may be unavailable to take the place of cold deck values.

(5) **Random imputation**

First the imputation classes are specified. Suppose for the hth imputation class n_h is the epsem sample size out of which r_h are respondents and $m_h = n_h - r_h$ are nonrespondents. Although $m = \sum_h m_h$ should be less than $r = \sum_h r_h$, the overall nonresponse rate $\frac{m}{n}$ (writing $n = \sum_h n_h$) being required to be substantially less than $\frac{1}{2}$ for general credibility and acceptability of the survey results, for a particular class h, it is quite possible that m_h may exceed r_h. Keeping this in mind, let for each h two integers k_h and t_h be chosen such that $m_h = k_h r_h + t_h$ ($k_h, t_h \geq 0$, taking $k_h = 0$ if $m_h < r_h$). Then, an SRSWOR of t_h is chosen out of the r_h respondents to serve as donors for the m_h missing records $(k_h + 2)$ times each and the remaining $(r_h - t_h)$ respondents serving as donors $(k_h + 1)$ times each. Further improvements of this random imputation procedure are available, leading to more complexities but possibly improved efficacies. Performances of this procedure may be examined with considerably complex analysis.

(6) **Flexible matching imputation**

This is a modification of hot deck practiced in the U.S. Bureau of the Census. Here, on the basis of data on numerous control variables considered in a hierarchical pattern in order of importance, for each recipient a suitable matching donor is determined, and in such determinations stringencies are avoided by dropping some of the control variables in the lower rungs of the hierarchy if found necessary to create a good match.

(7) **Distance function matching**

After creating imputation classes on the basis of control variables while fixing up donor–recipient matching, some ambiguities are required to be resolved on the borders of consecutive classes. For a smooth resolution the closeness of a match is often assessed in terms of a distance function. Different measures of distance, including MAHALANOBIS distance in case of availability of multiple control variables, and also those based on transformations including ranks, logarithmic transforms, etc., are tried in finding good neighbors or, if possible, nearest neighbors in picking up right donors for recipients. FORD (1976) and SANDE (1979) are appropriate references to throw further light on this method of imputation.

(8) **Regression imputation**

Suppose $x_1, \ldots, x_t$ are control variables with values available on both the respondents and nonrespondents, the potential donors and recipients respectively, while y is the variable of interest with values available only for the respondents. Using y and $x_j (j = 1, \ldots, t)$ values on the respondents is then established a regression line, which is utilized in obtaining predicted values on y for nonrespondents corresponding to each nonrespondent's x_j value. The predicted value is then usable for imputation either by itself or with a random error component added to it. If the control variables are all qualitative then log-linear or logistic models are often postulated in deriving the predicted values. If both qualitative and quantitative variables are available, then the former are often replaced by dummy variables in obtaining a right regression function. For alternatives and further discussions, one should consult FORD, KLEWENO and TORTORA (1980) and KALTON (1983b).

(9) **Multiple imputations**

While applying any one of several available imputation techniques, one must be aware that each imputed value is fake, as it cannot be claimed to be the real value for a missing one. Imputation cannot create any information that is really absent. So, it is useful to obtain repeated imputed values for each missing record by applying the same imputation techniques several, $c(> 1)$ times, and also by applying different imputation techniques repeatedly to compare among the resulting final estimates using the imputed values for satisfaction about their usefulness. RUBIN (1976, 1977, 1978, 1983) is an outstanding advocate for trying multiple imputed values in examining the performances of one or more of the available imputation techniques in any given context. Multiple imputation facilitates variance estimation, extending the technique of subsampling replication variance estimation procedure suitably adaptable in this context. For example, if z is any statistic obtained on the basis of multiple imputations replicated $C(> 1)$ times, z_j being its value for the jth replicate ($j = 1, \ldots, C$), $\bar{z} = \frac{1}{C}\sum_{j=1}^{C} z_j$, and $\widehat{v}_j$ is an estimated variance of z_j, then RUBIN's (1979) formula for estimating the variance of $\bar{z}$ is

$$v(\bar{z}) = \frac{1}{C}\sum_{1}^{C}\widehat{v}_j + \frac{1}{C-1}\sum_{1}^{C}(z_j - \bar{z})^2$$

For further details, one should consult RUBIN (1983) and KALTON (1983b).

(10) **Repeated replication imputation**

KISH (cf. KALTON, 1983b) recommends a variation but an analogue of multiple imputation technique that consists of splitting the sample into two or more parts, as in interpenetrating or replicated sampling, each part containing both respondents and nonrespondents, the response rates in the two or more such parts being usually different. A method is then applied using suitable weights, taking account of these differential response rates in the parts so that the bias due to nonresponse may be reduced when the donors are appropriately sampled in the two or more parts of the sample. In RUBIN's multiple imputation, donor values are duplicated to compensate for nonresponse and the process is then replicated. In KISH's repeated replication technique, first the sample is replicated and then in each replicate there is duplication of donor values to compensate for nonresponse. The latter procedure involves selection of donors without replacement and hence is likely to yield lower variances than the former, which involves selection of donors with replacement.

Epilogue

This book is, of course, not a suitable substitute for a well-chosen sample of published materials from the entire literature on theory and methods of survey sampling. In fact, a careful reader of the contents of even the limited bibliography we have annexed must be infinitely better equipped with the message we intend to convey than one depending exclusively on it. Yet, we claim it justifies itself because of its restricted size designed for rapid communication.

Requirements in a design- or, randomization- or, briefly, p-based approach toward estimating a total Y by a statistic t_p based on a sample s chosen with probability $p(s)$ are the following. (a) The bias $B_p(t_p)$ should be absent, or at least numerically small, (b) the variance $V_p(t_p)$ as well as the mean square error $M_p(t_p)$ should be small, and (c) a suitable estimator $v_p(t_p)$ for $V_p(t_p)$ should be available. One may use the standardized estimator (SZE) $(t_p - Y)/\sqrt{v_p(t_p)}$ to construct a confidence interval of a limited length covering the unknown Y with a preassigned nominal confidence coefficient $(1 - \alpha)$, close to 1, which is the coverage probability calculated in terms of $p(s)$. If the exact magnitude of its bias cannot be controlled,

t_p should at least be consistent, or at least its asymptotic p bias should be small.

Here the concept of asymptotics is not unique. We mentioned briefly one approach due to BREWER (1979). But we did not discuss one due to FULLER and ISAKI (1981) and ISAKI and FULLER (1982), which considers nested sequences of finite populations $U_k(U_k \subset U_{k'}, k < k')$ of increasing sizes $N_k(N_k < N_{k'}, k < k')$ from which independent samples s_k of sizes $n_k(< n_{k'}, k < k')$ are drawn according to sequences p_k of designs.

The SZE mentioned above is required to converge in law to the standardized normal deviate τ. The inference made with this approach is regarded as robust in the sense that it is valid irrespective of how the coordinates of $\underline{Y} = (Y_1, \ldots, Y_N)'$ are distributed of which Y is the total. The sampled and unsampled portions of the population are conceptually linked through hypothetically repeatable realization of samples. So the selection probability of a sample out of all speculatively possible samples constitutes the only basis for any inference.

In the p-based approach the emphasis is on the property of the sampling strategy specified with reference to the hypothetical p distribution of the estimators, rather than on how good or bad the sample actually drawn is. In the predictive model-based (m-based, in brief) approach, however, inference is conditional on the realized sample, which is an ancillary statistic. The speculation is on how the underlying population vector $\underline{Y} = (Y_1, \ldots, Y_N)'$ is generated through an unknown process of a random mechanism. In the light of available background information, a probability distribution for $\underline{Y}$ is postulated within a reasonable class, called a superpopulation model. Under a model, M, a predictor t_m for Y is adopted that is m unbiased, that is, $E_m(t_m - Y) = 0$ for every sample such that $V_m(t_m - Y)$ is minimum among m-unbiased predictors that are linear in the sampled Y_i's.

A design, however, is chosen consistently within one's resources such that $E_p V_m(t_m - Y)$ is minimal. An optimal design here turns out purposive, that is, nonrandom.

To complete the inference, one needs an estimator v_m for $V_m(t_m - Y)$ and an SZE of the form $(t_m - Y)/\sqrt{v_m}$, which again is required to converge in law to τ. As a result, a confidence interval for Y may be set up with a nominal coverage probability calculated with respect to speculated unanswered questions about the performances of t_m, v_m and $(t_m - Y)/\sqrt{v_m}$ when the postulated model is incorrect. If a correct model is M_0, it is not easy to speculate on the m bias of t_m

$$E_{m_0}(t_m - Y) = B_{m_0}(t_m),$$

the m MSE of t_m

$$E_{m_0}(t_m - Y)^2 = M_{m_0}(t_m)$$

the m bias of v_m

$$E_{m_0}[v_m - M_{m_0}(t_m)] = B_{m_0}(v_m),$$

and the distribution of $(t_m - Y)/\sqrt{v_m}$ when $\underline{Y}$ is generated according to M_0. So, the question of robustness is extremely crucial here.

One approach to retain m unbiasedness of t_m in case of modest departure from a postulated model is to adjust the sampling design. The concept of balanced sampling that demands equating sample and population moments of an auxiliary x variable is very important in this context, as emphasized by ROYALL and his colleagues. They also demonstrate the need for alternatives to v_m as m variance estimators that retain m unbiasedness and preserve asymptotic normality of revised SZEs. A net beneficial impact of this approach on survey sampling theory and practice has been that some classical p-based strategies like ratio and regression estimators with or without stratification, weighted differentially across the strata, have been confirmed to be serviceable predictors and, more importantly, alternative variance estimators for several such common estimation procedures for total have emerged.

A further important outcome is the realization that a re-evaluation of p-based procedures is necessary and useful in terms of their performances, not over hypothetical averaging over all possible samples, but through their conditional behavior averaging over only samples sharing in common some discernible features with those in the sample at hand.

ROYALL, the chief promoter of predictive methodology in survey sampling, and his colleagues CUMBERLAND and EBERHARDT, have demonstrated that $\bar{x}$-dependent variation of variance estimators of ratio and regression estimators is a behavior worthy of attention that is not revealed if one blindly follows the classical p-based procedures. Inspired by this demonstration WU, DENG, SÄRNDAL, KOTT, and others have derived useful alternative variance estimators, keeping eyes to their conditional behaviors. HOLT and SMITH (1979) have emphasized how in poststratified sampling the observed sample configuration $\underline{n} = (n_1, \ldots, n_L)$ for the given L post-strata should be used in a variance estimator rather than averaging over it, and then its variation conditional on $\underline{n}$ and how it is useful to set up conditional confidence intervals should be studied. J. N. K. RAO (1985) has further stressed how efficacious is conditional inference in survey sampling, but also illustrated several associated difficulties. GODAMBE (1986), SÄRNDAL, SWENSSON and WRETMAN (1989), and KOTT (1990) have also given new variance estimators with good design- and model-based properties. SÄRNDAL and HIDIROGLOU (1989) recommended setting up confidence intervals with preassigned conditional coverage probabilities that are maintained unconditionally and have given specific recipes with demonstrated serviceability.

Followers of HANSEN, MADOW and TEPPING (1983) would agree to live with model-based predictors provided, in case of large samples, they have good design-based properties. Especially if a t_m has small $|B_p(t_m)|$ and hence, hopefully, also a controlled $M_p(t_m)$, then it may be admitted as a robust procedure. BREWER (1979) (a) recommended that to avoid exclusive model dependence t_m need not be chosen as the BLUP and (b) discouraged purposive sampling. Instead he based his t_m on a design to invest it with good design properties. At least the limiting value of $|B_p(t_m)|$ for large samples should be zero. A preferred t_m is one for which the lower bound of the limiting value of $E_m E_p(t_m - Y)^2$ is attained, and the right design is one for which this lower bound is minimized. SÄRNDAL (1980, 1981, 1982, 1984, 1985) has alternative recommendations in favor of what he called the GREG predictors, which are robust in the sense of being asymptotically design unbiased (ADU).

WRIGHT (1983) introduced the wider class of QR predictors covering both linear predictors (LPRE) including BREWER's. GREG, and SÄRNDAL and WRIGHT (1984) examine their ADU properties. MONTANARI (1987) enlarges this class, further accommodating correlated residuals. LITTLE (1983) considers GREG predictors inferior to LPRE and shows that the latter are ADU and ADC provided they originate from a modeled regression curve with a non-zero intercept term for each of a number of identifiable groups into which the population is divisible. This leads to expensive strategies demanding group-wise estimation of each intercept term. An adaption of JAMES-STEIN procedures as empirical Bayes estimators, which involve borrowing strength across the groups with unrepresented or underrepresented groups is, however, recommended in case one cannot afford adequate group-wise sampling.

An accredited merit of this approach is that a predictor is good if the underlying model is correct, but is nevertheless robust in case the model is faulty because it is ADU or ADC. But a criticism against it is that its model-based property is conditional on the chosen sample, while its asymptotic design property is unconditional and based on speculation over all possible samples. For a better design-based justification a procedure should fare well conditionally when the reference set for the repeated sampling is a proper but meaningful subset of all possible samples. For example, averaging should be over a set of samples sharing certain recognizable common features of the sample at hand. SÄRNDAL and HIDIROGLOU (1989), however, have shown that GREG predictors and some modified ones adapted from them have good conditional design-based properties.

Advancing conditional arguments, ROBINSON (1987) has proposed a conditional bias-corrected modification to a ratio estimator of Y in case $\overline{X}$ is known, given by

$$t_d = X\left(\frac{\overline{y}}{\overline{x}} + \left(\frac{\overline{y}}{\overline{x}} - b\right)\left(1 - \frac{\overline{X}}{\overline{x}}\right)\right)$$

where

$$b = \sum_s (Y_i - \overline{y})(X_i - \overline{x}) / \sum_s (X_i - \overline{x})^2$$

postulating asymptotic bivariate normality for the joint distribution of $(\overline{x}, \overline{y})$ with an approximate variance estimator as

$$v_2 = \left(\frac{\overline{X}}{\overline{x}}\right)^2 v_0, \; v_0 = \frac{1-f}{n-1} \sum_s \left(Y_i - \frac{\overline{y}}{\overline{x}} X_i\right)^2$$

Asymptotics have been effectively utilized in the survey sampling context by KREWSKI and RAO (1981), who have established asymptotic normality of nonlinear statistics given by (a) linearization, (b) BRR, and (c) jackknife methods and consistency of the corresponding variance estimators when they are based on large numbers of strata, although with modest rates of sampling of psus within strata. As their first-order analysis proves inconclusive to arrange these three procedures in order of merit, RAO and WU (1985) resort to second-order analysis to derive additional results.

Earlier comparative studies of these procedures due, for example, to KISH and FRANKEL (1970) were exclusively empirical. Incidentally, MCCARTHY (1969) restricted BRR with two sampled units per stratum, while GURNEY and JEWETT (1975) extended allowing more but common per stratum sample size provided it is a prime number. KREWSKI (1978) has examined stabilities of BRR-based variance estimation.

What now transpires as a palpable consensus among sampling experts is that superpopulation modeling cannot be ruled out from sampling practice. It is useful in adopting a sampling strategy, but the question is whether the inference should be based on (a) the model ignoring the design, (b) the speculation over repeated sampling out of all possible samples, (c) the speculation over repeated sampling out of a meaningful proper subset of all possible samples, (d) the speculation over repeated sampling in either of these two ways and also over realization of the population vector in the modeled way.

A model, of course, is a recognized necessity (a) in the presence of nonresponse and (b) in inference concerning small domain characteristics that needs borrowing strength, implicity or explicitly postulating similarity across domains with inadequate sample representation. But, in other situations, its utility is controversial. Even if one adopts a model, inference procedure must have an built-in protective arrangement to remain valid even in case its postulation is at fault. We have mentioned a few robustness preserving techniques. We may also add that sensitivity analyses to validate a postulated model for the finite population vector of variate values through a consistency check with the realized survey data are impracticable in large-scale surveys. More information is available from RAO (1971), GODAMBE (1982), CHAUDHURI and VOS (1988), SMITH (1976, 1984), KALTON (1983a), IACHAN (1984), CUMBERLAND and ROYALL (1981), VALLIANT (1987a, 1987b), RAO and BELLHOUSE (1989), ROYALL and PFEFFERMANN (1982), SCOTT (1977), SCOTT, BREWER and HO (1978), and the references cited therein.

The generalized regression estimators of CSW (1976) are the pioneering illustrations of the outcomes of the model-assisted approach. Their forms are motivated by an underlying regression model, for example,

$$y_i = \beta x_i + \in_i$$

with β as an unknown slope parameter, x_i's as known positive numbers, and $\in_i$'s as unknown random errors.

In estimating $Y = \Sigma y_i = \beta X + \Sigma \in_i$ one is motivated to estimate β by

$$b_Q = \frac{\Sigma y_i x_i Q_i I_{si}}{\Sigma x_i^2 Q_i I_{si}}$$

with Q_i as an estimator for $\frac{1}{V_m(\in_i)}$.

This motives the choice of

$$t_g = \Sigma \frac{y_i}{\pi_i} I_{si} + b_Q \left(X - \Sigma \frac{x_i}{\pi_i} I_{si} \right)$$

or of

$$t_{gb} = \Sigma y_i b_{si} I_{si} + b_Q (X - \Sigma x_i b_{si} I_{si}) .$$

A t_g or t_{gb} is privileged to have the purely design-based property of being an ADU as well as an ADC estimator for Y for any choice of Q_i as a positive number. However, a right choice of Q_i is needed in rendering t_g or t_{gb} close to Y along with an estimated measure of its error in repeated sampling from $U = (1, \ldots, i, \ldots, N)$ under control.

An alternative purely design-based motivation for the introduction of t_g or t_{gb} is also available, called the **calibration approach** thanks to the intiative taken by ZIESCHANG (1990), and DEVILLE and SÄRNDAL (1992), with plenty of follow-up activities as well.

The GREG estimator t_g for Y is a modification of a basic estimator (HORVITZ–THOMPSON, HT, 1952)

$$t_H = \Sigma \frac{y_i}{\pi_i} I_{si}.$$

Writing $a_i = \frac{1}{\pi_i}$ and supposing positive numbers x_i's are available, let us revise the initial weights a_i for y_i by way of a possible improvement in the following possible ways:

(a) The revised weights w_i's are to be chosen such that
(b) they satisfy the side conditions, better known as calibration constants or calibration equations

$$\Sigma w_i x_i I_{si} = \Sigma x_i$$

and

(c) that w_i's are close to a_i's is terms of the minimized distance to be measured by

(d) $$\Sigma [c_i(w_i - a_i)^2/a_i] I_{si}$$

with suitably chosen positive constants c_i's.
The resulting choice of w_i's is

$$g_{si} = 1 + \left(X - \Sigma \frac{x_i}{\pi_i} I_{si}\right) \frac{x_i/(c_i a_i)}{\Sigma (x_i^2/c_i) I_{si}}, i \in s.$$

The resulting estimator for Y, namely,

$$\Sigma y_i a_i g_{si} I_{si}$$

coincides with t_g on choosing $c_i = \frac{1}{Q_i}, i \in s$. Then the purely design-based t_g is the same as the model-assisted GREG predictor for Y expressing t_g in the form

$$\begin{aligned} t_g &= X b_Q + \Sigma \frac{y_i - b_Q x_i}{\pi_i} I_{si} \\ &= X b_Q + \Sigma \frac{e_i}{\pi_i} I_{si} \end{aligned}$$

Calling $e_i = y_i - b_Q x_i$ the residual, we may recall that it is a special case of the QR predictors for Y introduced by WRIGHT (1983), namely,

$$t_{QR} = Xb_Q + \Sigma r_i b_i I_{si}$$

with $r_i(\geq 0)$ chosen as certain non-negative constants free of $\underline{Y} = (y_1, \ldots, y_i, \ldots, y_N)$.

ROYALL's (1970) predictor for Y is of the form

$$\begin{aligned} t_{R0} &= \Sigma y_i I_{si} + b_Q(X - \Sigma x_i I_{si}) \\ &= Xb_Q + \Sigma e_i I_{si}. \end{aligned}$$

Thus the choices $r_i = \frac{1}{\pi_i}, 1$, respectively, yield from t_{QR} the GREG predictor and ROYALL's predictors. For the choice $r_i = 0$ in t_{QR} one gets the projective estimator

$$t_{PR0} = Xb_Q \text{ for } Y.$$

It is possible also to establish t_{QR} as a calibration estimator.

If t_{R0} coincides with t_{PR0} for a specific choice of Q_i, it is called a cosmetic predictor or estimator. One possible example for it is the ratio estimator or predictor namely

$$t_R = X\frac{\Sigma y_i I_{si}}{\Sigma x_i I_{si}}.$$

A QR is called a restricted QR predictor t_{RQR} if some restrictions are imposed on the possible magnitudes allowed for Q_i and r_i's. For a calibration estimator, sometimes the assignable weights w_i's are restricted or limited to certain preassigned ranges like $L_i < w_i < U_i$, especially $w_i \geq 0$. Then they are called limited calibration estimators. In the recent volumes of *Survey Methodology*, many relevant illustrations are available. For the sake of simplicity, we have illustrated the case of only a single auxiliary variable x, but the literature covers several of them.

An advantage of this interpretation of a GREG estimator or predictor as a calibration estimator is that it gets recognized as a robust estimator as it is totally model free, not only for large sample sizes in an asymptotic sense. Its ADU or ADC property alone is not its only guarantee to be robust.

In the finite population context, CHAMBERS (1986) pointed out the need for outler-robust estimators, and prior to him BARNETT and LEWIS (1994) also discuss the problem with outliers in survey sampling, suggesting ways and means of tackling them.

SÄRNDAL (1996) made an epoch-making recommendation of employing procedures that bypass the need to include the cross-product terms in the quadratic forms in which variance or mean square error estimators for linear estimators for finite population totals are expressed covering HORVITZ–THOMPSON and generalized regression estimators. The prime need for this is that exact formulae for π_{ij} for many sampling schemes are hard to develop. They occur in too many cross-product terms destabilizing the magnitudes of the variance or MSE estimators for large- and moderate-sized samples. He prescribes the use of Poisson sampling or its special case, Bernoulli sampling, for which $\pi_{ij} = \pi_i \pi_j$ as noted by HÁJEK (1964, 1981). His second prescription is to employ approximations for the variance or MSE estimators that are expressible in terms of squared residuals with positive multipliers avoiding the cross-product terms. He has shown that stratified simple random sampling (STSRS) or stratified Bernoulli sampling (STBE) employing GREG estimators in suitable forms yields quite efficient procedures. DEVILLE (1999), BREWER (1999a, 2000), and BREWER and GREGOIRE (2000) also propagate the utility of this approach, especially by approximating π_{ij}'s in terms of π_i's with suitable corrective terms.

For sampling schemes with sample sizes fixed at a number, n, BREWER (2000) expresses

$$V(t_H) = \Sigma y_i^2 \left(\frac{1-\pi_i}{\pi_i}\right) + \sum\sum_{i \neq j} y_i y_j \left(\frac{\pi_{ij} - \pi_i \pi_j}{\pi_i \pi_j}\right)$$

as

$$V(t_H) = \Sigma \pi_i (1-\pi_i) \left(\frac{y_i}{\pi_i} - \frac{Y}{n}\right)^2 + \sum\sum_{i \neq j} (\pi_{ij} - \pi_i \pi_j) \left(\frac{y_i}{\pi_i} - \frac{Y}{n}\right) \left(\frac{y_j}{\pi_j} - \frac{Y}{n}\right),$$

approximates π_{ij}, for example, by

$$\pi_{ij}(B) = \pi_i \pi_j \left(\frac{c_i + c_j}{2}\right)$$

with c_i chosen in (0, 1), approximates $V(t_H)$ by

$$V_B(t_H) = \Sigma \pi_i (1 - c_i \pi_i) \left(\frac{y_i}{\pi_i} - \frac{Y}{n}\right)^2$$

and estimates it by

$$v_B(t_H) = \Sigma \left(\frac{1}{c_i} - \pi_i\right) \left(\frac{y_i}{\pi_i} - \frac{t_H}{n}\right)^2 I_{si}$$

PAL (2003) has generalized BREWER's (2000) form of $V(t_H)$ to

$$V(t_H) = \Sigma \pi_i (1 - \pi_i) \left(\frac{y_i}{\pi_i} - \frac{Y}{\nu}\right)^2 + \sum\sum_{i \neq j} (\pi_{ij} - \pi_i \pi_j) \left(\frac{y_i}{\pi_i} - \frac{Y}{\nu}\right) \left(\frac{y_j}{\pi_j} - \frac{Y}{\nu}\right) - Y^2 \left(1 - \frac{1}{\nu} + \frac{1}{\nu^2} \sum\sum_{i \neq j} \pi_{ij}\right) + \frac{2Y}{\nu} \Sigma \frac{y_i}{\pi_i} \left(\sum_{j \neq i} \pi_{ij}\right)$$

which is correct for any number of distinct units $\nu(s)$ for a sample s with $\nu = E_p(\nu(s))$.

Thus, with BREWER's (2000) approximation for π_{ij} as given earlier $V(t_H)$ approximates to

$$V_{AB}(t_H) = \Sigma y_i^2 \left(\frac{1 - \pi_i}{\pi_i}\right) + \Sigma \pi_i^2 (c_i - 1) \left(\frac{y_i}{\pi_i} - \frac{Y}{\nu}\right)^2$$

for which an estimator is

$$v_{AB}(t_H) = \Sigma y_i^2 \frac{1 - \pi_i}{\pi_i} \frac{I_{si}}{\pi_i} + \Sigma \pi_i \left(1 - \frac{1}{c_i}\right) \left(\frac{y_i}{\pi_i} - \frac{t_H}{\nu}\right)^2 I_{si}$$

Poisson's sampling scheme needs no such approximations but is handicapped because $\nu(s)$ for it varies over its entire range $(0, 1, \ldots, N - 1, N)$, which is undesirable. To avoid this, GROSENBAUGH's (1965) $3P$ sampling, OGUS and CLARK's (1971) modified Poisson sampling, further discussed by

BREWER, EARLY and JOYCE (1972), and BREWER, EARLY and HANIF's (1984), use of collocated sampling, and OHLSSON's (1995), use of permanent random numbers (PRN) to effect co-ordination in rotation vis-a-vis Poisson sampling, are all important developments receiving attention over a protracted time period.

In modified Poisson sampling (MPS) one has to repeat the Poisson scheme each time it culminates in having $\nu(s) = 0$ with revised selection probabilities to retain π_i in tact. CHAUDHURI and VOS (1988, p. 198) have clarified that for MPS one has

$$\pi_{ij} = \pi_i \pi_j (1 - P_0)$$

where $P_0 = Prob[\nu(s) = 0]$ derivable as a solution of

$$\prod_{i=1}^{N} [1 - \pi_i(1 - P_0)] = P_0$$

because $\pi_i(1 - P_0)$ is the revised selection probability of i for this MPS.

For MPS, $V(t_H)$ turns out to be

$$V(t_H) = \Sigma(1 - \pi_i)\frac{y_i^2}{\pi_i} - P_0(Y^2 - \Sigma y_i^2)$$

with an unbiased estimator as

$$v(t_H) = \Sigma(1 - \pi_i)\frac{y_i^2}{\pi_i}\frac{I_{si}}{\pi_i} - \frac{P_0}{1 - P_0}\left(t_H^2 - \Sigma\frac{y_i^2}{\pi_i}\frac{I_{si}}{\pi_i}\right)$$

An alternative approach is to employ original Poisson sampling combined with the estimator

$$t_{RH} = \frac{\nu}{\nu(s)} t_H = \frac{\nu}{\nu(s)} \Sigma y_i \frac{I_{si}}{\pi_i} \quad \text{if} \quad \nu(s) \neq 0$$

with its MSE estimators as

$$\begin{aligned} m_1 &= \Sigma\left(\frac{1 - \pi_i}{\pi_i}\right)\left(y_i - \frac{\pi_i}{\nu(s)} t_H\right)^2 \frac{I_{si}}{\pi_i} \\ &= 0, \quad \text{if } \nu(s) = 0 \end{aligned}$$

or

$$m_2 = \left(\frac{\nu}{\nu(s)}\right)^2 m_1.$$

For any general sampling scheme, STEHMAN and OVERTON (1994) use two approximations

$$\pi_{ij}(1) = \frac{(n-1)\pi_i\pi_j}{n-(\pi_i+\pi_j)/2}, \quad \pi_{ij}(2) = \frac{(n-1)\pi_i\pi_j}{n-\pi_i-\pi_j+\frac{1}{n}\Sigma\pi_i^2}$$

with the compulson that $\pi_i < 1\ \forall i$.

For circular systematic sampling (CSS) with probabilities proportional to sizes (PPS) that are positive integers x_i with the total X, we know from MURTHY (1957) that the execution steps are the following.

Let $k = [\frac{X}{n}]$ and R be a random integer chosen out of $1, 2, \ldots, X$. Then let

$$a_r = (R+kj)mod(X),\ j = 0, 1, \ldots, n-1,$$

$C_i = \sum_{j=0}^{i} x_j$. Then the sample consists of the unit N if $a_r = 0$ and of i if

$$C_{i-1} < a_r \leq C_i,\ \text{taking}\ C_0 = 0.$$

For this scheme, the intended sample size n may not be realized unless $np_i < 1\forall i$, writing $p_i = \frac{x_i}{X}$. Also, $\pi_i = \frac{1}{X}$ (number of samples with i), $\pi_{ij} = \frac{1}{X}$ (number of samples with i and j).

But π_{ij} turns out zero for many i, j's $(i \neq j)$. CHAUDHURI and PAL (2003) have shown that if, instead of this fixed interval equal to k CSSPPS, one employs its revised random interval k chosen at random out of $1, 2, \ldots, X-1$ form, then $\pi_{ij} > 0\forall i, j(i \neq j)$.

In order to avoid this shortcoming of CSSPPS that "π_{ij} equals zero for many $i \neq j$", rendering nonavailability of an unbiased estimator for the variance of a linear estimator for Y, HARTLEY and RAO (1962) gave their random CSSPPS scheme where CSSPPS method is applied with a prior random permutation of the units of $U = (1, \ldots, i, \ldots, N)$. For this scheme, provided $np_i < 1\forall i$, the intended sample size n is realized,

$\pi_i = np_i$ and also

$$\pi_{ij} = \left(\frac{n-1}{n}\right)\pi_i\pi_j + \left(\frac{n-1}{n^2}\right)\left(\pi_i^2\pi_j + \pi_i\pi_j^2\right)$$

$$- \left(\frac{n-1}{n^3}\right)\pi_i\pi_j\,\Sigma\pi_i^2 + \frac{2(n-1)}{n^3}\left(\pi_i^3\pi_i + \pi_i\pi_j^3 + \pi_i^2\pi_j^2\right)$$

$$- \frac{3(n-1)}{n^4}\left(\pi_i^2\pi_j + \pi_i\pi_j^2\right)\Sigma\pi_i^2 + \frac{3(n-1)}{n^5}\pi_i\pi_j\left(\Sigma\pi_i^2\right)^2$$

$$- \frac{2(n-1)}{n^4}\pi_i\pi_j\,\Sigma\pi_i^3 > 0\ \forall i \neq j$$

Let us now briefly discuss concepts of coordination in rotation sampling and of permanent random number (PRN) technique in sample selection.

If sampling needs to be repeated from the same population or essentially the same population subject to incidences of deaths, that is, dropouts, and of births, that is, addition of units, then in estimation of a population total or mean, it seems necessary that some of the units in every sample should be retained for ascertainment of facts on one or more subsequent occasions too. This is called rotation in sampling. Thus rotational sampling involves a problem of coordination. If two samples have an overlap of units, then there is positive coordination and one needs to adopt a policy of maximizing or minimizing positive coordination. If there is no overlap, then there is negative coordination. A useful technique of retaining the essential properties of a basic sampling scheme involving rotation of units is to use PRNs for the units. OHLSSON (1995) has described PRN technques for SRSWOR Bernoulli and Poisson sampling schemes with rotations allowing birth and deaths in respect of an initial population. Details are omitted here.

We conclude this text by recounting in brief one of our latest innovative techniques of cluster sampling in a particular mode. While commissioned by UNICEF in 1998, Indian Statistical Institute (ISI) undertook a health survey in the villages of an Indian district. It was found useful to first take an SRSWOR of a kind of selection units called PHC, the primary health centers, a few of which are localized in proximity to a bigger unit called BPHC (big PHC) such that the villages are to be treated in a separate and territorially nearby PHC or a BPHC. The PHCs linked to a BPHC together form a cluster. The sampling scheme actually employed added purposively each BPHC to which an initially chosen PHC was linked. This is a version of cluster sampling attaching varying inclusion probabilities to the BPHCs in the district and thus allowing various choices of unbiased estimation procedures. A simpler possible two-stage sampling with BPHCs as the first-stage units and the PHCs linked to the BPHCs as the second-stage units was avoided with the expectation of achieving wider territorial coverage of the district's PHCs and BPHCs and hence of higher information contents and resulting increased accuracy in estimation. Details are given by CHAUDHURI and PAL (2003).

Appendix

Abbreviations Used in the References

AISM	*Annals of the Institute of Statistical Mathematics*
AJS	*Australian Journal of Statistics*
AMS	*The Annals of Mathematical Statistics*
ANZJS	*The Australian and New Zealand Journal of Statistics*
Appl. Stat.	*Applied Statistics*
APSPST	*Applied Probability, Stochastic Processes and Sampling Theory* (see MacNeill and Umphrey, eds. [1987])
AS	*The Annals of Statistics*
ASA	*The American Statistical Association*
BISI	*Bulletin of the International Statistical Institute*
Bk	*Biometrika*
Bms	*Biometrics*
CDSS	*Current Developments in Survey Sampling* (see Swain [2000])
CSAB	*Calcutta Statistical Association Bulletin*

CSTM	*Current Statistics Theory and Methods* (Abstract)
CTS	*Current Topics in Survey Sampling* (see Krewski, Platek, and Rao, eds. [1981])
CSA	*Communications in Statistics A*
FSI	*Foundations of Statistical Inference*, (see Godambe and Sprott [1971])
HBS	*Handbook of Statistics*, vol. 6, (see Krishnaiah and Rao, eds. [1988])
ISR	*International Statistical Review*
JASA	*Journal of the American Statistical Association*
JISA	*Journal of the Indian Statistical Association*
JISAS	*Journal of the Indian Society of Agricultural Statistics*
JOS	*Journal of Offical Statistics*
JRSS	*Journal of the Royal Statistical Society*
JSPI	*Journal of Statistical Planning and Inference*
JSR	*Journal of Statistical Research*
Mk	*Metrika*
N	*Nature*
NDSS	*New Developments in Survey Sampling*, (see Johnson and Smith, eds. [1969])
NPTAS	*New Perspectives in Theoretical and Applied Statistics*, (see Puri, Vilalane and Wertz, eds.[1987])
PJS	*Pakistan Journal of Statistics*
RISI	*Revue de Statistique Internationale*
Sā	*Sankhya*
SJS	*Scandinavian Journal of Statistics*
SM	*Sociological Methodology*
SSM	*Survey Sampling and Measurement*, (see Nanboodiri, ed.[1978])
St	*The Statistician*
SUM	*Survey Methodology*
SESA, NIDA	*Synthetic Estimates for Small Areas*, (see Steinberg, ed.[1979])

References

1. AGGARWAL, O. P. (1959): *Bayes and minimax procedures in sampling from finite and infinite populations, I.* AMS, *30*, 206–218.
2. ALTHAM, P. A. E. (1976): *Discrete variable analysis for individuals grouped into families.* Bk, *63*, 263–269.
3. ANDERSON, T. W. (1957): *Maximum likelihood estimates for a multivariate normal distribution when some observations are missing.* JASA, *52*, 200–203.
4. ARNAB, R. (1988): *Variance estimation in multi-stage sampling.* AJS, *30*, 107–111.
5. BARNETT, V. and LEWIS, T. (1994): *Outliers in statistical data.* Wiley & Sons, New York.
6. BARTHOLOMEW, D. J. (1961): *A method of allowing for "not-at-home" bias in sample surveys.* Appl. Stat. *10*, 52–59.
7. BASU, D. (1971): *An essay on the logical foundations of survey sampling, Part I.* In: FSI, 203–242.
8. BAYLESS, D. L. and RAO, J. N. K. (1970): *An empirical study of estimators and variance estimators in unequal probability sampling ($n = 3$ or 4).* JASA, *65*, 1645–1667.
9. BELLHOUSE, D. R. (1985): *Computing methods for variance estimation in complex surveys.* JOS, *1*, 323–330.
10. BELLHOUSE, D. R. (1988): *Systematic sampling.* In: HBS, 125–145.
11. BICKEL, P. J. and FREEDMAN, D. A. (1981): *Some asymptotic theory for the bootstrap.* AS, *9*, 1196–1217.
12. BICKEL, P. J. and LEHMANN, E. L. (1981): *A minimax property of the sample mean in finite populations.* AS, *9*, 1119–1122.
13. BIYANI, S. H. (1980): *On inadmissibility of the Yates–Grundy variance estimator in unequal probability sampling.* JASA, *75*, 709–712.

14. BOLFARINE, H. and ZACKS, S. (1992): *Prediction theory for finite populations*. Springer-Verlag, New York.

15. BRACKSTONE, G. J. and RAO, J. N. K. (1979): *An investigation of raking ratio estimators*. Sā, *41*, 97–114.

16. BREWER, K. R. W. (1963): *Ratio estimation and finite populations: Some results deducible from the assumption of an underlying stochastic process*. AJS, *5*, 93–105.

17. BREWER, K. R. W. (1979): *A class of robust sampling designs for large-scale surveys*. JASA, *74*, 911–915.

18. BREWER, K. R. W. (1990): *Review of unified theory and strategies of survey sampling by* CHAUDHURI, A. *and* VOS, J. W. E. In: JOS, *6*, 101–104.

19. BREWER, K. R. W. (1994): *Survey sampling inference: Some past perspectives and present prospects*. PJS, *10*, 15–30.

20. BREWER, K. R. W. (1999a): *Design-based or prediction inference? Stratified random vs. stratified balanced sampling*. ISR, *67*, 35–47.

21. BREWER, K. R. W. (1999b): *Cosmetic calibration with unequal probability sampling*. SUM, *25*(2), 205–212.

22. BREWER, K. R. W. (2000): *Deriving and estimating an approximate variance for the Horvitz–Thompson estimator using only first order inclusion-probabilities. Contributed to second international conference on establishment surveys*. Buffalo, NY, 17–21 (unpublished).

23. BREWER, K. R. W., EARLY, L. J. and HANIF, M. (1984): *Poisson, modified Poisson and collocated sampling*. JSPI, *10*, 15–30.

24. BREWER, K. R. W., EARLY, L. J. and JOYCE, S. F. (1972): *Selecting several samples from a single population*. AJS, *14*(3), 231–239.

25. BREWER, K. R. W. and GREGOIRE, T. G. (2000): *Estimators for use with Poisson sampling and related selection procedures*. Invited paper in second international conference on establishment surveys. Buffalo, NY, 17–21 (unpublished).

26. BREWER, K. R. W. and HANIF, M. (1983): *Sampling with unequal probabilities*. Springer-Verlag, New York.

27. BREWER, K. R. W., HANIF, M. and TAM, S. M. (1988): *How nearly can model-based prediction and design-based estimation be reconciled*. JASA, *83*, 128–132.

28. BREWER, K. R. W. and MELLOR, R. W. (1973): *The effect of sample structure on analytical surveys*. AJS, *15*, 145–152.

29. BRIER, S. E. (1980): *Analysis of contingency tables under cluster sampling*. Bk, *67*, 91–96.

30. CASSEL, C. M., SÄRNDAL, C. E. and WRETMAN, J. H. (1976): *Some results on generalized difference estimation and generalized regression estimation for finite populations*. Bk, *63*, 615–620.

31. CASSEL, C. M., SÄRNDAL, C. E. and WRETMAN, J. H. (1977): *Foundations of inference in survey sampling*. John Wiley & Sons, New York.

32. CASSEL, C. M., SÄRNDAL, C. E. and WRETMAN, J. H. (1983): *Some uses of statistical models in connection with the nonresponse problem*. In: IDSS, *3*, 143–170.

33. CHAMBERS, R. L. (1986): *Outlier robust finite population estimation*. JASA, *81*, 1063–1069.

34. CHAMBERS, R. L., DORFMAN, A. H. and WEHRLY, T. E. (1993): *Bias robust estimation infinite populations using nonparametric calibration*. JASA, *88*, 268–277.

35. CHAUDHURI, A. (1985): *On optimal and related strategies for sampling on two occasions with varying probabilities*. JISAS, *37*(1), 45–53.

36. CHAUDHURI, A. (1988): *Optimality of sampling strategies*. In: HBS, *6*, 47–96.

37. CHAUDHURI, A. (1992): *A note on estimating the variance of the regression estimator*. Bk, *79*, 217–218.

38. CHAUDHURI, A. (2000a): *Network, adaptive sampling*. CSAB, 237–253.

39. CHAUDHURI, A. (2000b): *Mean square error estimation in multistage and randomized response surveys*. In: CDSS, ed. Swain, A. K. P. C., Ulkal University, Bhubaneswar, 9–20.

40. CHAUDHURI, A. (2001): *Using randomized response from a complex survey to estimate a sensitive proportion in a dichotomous finite population.* JSPI, *94*, 37–42.

41. CHAUDHURI, A. and ADHIKARI, A. K. (1983): *On optimality of double sampling strategies with varying probabilites.* JSPI, *8*, 257–265.

42. CHAUDHURI, A. and ADHIKARI, A. K. (1985): *Some results on admissibility and uniform admissibility in double sampling.* JSPI, *12*, 199–202.

43. CHAUDHURI, A. and ADHIKARI, A. K. (1987): *On certain alternative IPNS schemes.* JISAS, *39*(2), 121–126.

44. CHAUDHURI, A., ADHIKARI, A. K. and DIHIDAR, S. (2000a): *Mean square error estimation in multi-stage sampling.* Mk, *52*, 115–131.

45. CHAUDHURI, A., ADHIKARI, A. K. and DIHIDAR, S. (2000b): *On alternative variance estimators in three-stage sampling.* PJS, *16*(3), 217–227.

46. CHAUDHURI, A. and ARNAB, R. (1979): *On the relative efficiencies of sampling strategies under a super population model.* Sā, *41*, 40–53.

47. CHAUDHURI, A. and ARNAB, R. (1982): *On unbiased variance-estimation with various multi-stage sampling strategies.* Sā, *44*, 92–101.

48. CHAUDHURI, A. and MITRA, J. (1992): *A note on two variance estimators for Rao-Hartley-Cochran estimator.* CSA, *21*, 3535–3543.

49. CHAUDHURI, A., BOSE, M. and GHOSH, J. K. (2003): *An application of adaptive sampling to estimate highly localized population segments.* In: JSPI.

50. CHAUDHURI, A. and MAITI, T. (1994): *Variance estimation in model assisted survey sampling.* CSA, *23*, 1203–1214.

51. CHAUDHURI, A. and MAITI, T. (1995): *On the regression adjustment to* RAO-HARTLEY-COCHRAN *estimator.* JSR, *29*(1), 71–78.

52. CHAUDHURI, A. and MAITI, T. (1997): *Small domain estimation by borrowing strength across time and domain: a case study*. Comp. Stat. Simul. Comp., *26*(4), 1547–1558.

53. CHAUDHURI, A. and MITRA, J. (1996): *Setting confidence intervals by ratio estimator*. CSA, *25*(5), 1135–1148.

54. CHAUDHURI, A. and MUKERJEE, R. (1988): *Randomized response: Theory and techniques*. Marcel Dekker, New York.

55. CHAUDHURI, A. and PAL, S. (2002): *On certain alternative mean square error estimators in complex survey*. JSPI, *104*(2), 363–375.

56. CHAUDHURI, A. and PAL, S. (2003): *On a version of cluster sampling and its practical use*. JSPI, *113*(1), 25–34.

57. CHAUDHURI, A., ROY, D. and MAITI, T. (1996): *A note on competing variance estimators in randomized response surveys*. AJS, *38*, 35–42.

58. CHAUDHURI, A. and VOS, J. W. E. (1988): *Unified theory and strategies of survey sampling*. North-Holland Publishers, Amsterdam.

59. CHENG, C. S. and LI, K. C. (1983): *A minimax approach to sample surveys*. AS, *11*, 552–563.

60. COCHRAN, W. G. (1977): *Sampling techniques*. John Wiley & Sons, New York.

61. COX, B., BINDER, D., CHINNAPPA, B., CHRISTIANSON, A., COLLEGE, M. and KOTT, P., eds. (1995): *Business survey methods*. J. Wiley, Inc.

62. CRAMÉR, H. (1966): *Mathematical methods of statistics*. Princeton University Press, Princeton, NJ.

63. CUMBERLAND, W. G. and ROYALL, R. M. (1981): *Prediction models and unequal probability sampling*. JRSS, *43*, 353–367.

64. CUMBERLAND, W. G. and ROYALL, R. M. (1988): *Does simple random sampling provide adequate balance?* JRSS, *50*, 118–124.

65. DAS, M. N. (1982): *Systematic sampling without drawback*. Tech. Rep. 8206, ISI, Delhi.

66. DEMETS, D. and HALPERIN, M. (1977): *Estimation of a simple regression coefficient in samples arising from a sub-sampling procedure*. Bms, *33*, 47–56.

67. DEMING, W. E. (1956): *On simplification of sampling design through replication with equal probabilites and without stages*. JASA, *51*, 24–53.

68. DENG, L. Y. and WU, C. F. J. (1987): *Estimation of variance of the regression estimator*. JASA, *82*, 568–576.

69. DEVILLE, J. C. (1988): *Estimation linéaire et redressement sur informations auxiliaires d'enquêtes par sondages*. In: MONFORT and LAFFOND, eds: *Essais en l'honneur d'Edmont Malinvaud* Economica, 915–927.

70. DEVILLE, J. C. (1999): *Variance estimation for complex statistics and estimators: Linearization and residual techniques*. SUM, *25*(2), 193–203.

71. DEVILLE, J. C. and SÄRNDAL, C. E. (1992): *Calibration estimators in survey sampling*. JASA, *87*, 376–382.

72. DEVILLE, J. C., SÄRNDAL, C. E. and SAUTORY, O. (1993): *Generalized raking procedures in survey sampling*. JASA, *88*, 1013–1020.

73. DORFMAN, A. H. (1993): *A comparion of design-based and model-based estimator of the finite population distribution function*. AJS, *35*, 29–41.

74. DOSS, D. C., HARTLEY, H. O. and SOMAYAJULU, G. R. (1979): *An exact small sample theory for post-stratification*. JSPI, *3*, 235–247.

75. DUCHESNE, P. (1999): *Robust calibration estimators*. SUM, *25*(1), 43–56.

76. DURBIN, J. (1953): *Some results in sampling theory when the units are selected with unequal probabilities*. JRSS, *15*, 262–269.

77. EFRON, B. (1982): *The jackknife, the bootstrap and other resampling plans*. Soc. Ind. Appl. Math. CBMS. Nat. SC. Found. Monograph 38.

78. EL-BADRY, M. A. (1956): *A simple procedure for mailed questionnaires*. JASA, *51*, 209–227.

79. ERICKSEN, E. P. (1974): *A regression method for estimating population changes of local areas*. JASA, *69*, 867–875.

80. FAY, R. E. (1985): *A jackknifed chi-squared test for complex samples*. JASA, *80*, 148–157.

81. FAY, R. E. and HERRIOT, R. A. (1979): *Estimation of income from small places: An application of James-Stein procedures to census data*. JASA, *74*, 269–277.

82. FELLEGI, I. P. (1963): *Sampling with varying probabilities without replacement: Rotating and non-rotating samples*. JASA, *58*, 183–201.

83. FELLEGI, I. P. (1978): *Approximate tests of independence and goodness of fit based upon stratified multi-stage samples*. SUM, *4*, 29–56.

84. FELLEGI, I. P. (1980): *Approximate tests of independence and goodness of fit based on stratified multi-stage samples*. JASA, *75*, 261–268.

85. FIRTH, D. and BENNETT, K. E. (1998): *Robust models in probability sampling*. JRSS, *60*(1), 3–21.

86. FORD, B. L. (1976): *Missing data procedures: A comparative study*. PSSSASA, 324–329.

87. FORD, B. L., KLEWENO, D. G. and TORTORA, R. D. (1980): *The effects of procedures which impute for missing items: A simulation study using an agricultural survey*. In: CTS, 413–436.

88. FULLER, W. A. (1975): *Regression analysis for sample survey*. Sā, *37*, 117–132.

89. FULLER, W. A. (1981): *Comment on a paper by* ROYALL, R. M. *and* CUMBERLAND, W. G. JASA, *76*, 78–80.

90. FULLER, W. A. and ISAKI, C. T. (1981): *Survey design under superpopulation models*. In: CTS, 199–226.

91. FULLER, W. A., LONGHIN, M. and BAKER, H. (1994): *Regression weighting in the presence of nonresponse with application to the 1987–1988 Nationwide Food Consumption Survey*. SUM, *20*, 75–85.

92. GABLER, S. (1990): *Minimax solutions in sampling from finite populations*. Springer-Verlag, New York.

93. GABLER, S. and STENGER, H. (2000): *Minimax strategies in survey sampling*. JSPI, *90*, 305–321.

94. GAUTSCHI, W. (1957): *Some remarks on systematic sampling*. AMS, *28*, 385–394.

95. GHOSH, M. (1987): *On admissibility and uniform admissibility in finite population sampling*. In: APSPST, 197–213.

96. GHOSH, M. (1989): *Estimating functions in survey sampling*. Unpublished manuscript.

97. GHOSH, M. and LAHIRI, P. (1987): *Robust empirical Bayes estimation of means from stratified samples*. JASA, *82*, 1153–1162.

98. GHOSH, M. and LAHIRI, P. (1988): *Bayes and empirical Bayes analysis in multi-stage sampling*. In *Statistical Decision Theory and Related Topics IV*, Eds. Gupta, S. S. and Berger, G. O., Springer, New York, 195–212.

99. GHOSH, M. and MEEDEN, G. (1986): *Empirical Bayes estimation in finite population sampling*. JASA, *81*, 1058–1062.

100. GHOSH, M. and MEEDEN, G. (1997): *Bayesian methods for finite population sampling*. Chapman & Hall, London.

101. GODAMBE, V. P. (1955): *A unified theory of sampling from finite populations*. JRSS, *17*, 269–278.

102. GODAMBE, V. P. (1960a): *An admissible estimate for any sampling design*, Sā, *22*, 285–288.

103. GODAMBE, V. P. (1960b): *An optimum property of regular maximum likelihood estimation*. AMS, *31*, 1208–1212.

104. GODAMBE, V. P. (1982): *Estimation in survey sampling: Robustness and optimality (with discussion)*. JASA, *77*, 393–406.

105. GODAMBE, V. P. (1986): *Quasi-score function, quasi-observed Fisher information and conditioning in survey sampling*. Unpublished manuscript.

106. GODAMBE, V. P. (1995): *Estimation of parameters in survey sampling: Optimality*. Canadian Journal of Statistics, *23*, 227–243.

107. GODAMBE, V. P. and JOSHI, V. M. (1965): *Admissibility and Bayes estimation in sampling finite populations, I*. AMS, *36*, 1707–1722.

108. GODAMBE, V. P. and SPROTT, D. A., eds. (1971): *Foundations of statistical inference*. Holt, Rinehart, Winston, Toronto.

109. GODAMBE, V. P. and THOMPSON, M. E. (1977): *Robust near optimal estimation in survey practice*. BISI, *47*(3), 129–146.

110. GODAMBE, V. P. and THOMPSON, M. E. (1986a): *Parameters of super-population and survey population: Their relationships and estimation*. ISR, *54*, 127–138.

111. GODAMBE, V. P. and THOMPSON, M. E. (1986b): *Some optimality results in the presence of non-response*. SUM, *12*, 29–36.

112. GROSENBAUGH, L. R. (1965): *Three "p" sampling theory and program THRP for computer generation of selection criteria*. USDA Forest Service Research Paper, PSW, *21*, 53.

113. GROSS, S. (1980): *Media estimation in sampling surveys*. Proc. Sec. survey sampling methods, Amer. Stat. Assoc., 181–184.

114. GURNEY, M. and JEWETT, R. S. (1975): *Constructing orthogonal replications for variance estimation*. JASA, *70*, 819–821.

115. HÁJEK, J. (1959): *Optimum strategy and other problems in probability sampling*. CPM, *84*, 387–473.

116. HÁJEK, J. (1960): *Limit distributions in simple random sampling from a finite population*. Publication of the Hungarian Academy of Science, *5*, 361–374.

117. HÁJEK, J. (1964): *Asymptotic theory of rejective sampling with varying probability from a single population*. PSW, *21*, 53.

118. HÁJEK, J. (1971): *Comment on a paper by* BASU, D. In: FSI, 203–242.

119. HÁJEK, J. (1981): *Sampling from a finite population*. Marcel Dekker, New York.

120. HANSEN, M. H. and HURWITZ, W. N. (1943): *On the theory of sampling from finite populations*. AMS, *14*, 333–362.

121. HANSEN, M. H. and HURWITZ, W. N. (1946): *The problem of non-response in sample surveys*. JASA, *41*, 517–529.

122. HANSEN, M. H., HURWITZ, W. N. and MADOW, W. G. (1953): *Sample survey methods and theory*. Vol. I and Vol. II. Wiley, New York.

123. HANSEN, M. H., MADOW, W. G. and TEPPING, B. J. (1983): *An evaluation of model-dependent and probability-sampling inferences in sample surveys*. JASA, *78*, 776–807.

124. HANURAV, T. V. (1966): *Some aspects of unified sampling theory*. Sā, *28*, 175–204.

125. HARTLEY, H. O. (1946): *Discussion of papers by F. Yates*. JRSS, *109*, 37.

126. HARTLEY, H. O. (1962): *Multiple frame surveys*. PSSSASA, 203–206.

127. HARTLEY, H. O. (1974): *Multiple frame methodology and selected applications*. Sā, *37*, 99–118.

128. HARTLEY, H. O. (1981): *Estimation and design for non-sampling errors of survey*. In: CTS, 31–46.

129. HARTLEY, H. O. and RAO, J. N. K. (1962): *Sampling with unequal probabilities and without replacement*. In: ASM, *33*, 350–374.

130. HARTLEY, H. O. and RAO, J. N. K. (1978): *Estimation of non-sampling variance components in sample surveys*. In: SSM, 35–43.

131. HARTLEY, H. O. and ROSS, A. (1954): *Unbiased ratio estimators*. N *174*, 270–271.

132. HARTLEY, H. O. and SIELKEN, R. L. (1975): *A "super-population view-point" for finite population sampling*. Bms, *31*, 411–422.

133. HEGE, V. S. (1965): *Sampling designs which admit uniformly minimum variance unbiased estimators*. CSAB, *14*, 160–162.

134. HEILBRON, D. C. (1978): *Comparison of estimators of the variance of systematic sampling*, Bk, *65*, 429–433.

135. HIDIROGLOU, M. A. and RAO, J. N. K. (1987): *Chi-squared tests with categorical data from complex surveys I, II*. JOS, *3*, 117–132, 133–140.

136. HIDIROGLOU, M. A. and SRINATH, K. P. (1981): *Some estimators of the population total from simple random samples containing large units*. JASA, *76*, 690–695.

137. HO, E. W. H. (1980): *Model-unbiasedness and the Horvitz-Thompson estimator in finite population sampling*. AJS, *22*, 218–225.

138. HOLT, D. and SCOTT, A. J. (1981): *Regression analysis using survey data*. St, *30*, 169–178.

139. HOLT, D., SCOTT, A. J. and EWINGS, P. D. (1980): *Chi-squared tests with survey data*. JRSS, A, *143*, 303–320.

140. HOLT, D. and SMITH, T. M. F. (1976): *The design of surveys for planning purposes*, AJS, *18*, 37–44.

141. HOLT, D. and SMITH, T. M. F. (1979): *Post-stratification*. JRSS, A, *142*, 33–46.

142. HOLT, D., SMITH, T. M. F. and WINTER, P. D. (1980): *Regression analysis of data from complex surveys*, JRSS, A, *143*, 474–487.

143. HORVITZ, D. G. and THOMPSON, D. J. (1952): *A generalization of sampling without replacement from a universe*. JASA, *47*, 663–685.

144. IACHAN, R. (1982): *Systematic sampling: A critical review*. ISR, *50*, 293–303.

145. IACHAN, R. (1983): *Measurement errors in surveys: A review*. *CSTM*, *12*, 2273–2281.

146. IACHAN, R. (1984): *Sampling strategies, robustness and efficiency: The state of the art*. ISR, *52*, 209–218.

147. ISAKI, C. T. and FULLER, W. A. (1982): *Survey design under the regression superpopulation model*. JASA, *77*, 89–96.

148. JAMES, W. and STEIN, C. (1961): *Estimation with quadratic loss*. Proc. 4th Berkeley Symposium on Math. Stat. Calif. Press. 361–379.

149. JESSEN, R. J. (1969): *Some methods of probability non-replacement sampling*. JASA, *64*, 175–193.

150. JOHNSON, N. L. and SMITH, H., Jr., eds. (1969): *New developments in survey sampling*. Wiley Interscience, New York.

151. JÖNRUP, H. and RENNERMALM, B. (1976): *Regression analysis in samples from finite populations*. SJS, *3*, 33–37.

152. KALTON, G. (1983a): *Models in the practice of survey sampling*. ISR, *51*, 175–188.

153. KALTON, G. (1983b): *Compensating for missing survey data.*

154. KEMPTHORNE, O. (1969): *Some remarks on statistical inference in finite sampling.* In: NDSS, 671–695.

155. KEYFITZ, N (1957): *Estimates of sampling variance when two units are selected from each stratum.* JASA, *52*, 503–510.

156. KISH, L. (1965): *Survey sampling.* John Wiley, New York.

157. KISH, L. and FRANKEL, M. R. (1970): *Balanced repeated replications for standard errors.* JASA, *65*, 1071–1094.

158. KISH, L. and FRANKEL, M. R. (1974): *Inference from complex samples (with discussion).* JRSS, *36*, 1–37.

159. KONJIN, H. S. (1962): *Regression analysis in sample surveys.* JASA, *57*, 590–606.

160. KOOP, J. C. (1967): *Replicated (or interpenetrating) samples of unequal sizes.* AMS, *38*, 1142–1147.

161. KOOP, J. C. (1971): *On splitting a systematic sample for variance estimation.* AMS, *42*, 1084–1087.

162. KOTT, P. S. (1990): *Estimating the conditional variance of a design consistent regression estimator.* JSPI, *24*, 287–296.

163. KREWSKI, D. (1978): *On the stability of some replication variance estimators in the linear case.* JSPI, *2*, 45–51.

164. KREWSKI, D., PLATEK, R., and RAO, J. N. K., eds. (1981): *Current topics in survey sampling.* Academic Press, New York.

165. KREWSKI, D. and RAO, J. N. K. (1981): *Inference from stratified samples: Properties of the linearization, jackknife and balanced repeated replication methods.* AS, *9*, 1010–1019.

166. KRISHNAIAH, P. R., and RAO, C. R., eds. (1988): *Handbook of statistics*, Vol. 6. North-Holland, Amsterdam.

167. KRÖGER, H., SÄRNDAL, C. E. and TEIKARI, I. (1999): *Poission mixture sampling: A family of designs for coordinated selection using permanent random numbers.* SUM, *25*, 3–11.

168. KUMAR, S., GUPTA, V. K. and AGARWAL, S. K. (1985): *On variance estimation in unequal probability sampling.* AJS, *27*, 195–201.

169. LAHIRI, D. B. (1951): *A method of sample selection providing unbiased ratio estimators*. BISI, *33*(2), 33–140.

170. LANKE, J. (1975): *Some contributions to the theory of survey sampling*. Unpublished Ph.D. THESIS, University of Lund, Sweden.

171. LITTLE, R. J. A. (1983): *Estimating a finite population mean from unequal probability samples*. JASA, *78*, 596–604.

172. MACNEILL, I. B. and UMPHREY, G. J., eds. (1987): *Applied probability, stochastic processes and sampling theory*. Reidell, Dordrecht.

173. MADOW, W. G., NISSELSON, H. and OLKIN, I., eds. (1983): *Incomplete data in sample surveys, vol. 1*, Academic Press, New York.

174. MADOW, W. G., and OLKIN, I., eds. (1983): *Incomplete data in sample surveys, vol. 3*, Academic Press, New York.

175. MADOW, W. G., OLKIN, I. and RUBIN, D. B., eds. (1983): *Incomplete data in sample surveys, vol. 2*, Academic Press, New York.

176. MAHALANOBIS, P. C. (1946): *Recent experiments in statistical sampling in the Indian Statistical Institute*. JRSS, *109*, 325–378.

177. MCCARTHY, P. J. (1969): *Pseudo-replication: Half-samples*. RISI., *37*, 239–264.

178. MCCARTHY, P. J. and SNOWDEN, C. B.(1985): *The bootstrap and finite population sampling*.

179. MEINHOLD, R. J. and SINGPURWALLA, N. D. (1983): *Understanding the Kalman filter*. ASA, *37*, 123–127.

180. MIDZUNO, H. (1952): *On the sampling system with probabilities proportionate to sum of sizes*. AISM, *3*, 99–107.

181. MONTANARI, G. E. (1987): *Post-sampling efficient QR-prediction in large-sample surveys*. ISR, *55*, 191–202.

182. MUKERJEE, R. and CHAUDHURI, A. (1990): *Asymptotic optimality of double sampling plans employing generalized regression estimators*. JSPI, *26*, 173–183.

183. MUKERJEE, R. and SENGUPTA, S. (1989): *Optimal estimation of finite population total under a general correlated model*. Bk, *76*, 789–794.

184. MUKHOPADHYAY, P. (1998): *Small area estimation in survey sampling*. Norasa Publishing House, New Delhi.

185. MURTHY, M. N. (1957): *Ordered and unordered estimators in sampling without replacement*. Sā, *18*, 379–390.

186. MURTHY, M. N. (1977): *Sampling theory and methods*. Stat. Pub. Soc., Calcutta.

187. MURTHY, M. N. (1983): *A framework for studying incomplete data with a reference to the experience in some countries in Asia and the Pacific*. In: IDSS, *3*, 7–24.

188. NAMBOODIRI, N. K., ed. (1978): *Survey sampling and measurement*. Academic Press, New York.

189. NATHAN, G. (1988): *Inference based on data from complex sample designs*. In: HBS, *6*, 247–266.

190. NATHAN, G. and HOLT, D. (1980): *The effect of survey design on regression analysis*. JRSS, *42*, 377–386.

191. NEYMAN, J. (1934): *On the two different aspects of the representative method: The method of stratified sampling and the method of purposive selection*. JRSS, *97*, 558–625.

192. OGUS, J. K. and CLARK, D. F. (1971): *A report on methodology*. Tech. Report No. 24, U.S. Bureau of the Census, Washington, DC, *77*, 436–438.

193. OH, H. L. and SCHEUREN, F. J. (1983): *Weighting adjustment for unit non-response*. In: IDSS, *2*, 143–184.

194. OHLSSON, E. (1989): *Variance estimation in the* RAO-HARTLEY-COCHRAN *procedure*. Sā, *51*, 348–361.

195. OHLSSON, E. (1995): *Coordinating samples using permanent random numbers*. In: 153–169.

196. PAL S. (2002): *Contributions to emerging techniques in survey sampling*. Unpublished Ph.D., thesis of ISI, Kolkata, India.

197. PEREIRA, C. A. and RODRIGUES, J. (1983): *Robust linear prediction in finite populations*. ISR, *51*, 293–300.

198. PFEFFERMANN, D. (1984): *A note on large sample properties of balanced samples*. JRSS, *46*, 38–41.

199. PFEFFERMANN, D. and HOLMES, D. J. (1985): *Robustness considerations in the choice of a method of inference for regression analysis of survey data*. JRSS, *148*, 268–278.

200. PFEFFERMANN, D. and NATHAN, G. (1981): *Regression analysis of data from a clustered sample*. JASA, *76*, 681–689.

201. PFEFFERMANN, D. and SMITH, T. M. F. (1985): *Regression models for grouped populations on cross-section surveys*. ISR, *53*, 37–59.

202. POLITZ, A. and SIMMONS, W. (1949): *I. An attempt to get the "not at homes" into the sample without callbacks. II. Further theoretical considerations regarding the plan for eliminating callbacks*. JASA, *44*, 9–31.

203. POLITZ, A. and SIMMONS, W. (1950): *Note on an attempt to get the "not at homes" into the sample without callbacks*. JASA, *45*, 136–137.

204. PORTER, R. D. (1973): *On the use of survey sample weights in the linear model*. AESM, *212*, 141–158.

205. PRASAD, N. G. N. (1988): *Small area estimation and measurement of response error variance in surveys*. Unpublished Ph.D. THESIS, Carleton University, Ottawa, Canada.

206. PRASAD, N. G. N. and RAO, J. N. K. (1990): *The estimation of the mean squared error of small area estimators*. JASA, *85*, 163–171.

207. PURI, M. L., VILALANE, J. P., and WERTZ, W., eds. (1987): *New perspectives in theoretical and applied statistics*. John Wiley & Sons, New York.

208. QUENOUILLE, M. H. (1949): *Approximate tests of correlation in time-series*. JRSS, *11*, 68–84.

209. RAJ, D. (1956): *Some estimators in sampling with varying probabilities without replacement*. JASA, *51*, 269–284.

210. RAJ, D. (1968): *Sampling theory*. McGraw-Hill, New York.

211. RAO, C. R. (1971): *Some aspects of statistical inference in problems of sampling from finite populations*. In: FSI, 177–202.

212. RAO, J. N. K. (1968): *Some small sample results in ratio and regression estimation*. JISA, *6*, 160–168.

213. RAO, J. N. K. (1969): *Ratio and regression estimators*. In: NDSS, 213–234.

214. RAO, J. N. K. (1971): *Some thoughts on the foundations of survey sampling*. JISAS, *23*(2), 69–82.

215. RAO, J. N. K. (1973): *On double sampling for stratification and analytical surveys*. Bk, *60*, 125–133.

216. RAO, J. N. K. (1975a): *Unbiased variance estimation for multistage designs*. Sā, *37*, 133–139.

217. RAO, J. N. K. (1975b): *Analytic studies of sample survey data*. SUM, *1*, 1–76.

218. RAO, J. N. K. (1979): *On deriving mean square errors and other non-negative unbiased estimators in finite population sampling*. JISA, *17*, 125–136.

219. RAO, J. N. K. (1985): *Conditional inference in survey sampling*. SUM, *11*, 15–31.

220. RAO, J. N. K. (1986): *Ratio estimators*. In: 7, 639–646.

221. RAO, J. N. K. (1987): *Analysis of categorical data from sample surveys*. In: NPTA, 45–60.

222. RAO, J. N. K. (1988): *Variance estimation in sample surveys*. In: HBS, *6*, 427–447.

223. RAO, J. N. K. and WU, C. F. J. (1988): *Resampling inference with complex survey data*. JASA, *80*, 620–630.

224. RAO, J. N. K. (1994): *Estimating totals and distribution functions using auxiliary information at the estimation stage*. JOS, *10*, 153–165.

225. RAO, J. N. K. (2002): *Small area estimation*. John Wiley, New York.

226. RAO, J. N. K. and BAYLESS, D. L. (1969): *An empirical study of estimators and variance estimators in unequal probability sampling of two units per stratum*. JASA, *64*, 540–549.

227. RAO, J. N. K. and BELLHOUSE, D. R. (1978): *Optimal estimation of a finite population mean under generalized random permutation models*. JSPI, *2*, 125–141.

228. RAO, J. N. K. and BELLHOUSE, D. R. (1989): *The history and development of the theoretical foundations of survey based estimation and statistical analysis*. Unpublished Manuscript.

229. RAO, J. N. K., HARTLEY, H. O. and COCHRAN, W. G. (1962): *On a simple procedure of unequal probability sampling without replacement*. JRSS, *24*, 482–491.

230. RAO, J. N. K. and NIGAM, A. K. (1990): *Optimal controlled sampling designs*. Bk, *77*(4), 807–814.

231. RAO, J. N. K. and SCOTT, A. J. (1979): *Chi-squared tests for analysis of categorical data from complex surveys*. PSRMASA, 58–66.

232. RAO, J. N. K. and SCOTT, A. J. (1981): *The analysis of categorical data from complex sample surveys: Chi-squared tests for goodness of fit and independence in two-way tables*. JASA, *76*, 221–230.

233. RAO, J. N. K. and SCOTT, A. J. (1984): *On chi-squared tests for multi-way tables with cell proportions estimated from survey data*. AS, *12*, 46–60.

234. RAO, J. N. K. and SCOTT, A. J. (1987): *On simple adjustments to chi-square tests with sample survey data*. AS, *15*, 385–397.

235. RAO, J. N. K. and THOMAS, D. R. (1988): *The analysis of cross-classified categorical data from complex sample surveys*. SM, *18*, 213–269.

236. RAO, J. N. K. and VIJAYAN, K. (1977): *On estimating the variance in sampling with probability proportional to aggregate size*. JASA, *80*, 620–630.

237. RAO, J. N. K. and WU, C. F. J. (1985): *Inference from stratified samples: Second-order analysis of three methods for non-linear statistics*. JASA, *80*, 620–630.

238. RAO, J. N. K. and WU, C. F. J. (1988): *Resampling inference with complex survey data*. JASA, *83*, 231–241.

239. RAO, P. S. R. S. (1983): *Randomization approach*. In: IDSS, 97–105.

240. RAO, P. S. R. S. (1988): *Ratio and regression estimators*. In: HBS, *6*, 449–468.

241. RAO, P. S. R. S. and RAO, J. N. K. (1971): *Small sample results for ratio estimators*. Bk, *58*, 625–630.

242. RAO, T. J. (1984): *Some aspects of random permutation models in finite population sampling theory*. Mk, *31*, 25–32.

243. RAY, S. and DAS, M. N. (1997): *Circular systematic sampling with drawback*. JISAS, *50*(1), 70–74.

244. RÉNYI, A., (1966): *Wahrscheinlichkeitsrechnung: mit einem Anhang über Informationstheorie*. 2. Aufl., Berlin (unpublished).

245. ROBERTS, G., RAO, J. N. K. and KUMAR, S. (1987): *Logistic regression analysis of sample survey data*. Bk, *74*, 1–12.

246. ROBINSON, J. (1987): *Conditioning ratio estimates under simple random sampling*. JASA, *82*, 826–831.

247. ROBINSON, P. M. and SÄRNDAL, C. E. (1983): *Asymptotic properties of the generalized regression estimator in probability sampling*. Sā, *45*, 240–248.

248. RODRIGUES, J. (1984): *Robust estimation and finite population*. PMS, *4*, 197–207.

249. ROY, A. S. and SINGH, M. P. (1973): *Interpenetrating subsamples with and without replacement*. Mk, *20*, 230–239.

250. ROYALL, R. M. (1970): *On finite population sampling theory under certain linear regression models*. Bk, *57*, 377–387.

251. ROYALL, R. M. (1971): *Linear regression models in finite population sampling theory*. In: FSI, 259–279.

252. ROYALL, R. M. (1979): *Prediction models in small area estimation*. In: SESA, NIDA, 63–87.

253. ROYALL, R. M. (1988): *The prediction approach to sampling theory*. In: SBH, *6*, 399–413.

254. ROYALL, R. M. (1992): *Robustness and optimal design under prediction models for finite populations*. SUM, *18*, 179–185.

255. ROYALL, R. M. and CUMBERLAND, W. G. (1978a): *Variance estimation in finite population sampling*. JASA, *73*, 351–358.

256. ROYALL, R. M. and CUMBERLAND, W. G. (1978b): *An empirical study of prediction theory in finite population sampling: Simple random sampling and the ratio estimator*. In: SSM, 293–309.

257. ROYALL, R. M. and CUMBERLAND, W. G. (1981a): *An empirical study of the ratio estimator and estimators of its variance*. JASA, *76*, 66–77.

258. ROYALL, R. M. and CUMBERLAND, W. G. (1981b): *The finite population linear regression estimator and estimators of its variance—An empirical study*. JASA, *76*, 924–930.

259. ROYALL, R. M. and CUMBERLAND, W. G. (1985): *Conditional coverage properties of finite population confidence intervals*. JASA, *80*, 355–359.

260. ROYALL, R. M. and EBERHARDT, K. R. (1975): *Variance estimators for the ratio estimator*. Sā, *37*, 43–52.

261. ROYALL, R. M. and HERSON, J. (1973): *Robust estimation in finite populations I, II*. JASA, *68*, 880–889, 890–893.

262. ROYALL, R. M. and PFEFFERMANN, D. (1982): *Balanced samples and robust Bayesian inference in finite population sampling*. Bk, *69*, 404–409.

263. RUBIN, D. B. (1976): *Inference and missing data*. Bk, *63*, 581–592.

264. RUBIN, D. B. (1977): *Formalizing subjective notions about the effect of non-respondents in sample surveys*. JASA, *72*, 538–543.

265. RUBIN, D. B. (1978): *Multiple imputations in sample surveys—A phenomenological Bayesian approach to non-response (with discussion and reply)*. Proc. Survey Research Methods Sec of Amer. Stat. Assoc., 20–34. Also in *Imputation and Editing of Faulty or Missing Survey Data*, US Dept. of Commerce, Bureau of the Census, 10–18.

266. RUBIN, D. B. (1979): *Illustrating the use of multiple imputations to handle non-response in sample surveys*. BISI,

267. RUBIN, D. B. (1983): *Conceptual issues in the presence of nonresponse*. In: IDSS, 123–142.

268. SALEHI, M. N. and SEBER, G. A. F. (1997): *Adaptive cluster sampling with networks selected without replacement*. BK, *84*, 209–219.

269. SALEHI, M. N. and SEBER, G. A. F. (2002): *Unbiased estimators for restricted adaptive cluster sampling*. ANZJS, *44*(1), 63–74.

270. SANDE, I. G. (1983): *Hot-deck imputation procedures*. In: IDSS, 339–349.

271. SÄRNDAL, C. E. (1980): *On π-inverse weighting versus best linear weighting in probability sampling*. Bk, *67*, 639–650.

272. SÄRNDAL, C. E. (1981): *Frameworks for inference in survey sampling with applications to small area estimation and adjustment for non-response*. BISI, *49*(1), 494–513.

273. SÄRNDAL, C. E. (1982): *Implications of survey design for generalized regression estimation of linear functions*. JSPI, *7*, 155–170.

274. SÄRNDAL, C. E. (1984): *Design-consistent versus model-dependent estimation for samll domains*. JASA, *79*, 624–631.

275. SÄRNDAL, C. E. (1985): *How survey methodologists communicate*. JOS, *1*, 49–63.

276. SÄRNDAL, C. E. (1996): *Efficient estimators with simple variance in unequal probability sampling*. JASA, *91*, 1289–1300.

277. SÄRNDAL, C. E. and HIDIROGLOU, M. A. (1989): *Small domain estimation: A conditional analysis*. JASA, *84*, 266–275.

278. SÄRNDAL, C. E. and HUI, T. K. (1981): *Estimation for non-response situations: To what extent must we rely on models?* In: CTS, 227–246.

279. SÄRNDAL, C. E., SWENSSON, B. and WRETMAN, J. H. (1992): *Model assisted survey sampling*. Springer-Verlag, New York.

280. SÄRNDAL, C. E. and WRIGHT, R. L. (1984): *Cosmetic form of estimators in survey sampling*. SJS, *11*, 146–156.

281. SATTERTHWAITE, F. E. (1946): *An approximate distribution of estimates of variance components*. Bms, *2*, 110–114.

282. SAXENA, B. C., NARAIN, P. and SRIVASTAVA, A. K. (1984): *Multiple frame surveys in two stage sampling*. Sā, *46*, 75–82.

283. SCOTT, A. J. (1977): *On the problem of randomization in survey sampling*. Sā, *39*, 1–9.

284. SCOTT, A. J., BREWER, K. R. W. and HO, E. W. H. (1978): *Finite population sampling and robust estimation*. JASA, *73*, 359–361.

285. SCOTT, A. J. and HOLT, D. (1982): *The effect of two-stage sampling on ordinary least squares theory*. JASA, *77*, 848–854.

286. SCOTT, A. J. and RAO, J. N. K. (1981): *Chi-squared tests for contingency tables with proportions estimated from survey data*. In: CTS, 247–265.

287. SCOTT, A. J. and SMITH, T. M. F. (1969): *Estimation in multi-stage surveys*. JASA, *64*, 830–840.

288. SCOTT, A. J. and SMITH, T. M. F. (1975): *Minimax designs for sample surveys*. Bk, *62*, 353–357.

289. SEN, A. R. (1953): *On the estimator of the variance in sampling with varying probabilities*. JISAS, *5*(2), 119–127.

290. SHAH, B. V., HOLT, M. M. and FOLSOM, R. E. (1977): *Inference about regression models from sample survey data*. BISI, *41*(3), 43–57.

291. SILVA, P. L. D. N., and SKINNER, C. J. (1995): *Estimating distribution functions with auxiliary information using poststratification*. JOS, *11*, 277–294.

292. SILVA, P. L. D. N., and SKINNER, C. J. (1997): *Variable selection for regression estimation in finite populations*. SUM, *23*, 23–32.

293. SINGH, A. C. and MOHL, C. A. (1996): *Understanding calibration estimators in survey sampling*. SUM, *22*, 107–115.

294. SINGH, D. and SINGH, P. (1977): *New systematic sampling*. JSPI, *1*, 163–177.

295. SIRKEN, M. G. (1983): *Handling missing data by network sampling*. In: IDSS, *2*, 81–90.

296. SITTER, R. R. (1992a): *A resampling procedure for complex survey data*. In: JASA, *87*, 755–765.

297. SITTER, R. R. (1992b): *Comparing three bootstrap methods for survey data*. Can. J. Stat. *20*, 133–154.

298. SKINNER, C. J. and RAO, J. N. K. (1996): *Estimation in dual frame surveys with complex designs*. JASA, *91*, 349–356.

299. SMITH, T. M. F. (1976): *The foundations of survey sampling: a review*. JRSS, *139*, 183–195.

300. SMITH, T. M. F. (1981): *Regression analysis for complex surveys*. In: CTS, 267–292.

301. SMITH, T. M. F. (1984): *Present position and potential developments: Some personal views: Sample surveys*. JRSS, *147*, 208–221.

302. SOLOMON, H. and STEPHENS, M. A. (1977): *Distribution of a sum of weighted chi-square variables*. JASA, *72*, 881–885.

303. SRINATH, K. P. (1971): *Multi-phase sampling in non-response problems*. JASA, *66*, 583–589.

304. SRINATH, K. P. und HIDIROGLOU, M. A. (1980): *Estimation of variance in multi-stage sampling*. Mk, *27*, 121–125.

305. STEHMAN, S. V. and OVERTON, W. S. (1994): *Comparison of variance estimators of the* HORVITZ–THOMPSON *estimator for randomized variable systematic sampling*. JASA, *89*, 30–43.

306. STEINBERG, J., ed., (1979): *Synthetic estimates for small areas* (Monograph 24). National Institute on Drug Abuse, Washington, D.C.

307. STENGER, H. (1986): *Stichproben*. Physica-Verlag, Heidelberg.

308. STENGER, H. (1988): *Asymptotic expansion of the minimax value in survey sampling*. Mk, *35*, 77–92.

309. STENGER, H. (1989): *Asymptotic analysis of minimax strategies in survey sampling*. AS, *17*, 1301–1314.

310. STENGER, H. (1990): *Asymptotic minimaxity of the ratio strategy*. Bk, *77*, 389–395.

311. STENGER, H. and GABLER, S. (1996): *A minimax property of* LAHIRI-MIDZUNO-SEN'*s sampling scheme*. Mk, *43*, 213–220.

312. STUKEL, D., HIDIROGLOU, M. A. and SÄRNDAL, C. E. (1996): *Variance estimation for calibration estimators: A comparison of jackknifing versus Taylor series linearization*. SUM, *22*, 117–125.

313. SUKHATME, P. V. (1954): *Sampling theory of surveys with applications*. Asia Publication House, London.

314. SUNDBERG, R. (1994): *Precision estimation in sample survey inference: A criterion for choice between various estimators*. Bk, *81*, 157–172.

315. SUNTER, A. B. (1986): *Implicit longitudinal sampling from adminstrative files: A useful technique*. JOS, *2*, 161–168.

316. SWAIN, A. K. P. C., ed. (2000): *Current developments in survey sampling*. Ulkal University, Bhubaneswar.

317. TALLIS, G. M. (1978): *Note on robust estimation in finite populations*. Sā, *40*, 136–138.

318. TAM, S. M. (1984): *Optimal estimation in survey sampling under a regression super-population model*. Bk, *71*, 645–647.

319. TAM, S. M. (1986): *Characterization of best model-based predictor in survey sampling*. Bk, *3*, 232–235.

320. TAM, S. M. (1988a): *Some results on robust estimation in finite population sampling*. JASA, *83*, 242–248.

321. TAM, S. M. (1988b): *Asymtotically design-unbiased prediction in survey sampling*. Bk, *75*, 175–177.

322. THOMAS, D. R. and RAO, J. N. K. (1987): *Small-sample comparisons of level and power for simple goodness-of-fit statistics under cluster sampling*. JASA, *82*, 630–636.

323. THOMPSON, M. E. (1971): *Discussion of a paper by* RAO, *C. R.*. In: FSI, 196–198.

324. THOMPSON, M. E. (1997): *Theory of sample surveys*. Chapman & Hall, London.

325. THOMPSON, S. K. (1992): *Sampling*. John Wiley & Sons, New York.

326. THOMPSON, S. K. and SEBER, G. A. F. (1996): *Adaptive sampling*. John Wiley & Sons, New York.

327. THOMSEN, I. (1973): *A note on the efficiency of weighting subclass means to reduce the effects of non-response when analyzing survey data*. ST, *4*, 278–283.

328. THOMSEN, I. (1978): *Design and estimation problems when estimating a regression coefficient from survey data*. Mk, *25*, 27–35.

329. THOMSEN, I. and SIRING, E. (1983): *On the causes and effects of non-response; Norwegian experiences*. In: IDSS, *3*, 25–29.

330. TORNQVIST, L. (1963): *The theory of replicated systematic cluster sampling with random start*. RISI, *31*, 11–23.

331. TUKEY, J. W. (1958): *Bias and confidence in not-quite large samples [abstract]*. AMS, *29*, 614.

332. VALLIANT, R. (1987a): *Conditional properties of some estimators in stratified sampling*. JASA, *82*, 509–519.

333. VALLIANT, R. (1987b): *Some prediction properties of balanced half-samples variance estimators in single-stage sampling*. JRSS, *49*, 68–81.

334. VALLIANT, R., DORFMAN, A. H. and ROYALL, R. M. (2000): *Finite population sampling and inference, a prediction approach*. John Wiley, New York.

335. WARNER, S. L. (1965): *RR: A survey technique for eliminating evasive answer bias*. JASA, *60*, 63–69.

336. WOLTER, K. M. (1984): *An investigation of some estimators of variance for systematic sampling*. JASA, *79*, 781–790.

337. WOLTER, K. M. (1985): *Introduction to variance estimation*. Springer-Verlag, New York.

338. WOODRUFF, R. S. (1971): *A simple method for approximating the variance of a complicated estimate*. JASA, *66*, 411–414.

339. WRIGHT, R. L. (1983): *Finite population sampling with multivariate auxiliary information*. JASA, *78*, 879–884.

340. WU, C. F. J. (1982): *Estimation of variance of the ratio estimator*. Bk, *69*, 183–189.

341. WU, C. F. J. (1984): *Estimation in systematic sampling with supplementary observations*. Sā, *46*, 306–315.

342. WU, C. F. J. and DENG, L. Y. (1983): *Estimation of variance of the ratio estimator: An empirical study*. In: Scientific Inference, Data Analysis and Robustness (G.E.P. Box et al., eds.), 245–277, Academic Press.

343. YATES, F. (1949): *Sampling methods for censuses and surveys*. Charles Griffin & Co., London.

344. YATES, F. and GRUNDY, P. M. (1953): *Selection without replacement from within strata with probability proportional to size*. JRSS, *15*, 253–261.

345. ZIESCHANG, K. D. (1990): *Simple weighting methods and estimation of totals in the consumer expenditure survey*. JASA, *85*, 986–1001.

346. ZINGER, A. (1980): *Variance estimation in partially systematic sampling*. JASA, *75*, 206–211.

List of Abbreviations, Special Notations, and Symbols

ADC	asymptotically design consistent	5.2	103
ADU	asymptotically design unbiased	5.2	103
BE	Bayes estimator	4.2.1	94
BLU	best linear unbiased	4.1.1	80
BLUE	best linear unbiased estimator	3.3.1	63
BLUP	best linear unbiased predictor	4.1.1	80
BRR	balanced repeated replication	11.2.1	265
CSW	Cassel-Särndal-Wretman	3.2.6	55
CV	coefficient of variation	7.1	135
deff	design effect	11.1.1	253
df	degrees of freedom	7	133
DR	direct response	12	275
EBE	empirical Bayes estimator	4.2.1	94
epsem	equiprobability selection methods	9	201
fsu	first-stage unit	8.1	176
GDE	generalized difference estimator	6.1.1	111
GLS	generalized least squares	11.2.2	266
GLSE	generalized least squares estimator	11.2.2	266
GREG	generalized regression	2.1	32
HH	Hansen–Hurwitz	2.2	13
HHE	Hansen–Hurwitz estimator	2.2	13
HL	homogeneous linear	1.2	3
HLU	homogeneous linear unbiased	1.2	4
HLUE	homogeneous linear unbiased estimator	3.1.1	35

HRE	Hartely–Ross estimator	2.4.7	29
HT	Horvitz–Thompson	1.2	4
HTE	Horvitz–Thompson estimator	1.2	4
IPF	iterated proportional fitting	13.4	308
IPNS	interpenetrating network of subsampling	9.3	208
IPPS	inclusion probability proportional to size	3.2.5	53
JSE	James–Stein estimator	4.2.2	94
L	linear	1.2	3
LMS	Lahiri-Midzuno-Sen	2.2	13
LPRE	linear predictor	6.1	113
LSE	least squares estimator	3.3.1	63
LU	linear unbiased	3.1.1	36
LUE	linear unbiased estimator	3.1.1	36
MLE	maximum likelihood estimator	3.3.1	162
MSE	mean square error	1.2	4
$\mathcal{M}_1$		3.2.2	46
$\mathcal{M}_2, \mathcal{M}_{2\gamma}$		3.2.5	54
$\mathcal{M}_{0\gamma}, \mathcal{M}_{1\gamma}, \mathcal{M}_{j\gamma}$		7.3	155
n(s)	sample size	1.2	2
NUCD	non-unicluster design	3.1.1	36
OLSE	ordinary least squares	11.2.2	268
$p_n, p_{n\mu}, p_{n\sigma}, p_{nx}$		3.7	52
πPS		3.2.5	53
PPS	probability proportional to size	7.5	171
PPSWOR	probability proportional to size without replacement	2.4.6	26
PPSWR	probability proportional to size with replacement	2.2	14
psu	primary stage unit	8.1	176
RHC	Rao-Hartley-Cochran	7.4	165
RHCE	Rao-Hartley-Cochran estimator	7.4	165

RR	randomized response	12	275
s	effective sample size	1.2	2
SDE	symmetrized Des Raj estimator	2.4.6	29
SL	significance level	11.1.1	254
SPRO	simple projection	6.1	113
SRSWOR	simple random sampling without replacement	1.2	3
SRSWR	simple random sampling with replacement	1.2	4
$\bar{t}$	Horvitz–Thompson estimator	2.4.4	23
t_{QR}	QR predictor	6.1.3	118
UCD	unicluster design	3.1.1	36
UE	unbiased estimator	1.2	4
UMV	uniformly minimum variance unbiased estimator	3.1.1	33
UMVUE	uniformly minimum variance unbiased estimator	3.1.1	33
WOR	without replacement	1.2	3
WR	with replacement	1.2	3

Author Index

Subject Index

Printed in the United States
by Baker & Taylor Publisher Services